Comprehensive Quality by Design for
Pharmaceutical Product Development
and Manufacture

Comprehensive Quality by Design for Pharmaceutical Product Development and Manufacture

Edited by Gintaras V. Reklaitis, Christine Seymour, and Salvador García-Munoz

Registered Office

John Wiley & Sons, Inc., 111 River Street, Hoboken, NJ 07030, USA

Editorial Office

111 River Street, Hoboken, NJ 07030, USA

For details of our global editorial offices, customer services, and more information about Wiley products visit us at www.wiley.com.

Library of Congress Cataloguing-in-Publication Data

Names: Reklaitis, G. V., 1942– editor. | Seymour, Christine, 1967– editor. | García-Munoz, Salvador, 1971– editor.
Title: Comprehensive quality by design for pharmaceutical product development and manufacture / edited by Gintaras V. Reklaitis, Christine Seymour, Salvador García-Munoz.
Description: Hoboken, NJ : John Wiley & Sons, 2017. | Includes bibliographical references and index. |
Identifiers: LCCN 2017016418 (print) | LCCN 2017025889 (ebook) | ISBN 9781119356165 (pdf) | ISBN 9781119356172 (epub) | ISBN 9780470942376 (cloth)
Subjects: LCSH: Drugs–Design. | Pharmaceutical technology–Quality control.
Classification: LCC RS420 (ebook) | LCC RS420 .C653 2017 (print) | DDC 615.1/9–dc23
LC record available at https://lccn.loc.gov/2017016418

Cover Design: Wiley
Cover Images: (Background) © BeholdingEye/Gettyimages;
(Graph) From Chapter 2, Courtesy of Chatterjee, Moore and Nasr

Set in 10/12pt Warnock by SPi Global, Pondicherry, India

Printed in the United States of America

10 9 8 7 6 5 4 3 2 1

Contents

List of Contributors   *xiii*
Preface   *xix*

1   **Introduction**   *1*
    Christine Seymour and Gintaras V. Reklaitis
1.1   Quality by Design Overview   *1*
1.2   Pharmaceutical Industry   *2*
1.3   Quality by Design Details   *3*
1.4   Chapter Summaries   *4*
    References   *7*

2   **An Overview of the Role of Mathematical Models in Implementation of Quality by Design Paradigm for Drug Development and Manufacture**   *9*
    Sharmista Chatterjee, Christine M. V. Moore, and Moheb M. Nasr
2.1   Introduction   *9*
2.2   Overview of Models   *9*
2.3   Role of Models in QbD   *12*
2.3.1   CQA   *13*
2.3.2   Risk Assessment   *13*
2.3.3   Design Space   *14*
2.3.4   Control Strategy   *19*
2.4   General Scientific Considerations for Model Development   *20*
2.4.1   Models for Process Characterization   *21*
2.4.2   Models for Supporting Analytical Procedures   *22*
2.4.3   Models for Process Monitoring and Control   *22*
2.5   Scientific Considerations for Maintenance of Models   *22*
2.6   Conclusion   *23*
    References   *23*

3   **Role of Automatic Process Control in Quality by Design**   *25*
Mo Jiang, Nicholas C. S. Kee, Xing Yi Woo, Li May Goh, Joshua D. Tice,
Lifang Zhou, Reginald B. H. Tan, Charles F. Zukoski, Mitsuko Fujiwara,
Zoltan K. Nagy, Paul J. A. Kenis, and Richard D. Braatz
3.1   Introduction   *25*
3.2   Design of Robust Control Strategies   *31*
3.3   Some Example Applications of Automatic Feedback Control   *35*
3.4   The Role of Kinetics Modeling   *40*
3.5   Ideas for a Deeper QbD Approach   *42*
3.6   Summary   *44*
     Acknowledgments   *46*
     References   *47*

4   **Predictive Distributions for Constructing the ICH Q8**
     Design Space   *55*
John J. Peterson, Mohammad Yahyah, Kevin Lief, and Neil Hodnett
4.1   Introduction   *55*
4.2   Overlapping Means Approach   *56*
4.3   Predictive Distribution Approach   *59*
4.4   Examples   *61*
4.4.1   A Mechanistic Model Example   *62*
4.4.2   An Empirical Model Example   *64*
4.5   Summary and Discussion   *68*
     Acknowledgment   *69*
     References   *69*

5   **Design of Novel Integrated Pharmaceutical Processes:**
     A Model-Based Approach   *71*
Alicia Román-Martínez, John M. Woodley, and Rafiqul Gani
5.1   Introduction   *71*
5.2   Problem Description   *73*
5.2.1   Mathematical Formulation   *73*
5.2.2   Solution Approach   *75*
5.3   Methodology   *76*
5.3.1   Superstructure   *77*
5.3.2   Model Development   *78*
5.3.3   Decomposition Strategy   *79*
5.4   Application: Case Study   *80*
5.4.1   Stage 1: Problem Definition   *81*
5.4.2   Stage 2: Data/Information Collection/Analysis   *81*
5.4.3   Stage 3: Superstructure, Model Development, and Decomposition
     Strategy   *82*

5.4.4 Stage 4: Generation of Feasible Candidates and Screening *82*
5.4.5 Stage 5: Screening by Process Model *84*
5.4.6 Stage 6: Evaluation of the Feasible Options: Calculation of the Objective Function *88*
5.5 Conclusions *91*
References *91*

6 Methods and Tools for Design Space Identification in Pharmaceutical Development *95*
Fani Boukouvala, Fernando J. Muzzio, and Marianthi G. Ierapetritou
6.1 Introduction *95*
6.2 Design Space: A Multidisciplinary Concept *98*
6.3 Integration of Design Space and Control Strategy *102*
6.4 Case Studies *102*
6.4.1 Design Space of a Continuous Mixer: Use of Data-Driven-Based Approaches *102*
6.4.2 Roller Compaction Case Study: Integration of Control Strategy and Its Effects on the Design Space *107*
6.4.2.1 Deterministic Design Space *110*
6.4.2.2 Stochastic Design Space *112*
6.4.2.3 Effect of Control Strategies on the Design Space *113*
6.5 Conclusions *119*
Acknowledgment *120*
References *120*

7 Using Quality by Design Principles as a Guide for Designing a Process Control Strategy *125*
Christopher L. Burcham, Mark LaPack, Joseph R. Martinelli, and Neil McCracken
7.1 Introduction *125*
7.2 Chemical Sequence, Impurity Formation, and Control Strategy *130*
7.2.1 Chemical Sequence *130*
7.2.2 Impurity Formation *131*
7.2.3 Control Strategy *136*
7.3 Mass Transfer and Reaction Kinetics *140*
7.3.1 CO_2 Mass Transfer Model *140*
7.3.1.1 Determination of Henry's Law Constant *143*
7.3.1.2 Determination of the Mass Transfer Coefficient *145*
7.3.2 Reaction Kinetics *149*
7.3.2.1 Deprotection Reaction Kinetics *151*
7.3.2.2 Calculation of Dissolution Constants *157*
7.3.2.3 Coupling Reaction Kinetics *159*
7.4 Optimal Processing Conditions *165*

7.4.1   Use of Combined Models   *166*
7.4.2   Carbon Dioxide Removal Process Options   *167*
7.5   Predicted Product Quality under Varied Processing Conditions   *174*
7.5.1   Virtual Execution of PAR and Design Space Experiments   *175*
7.5.1.1 Process Parameters   *177*
7.5.2   Acceptable In Situ Values   *177*
7.5.3   PAR Simulation   *178*
7.5.4   Design Space Simulation: Interactions   *178*
7.5.5   Design Space Simulation: Screening Design Experiment
         and Multifactor Experiment Simulation and Data Analysis   *183*
7.5.6   Confirmation of the Design Space with Experiment   *186*
7.6   Conclusions   *186*
         Acknowledgments   *187*
         Notation   *187*
         Acronyms   *187*
         Symbols   *187*
         Notes   *189*
         References   *189*

8   **A Strategy for Tablet Active Film Coating Formulation
         Development Using a Content Uniformity Model and Quality
         by Design Principles**   *193*
         *Wei Chen, Jennifer Wang, Divyakant Desai, Shih-Ying Chang,
         San Kiang, and Olav Lyngberg*
8.1   Introduction   *193*
8.2   Content Uniformity Model Development   *197*
8.2.1   Principles of the Model   *198*
8.2.2   Total Residence Time and Fractional Residence Time   *199*
8.2.3   The RSD Model Derivation   *201*
8.2.4   Model Parameters and Their Measurements   *204*
8.2.4.1 Tablet Velocity   *205*
8.2.4.2 Tablet Number Density   *207*
8.2.4.3 Spray Zone Width   *208*
8.3   RSD Model Validation and Sensitivity Analysis for Model
         Parameters   *212*
8.3.1   Model Validation   *213*
8.3.2   Effect of Spray Zone Width on Content Uniformity   *215*
8.3.3   Effect of Tablet Velocity on Content Uniformity   *216*
8.3.4   Effect of Tablet Size on Content Uniformity   *217*
8.3.5   Effect of Pan Load on Content Uniformity   *217*
8.3.6   Effect of Coating Time on Content Uniformity   *218*
8.4   Model-Based Design Space Establishment for Tablet Active
         Film Coating   *219*

8.4.1   Establish a Model-Based Process Design Space at
         a Defined Scale   *220*
8.4.2   Model-Based Scale-Up   *226*
8.4.3   Model-Based Process Troubleshooting   *228*
8.5   Summary   *229*
         Notations   *230*
         References   *230*

**9   Quality by Design: Process Trajectory Development for a Dynamic
         Pharmaceutical Coprecipitation Process Based on an Integrated
         Real-Time Process Monitoring Strategy**   *235*
         Huiquan Wu and Mansoor A. Khan
9.1   Introduction   *235*
9.2   Experimental   *237*
9.2.1   Materials   *237*
9.2.2   Equipment and Instruments   *237*
9.3   Data Analysis Methods   *239*
9.3.1   PCA and Process Trajectory   *239*
9.3.2   Singular Points of a Signal   *239*
9.4   Results and Discussion   *240*
9.4.1   Using Offline NIR Measurement to Characterize the Naproxen–
         Eudragit L100 Binary Powder Mixing Process   *241*
9.4.2   Using In-Line NIR Spectroscopy to Monitor the Alcohol–Water
         Binary Liquid Mixing Process   *242*
9.4.3   Real-Time Integrated PAT Monitoring of the Dynamic
         Coprecipitation Process   *243*
9.4.4   3D Map of NIR Absorbance–Wavelength–Process Time (or Process
         Sample) of the Coprecipitation Process   *244*
9.4.5   Process Signature Identification   *245*
9.4.6   Online Turbidity Monitoring of the Process   *248*
9.5   Challenges and Opportunities for PCA-Based Data Analysis
         and Modeling in Pharmaceutical PAT and QbD Development   *250*
9.6   Conclusions   *252*
         Acknowledgments   *252*
         References   *253*

**10   Application of Advanced Simulation Tools for Establishing Process
         Design Spaces Within the Quality by Design Framework**   *257*
         Siegfried Adam, Daniele Suzzi, Gregor Toschkoff, and Johannes G. Khinast
10.1   Introduction   *257*
10.2   Computer Simulation-Based Process Characterization
         of a Pharmaceutical Blending Process   *261*
10.2.1   Background   *261*

10.2.2   Goals   *263*
10.2.3   Material and Methods   *264*
10.2.3.1   Application of QbD Concepts   *264*
10.2.3.2   Model and Numerical Simulation   *267*
10.2.3.3   Process Characterization Experimental Design   *268*
10.2.4   Results and Discussion   *272*
10.2.5   Conclusion   *276*
10.3   Characterization of a Tablet Coating Process via CFD Simulations   *276*
10.3.1   Introduction   *276*
10.3.2   Background   *278*
10.3.3   Methods   *280*
10.3.3.1   Model and Numerical Simulation   *281*
10.3.3.2   Simulation Design and Characterization   *284*
10.3.3.3   Potentially Critical Input Parameters   *286*
10.3.4   Results and Discussion   *287*
10.3.4.1   Time Development of Mean Thickness and RSD   *288*
10.3.4.2   Knowledge Space   *290*
10.3.5   Summary   *294*
10.4   Overall Conclusions   *294*
References   *295*

11   **Design Space Definition: A Case Study—Small Molecule Lyophilized Parenteral**   *301*
Linas Mockus, David LeBlond, Gintaras V. Reklaitis, Prabir K. Basu, Tim Paul, Nathan Pease, Steven L. Nail, and Mansoor A. Khan
11.1   Introduction   *301*
11.2   Case Study: Bayesian Treatment of Design Space for a Lyophilized Small Molecule Parenteral   *302*
11.2.1   Arrhenius Accelerated Stability Model with Covariates for a Pseudo-Zero-Order Degradation Process   *302*
11.2.2   Design Space Definition   *307*
11.3   Results   *307*
11.4   Conclusions   *311*
Appendix 11.A   Implementation Using WinBUGS and R   *311*
11.A.1   WinBUGS Model   *312*
11.A.2   Data Used for Analysis   *312*
11.A.3   Calling WinBUGS from R   *314*
11.A.4   Calculating the Predictive Posterior Probability of Meeting Shelf Life   *315*
Notation   *316*
Acknowledgments   *317*
References   *317*

12 **Enhanced Process Design and Control of a Multiple-Input Multiple-Output Granulation Process** *319*
Rohit Ramachandran
12.1 Introduction and Objectives *319*
12.2 Population Balance Model *320*
12.2.1 Compartmentalized Population Balance Model *322*
12.3 Simulation and Controllability Studies *323*
12.4 Identification of Existing "Optimal" Control-Loop Pairings *327*
12.4.1 Discarding n_1 *328*
12.4.2 Discarding n_2 *328*
12.4.3 Discarding n_3 *328*
12.4.4 Discarding n_4 *329*
12.4.5 Discussion *329*
12.5 Novel Process Design *330*
12.5.1 Identification of Kernels *331*
12.5.2 Proposed Design and Control Configuration *331*
12.6 Conclusions *335*
References *336*

13 **A Perspective on the Implementation of QbD on Manufacturing through Control System: The Fluidized Bed Dryer Control with MPC and NIR Spectroscopy Case** *339*
Leonel Quiñones, Luis Obregón, and Carlos Velázquez
13.1 Introduction *339*
13.2 Theory *340*
13.2.1 Fluidized Bed Dryers (FBDs) *340*
13.2.2 Process Control *341*
13.2.2.1 Proportional Integral Derivative (PID) Control *342*
13.2.2.2 Model Predictive Control (MPC) *342*
13.3 Materials and Methods *344*
13.3.1 Materials *344*
13.3.2 Equipment *344*
13.3.3 MPC Implementation *346*
13.4 Results and Discussion *348*
13.4.1 Process Model *348*
13.4.2 Control Performance with Nominal Process Parameters *349*
13.4.3 Control Performance with Non-nominal Model Parameters *352*
13.5 Continuous Fluidized Bed Drying *355*
13.6 Control Limitations *356*
13.7 Conclusions *357*
Acknowledgment *357*
References *357*

14   **Knowledge Management in Support of QbD**   *361*
           G. Joglekar, Gintaras V. Reklaitis, A. Giridhar, and Linas Mockus
14.1   Introduction   *361*
14.2   Knowledge Hierarchy   *363*
14.3   Review of Existing Software   *364*
14.4   Workflow-Based Framework   *365*
14.4.1   Scientific Workflows   *366*
14.4.2   Business Workflows   *367*
14.4.3   Comprehensive Workflow-Based Knowledge Management
           System   *368*
14.5   Drug Substance Case Study   *368*
14.5.1   Process Description   *368*
14.5.2   Workflow-Based Representation of the Semagacestat Study   *370*
14.5.3   Using Workflows   *373*
14.6   Design Space   *374*
14.6.1   Design Space Example   *374*
14.6.2   Systematic Approach to Determining Design Space   *375*
14.6.3   Workflow-Based Approach to Design Space Development   *375*
14.6.4   Drug Product Case Study   *378*
14.7   Technical Challenges   *382*
14.7.1   Human–Machine Interaction Design   *382*
14.7.2   Extraction of Operational Data   *383*
14.7.3   Collection of Tacit Knowledge   *383*
14.8   Conclusions   *384*
           References   *385*

           Index   *387*

List of Contributors

Siegfried Adam
Research Center Pharmaceutical
Engineering GmbH
Graz, Austria

Prabir K. Basu
QbD Consultant
Mt Prospect, IL, USA

Richard D. Braatz
Department of Chemical
Engineering
Massachusetts Institute of
Technology
Cambridge, MA
and
Department of Chemical and
Biomolecular Engineering
University of Illinois at
Urbana-Champaign
Urbana, IL, USA

Christopher L. Burcham
Small Molecule Design and
Development
Eli Lilly and Company
Indianapolis, IN, USA

Shih-Ying Chang
Drug Product Science and
Technology

Bristol-Myers Squibb Company
New Brunswick, NJ, USA

Sharmista Chatterjee
Office of New Drugs Quality
Assessment
Food and Drug Administration
Silver Spring, MD, USA

Wei Chen
Drug Product Science and
Technology
Bristol-Myers Squibb Company
New Brunswick, NJ, USA

Divyakant Desai
Drug Product Science and
Technology
Bristol-Myers Squibb Company
New Brunswick, NJ, USA

Mitsuko Fujiwara
Department of Chemical and
Biomolecular Engineering
University of Illinois at
Urbana-Champaign
Urbana, IL, USA

Rafiqul Gani
Department of Chemical and
Biochemical Engineering

Technical University of Denmark
Lyngby, Denmark

A. Giridhar
Davidson School of Chemical
Engineering
Purdue University
West Lafayette, IN, USA

Li May Goh
Department of Chemical and
Biomolecular Engineering
University of Illinois at
Urbana-Champaign
Urbana, IL, USA

Neil Hodnett
GlaxoSmithKline Pharmaceuticals
Brentford, UK

Marianthi G. Ierapetritou
Department of Chemical and
Biochemical Engineering
Rutgers University
Piscataway, NJ, USA

Fernando J. Muzzio
Department of Chemical and
Biochemical Engineering
Rutgers University
Piscataway, NJ, USA

Mo Jiang
Department of Chemical
Engineering
Massachusetts Institute of
Technology
Cambridge, MA, USA

G. Joglekar
Davidson School of Chemical
Engineering
Purdue University
West Lafayette, IN, USA

Nicholas C. S. Kee
Department of Chemical and
Biomolecular Engineering
University of Illinois at
Urbana-Champaign
Urbana, IL, USA
and
National University of Singapore,
Block E5
and
Institute of Chemical and
Engineering Sciences
Jurong Island, Singapore

Paul J. A. Kenis
Department of Chemical and
Biomolecular Engineering
University of Illinois at
Urbana-Champaign
Urbana, IL, USA

Mansoor A. Khan
Division of Product Quality Research
(DPQR, HFD-940)
OTR, Office of Pharmaceutical
Quality, Center for Drug Evaluation
and Research, Food and Drug
Administration
Silver Spring, MD, USA

Current affiliation
Rangel College of Pharmacy, Texas
A&M University Health Science Center
College Station, TX, USA

Johannes G. Khinast
Research Center Pharmaceutical
Engineering GmbH
and
Institute for Process and Particle
Engineering
Graz University of Technology
Graz, Austria

San Kiang
Drug Product Science and
Technology
Bristol-Myers Squibb Company
New Brunswick, NJ, USA

Mark LaPack
Mark LaPack & Associates
Consulting, LLC, Lafayette
IN, USA

David LeBlond
Applied Statistics consultant
Wadsworth, IL, USA

Kevin Lief
GlaxoSmithKline
Pharmaceuticals
Brentford, UK

Olav Lyngberg
Chemical Development, Research
and Development
Bristol-Myers Squibb Company
New Brunswick, NJ, USA

Joseph R. Martinelli
Small Molecule Design and
Development
Eli Lilly and Company
Indianapolis, IN, USA

Neil McCracken
Small Molecule Design and
Development
Eli Lilly and Company
Indianapolis, IN, USA
and
Bioproduct Research and
Development
Eli Lilly and Company
Indianapolis, IN, USA

Linas Mockus
Davidson School of Chemical
Engineering
Purdue University
West Lafayette, IN, USA

Christine M. V. Moore
Global CMC Policy
Merck, Inc.
Philadelphia, PA, USA

Fani Boukouvala
Department of Chemical and
Biochemical Engineering
Rutgers University
Piscataway, NJ, USA

Zoltan K. Nagy
School of Chemical Engineering
Purdue University, Forney Hall of
Chemical Engineering
West Lafayette, IN, USA

Steven L. Nail
Pharmaceutical Development
Baxter Pharmaceutical Solutions,
LLC
Bloomington, IN, USA

Moheb M. Nasr
GlaxoSmithKline
Pharmaceuticals
Brentford, UK

Luis Obregón
Pharmaceutical Engineering
Research laboratory, Chemical
Engineering Department
University of Puerto Rico at
Mayaguez
Mayaguez, Puerto Rico

Tim Paul
Pharmaceutical Development
Baxter Pharmaceutical
Solutions, LLC
Bloomington, IN, USA

Nathan Pease
Pharmaceutical Development
Baxter Pharmaceutical
Solutions, LLC
Bloomington, IN, USA

John J. Peterson
GlaxoSmithKline
Pharmaceuticals
Brentford, UK

Leonel Quiñones
Pharmaceutical Engineering
Research laboratory, Chemical
Engineering Department
University of Puerto Rico
at Mayaguez
Mayaguez, Puerto Rico

Rohit Ramachandran
Department of Chemical and
Biochemical Engineering
Rutgers, The State University
of New Jersey
Piscataway, NJ, USA

Gintaras V. Reklaitis
Davidson School of Chemical
Engineering
Purdue University
West Lafayette, IN, USA

Alicia Román-Martínez
Department of Chemical and
Biochemical Engineering
Technical University of Denmark
Lyngby, Denmark

and
Facultad de Ciencias Químicas
Universidad Autónoma de San
Luis Potosí
San Luis Potosí, Mexico

Christine Seymour
Global Regulatory Chemistry
and Manufacturing Controls
Pfizer Inc.
Groton, CT, USA

Daniele Suzzi
Research Center Pharmaceutical
Engineering GmbH
Graz, Austria

Reginald B. H. Tan
National University of Singapore,
Block E5
and
Institute of Chemical and
Engineering Sciences
Jurong Island, Singapore

Joshua D. Tice
Department of Chemical and
Biomolecular Engineering
University of Illinois at
Urbana-Champaign
Urbana, IL, USA

Gregor Toschkoff
Research Center Pharmaceutical
Engineering GmbH
Graz, Austria

Carlos Velázquez
Pharmaceutical Engineering
Research laboratory, Chemical
Engineering Department
University of Puerto Rico at
Mayaguez
Mayaguez, Puerto Rico

Jennifer Wang
Drug Product Science and Technology
Bristol-Myers Squibb Company
New Brunswick, NJ, USA

Xing Yi Woo
The Jackson Laboratory
Bar Harbor, ME, USA

John M. Woodley
Department of Chemical and
Biochemical Engineering
Technical University of Denmark
Lyngby, Denmark

Huiquan Wu
Division of Product Quality
Research (DPQR, HFD-940)
OTR, Office of Pharmaceutical
Quality, Center for Drug Evaluation
and Research, Food and Drug
Administration
Silver Spring, MD, USA

Current affiliation
Process Assessment Branch II,
Division of Process Assessment 1
Office of Process and Facilities,
Office of Pharmaceutical Quality,
Center for Drug Evaluation
and Research, Food and Drug
Administration
Silver Spring, MD, USA

Mohammad Yahyah
GlaxoSmithKline
Pharmaceuticals
Brentford, UK

Lifang Zhou
Department of Chemical
Engineering
Massachusetts Institute of
Technology
Cambridge, MA, USA

Charles F. Zukoski
Department of Chemical and
Biomolecular Engineering
University of Illinois at
Urbana-Champaign
Urbana, IL
and
Department of Chemical and
Biological Engineering
University at Buffalo
Buffalo, NY, USA

Preface

Quality by design (QbD) is a scientific and risk-based approach to pharmaceutical product development and manufacturing that is becoming firmly established in the pharmaceutical industry. This volume contains chapters covering the various tools and considerations that come into play when the QbD approach is employed. The contributions are based on presentations from the series of AIChE topical conferences on the theme Comprehensive Quality by Design (QbD). The AIChE QbD Symposium was first held in 2009, was continued for several years, and then in 2013 transitioned to sessions organized under a new AIChE entity, the Pharmaceutical Discovery, Development and Manufacturing (PD^{2}M) Forum. While the initial directions envisioned for QbD has evolved in the course of its adoption and adaption by the industry, the stimulus for innovation in development and manufacture that it provided has been very valuable. The chapters in this volume capture some of the evolution of QbD that has occurred.

The idea for compiling the perspectives of a number of prominent contributors to the development and application of QbD into a coherent volume has to be credited to Dr. Salvador Garcia-Munoz. His enthusiasm for the approach stimulated our collective recognition that QbD needs to be actively promulgated as appropriate not just for the leading organization in this industry but by the entire pharmaceutical industry. We appreciate the willingness of the authors of the chapters contained herein to share their views and experiences on QbD, the support of the AICHE Publications Committee in approving this project, and the patience of the publisher, Wiley, in dealing with the various delays that our poor time management skills brought about.

March 31, 2017

Gintaras V. Reklaitis
and Christine Seymour

1

Introduction

Christine Seymour[1] and Gintaras V. Reklaitis[2]

[1] *Global Regulatory Chemistry and Manufacturing Controls, Pfizer Inc., Groton, CT, USA*
[2] *Davidson School of Chemical Engineering, Purdue University, West Lafayette, IN, USA*

1.1 Quality by Design Overview

QbD emerged as a cultural change in the pharmaceutical industry, which promoted a scientific and risk-based approach to pharmaceutical product development and manufacturing. Historically, pharmaceutical development and manufacturing had emphasized checklist-based operations rather than scientific understanding. The high attrition rates of drug candidates during development and the high value of pharmaceutical products, along with extremely high regulatory burden, had led to business practices that minimized risk and restricted process changes and the implementation of new technology.

Traditionally, pharmaceutical process and product development utilized empirical and univariate experimentation and pharmaceutical processes operated at fixed process conditions with offline analytical testing (with a long feedback timeline) and end-product testing. In addition, it was typical of pharmaceutical companies to provide the regulatory agencies with minimal process and scientific information, and regulatory agencies responded with a wealth of detail queries.

QbD is a "systematic approach to pharmaceutical development and manufacturing that is based on science and quality risk management and begins with predefined objectives and emphasizes product and process understanding as well as process control" [1]. QbD emphasizes multivariable experimentation, design of experiments, process modeling, kinetics, thermodynamics, online analytical testing, and so on. In addition, it has become an improved regulatory paradigm in which the scientific understanding of the product and process has to be provided to the regulatory agencies. This new regulatory model is intended

Comprehensive Quality by Design for Pharmaceutical Product Development and Manufacture,
First Edition. Edited by Gintaras V. Reklaitis, Christine Seymour, and Salvador García-Munoz.
© 2017 American Institute of Chemical Engineers, Inc. Published 2017 by John Wiley & Sons, Inc.

to allow for higher transparency, higher quality, and the implementation of modern manufacturing techniques, such as continuous processing, as well as continuous improvement of commercial pharmaceutical processes.

1.2   Pharmaceutical Industry

The active pharmaceutical ingredient or drug substance is the active component of the pharmaceutical product, and typically small-molecule drug substances are produced by a multistep synthesis, which involves a sequence of chemical reactions followed by purification/isolation unit operations. Historically, drug substance pharmaceutical processes consisted of batch operations such as reactions, extraction, distillation, crystallization, filtration/centrifugation, drying, and milling.

The drug product is the pharmaceutical formulation that the patient receives and is often in the form of tablets or capsules; other common formulations are oral solutions, topical transdermal patches, and lyophiles or sterile solutions for injection. Historically, drug product processes also consisted of batch operations such as blending, granulation, drying, tableting, encapsulation, or filling depending on the final formulation.

The history of regulations [2, 3] shows an increase in regulatory control after catastrophes; one of the most tragic incidents in the United States was the elixir sulfanilamide incident of 1937 where diethylene glycol was used as a solvent in a pediatric cough syrup and resulted in more than a hundred deaths. This incident led to the Food, Drug, and Cosmetic Act, which increased the US Food and Drug Administration (FDA)'s authority to regulate drugs and required premarketing safety approval for new medications. Another tragic incident was the thalidomide disaster of 1961 where approximately 12 000 infants in over 50 countries were born with severe malformations. This incident led to the Kefauver–Harris Drug Amendment, which increased the FDA's authority to require safety and efficacy prior to marketing (and tighter controls of clinical trials). These and other incidents led to tighter regulations and, in the 1960s and 1970s, to a rapid increase in national pharmaceutical regulations; simultaneously, many pharmaceutical companies were also globalizing.

The global harmonization of pharmaceutical guidelines across the developed economics was initiated in 1990 through the International Council for Harmonization (ICH) of Technical Requirements for Pharmaceuticals for Human Use [4]. ICH is cosponsored by regulatory agencies and industrial organizations, as well as many observing organizations. These include the European Commission; the US FDA; Ministry of Health, Labour and Welfare of Japan; the European Federation of Pharmaceutical Industries and Associations; the Japan Pharmaceutical Manufacturers Association; the Pharmaceutical Research and Manufacturers of America; Health Canada;

Swissmedic; ANVISA of Brazil; Ministry of Food and Drug Safety of the Republic of Korea; the International Generic and Biosimilar Medicines Association; the World Self-Medication Industry; and the Biotechnology Innovation Organization. ICH has set a structure and process for the proposal, review, and implementation for efficacy, safety, and quality guidelines as well as dossier format requirements. The initial ICH guidelines set common structure stability, analytical methods, impurity control and drug substance, and drug product specifications requirements for drug substance and drug product, and the early ICH guidelines emphasized testing for quality.

1.3 Quality by Design Details

QbD was introduced through the "QbD tripartite" of ICH guidelines: ICH Q8 (R2) Pharmaceutical Development, ICH Q9 Quality Risk Management, and ICH Q10 Pharmaceutical Quality Systems. ICH Q8 Pharmaceutical Development describes the principles of QbD, outlines the key elements, and provides illustrative examples for pharmaceutical drug products. ICH Q9 Quality Risk Management offers a systematic process for the assessment, control, communication, and review of risks to the quality of the drug product. In addition, it states that "the evaluation of the risk to the quality should be based on scientific knowledge and ultimately linked to the protection of the patient" [5]. ICH Q10 Pharmaceutical Quality Systems describes a "comprehensive model for an effective pharmaceutical quality system that is based on International Standards Organization quality concepts and includes applicable Good Manufacturing Practices" [6]. The fourth QbD ICH guideline (considered the drug substance equivalent of ICH Q8) for enhanced active pharmaceutical ingredient synthesis and process understanding, Q11 Development and Manufacturing of Drug Substances (Chemical Entities and Biotechnological/Biological Entities), was approved in November 2012 [7]. The fifth QbD ICH guideline (ICH Q12), Technology and Regulatory Considerations for Pharmaceutical Product Lifecycle Management, is currently under development.

The QbD approach begins with Quality Target Product Profile, which is a prospective summary of the quality characteristics of the pharmaceutical product that ensures the desired quality, safety, and efficacy, and works backward through the drug product and drug substance processes establishing a holistic understanding of which attributes are linked to patients' requirements and functional relationships of these attributes.

The next step in QbD is a systematic approach to determine the aspects of the drug substance and drug product manufacturing processes that impact the Quality Target Product Profile. A risk assessment is conducted to identify the quality attributes and process parameters that could potentially impact product safety and/or efficacy, utilizing prior scientific knowledge gained from first principles, literature, and/or similar processes.

The output of the risk assessment is a development plan, in which multivariable experiments, kinetics, and/or modeling is typically utilized. The goal of the plan is to establish a holistic understanding of how attributes and parameters are functionally interrelated throughout the entire drug substance and drug product processes. The result is control strategy, which links parameters and attributes to the Quality Target Product Profile.

This approach provides a comprehensive understanding of the critical quality attributes, which is a "physical, chemical, biological, or microbiological property or characteristic that should be in an appropriate limit, range or distribution to ensure the desired product quality," and of how the process parameters are related to the quality attributes and how probable they can impact quality.

The enhanced understanding of products and processes, along with quality risk management, leads to a product control strategy, which might include a design space (that is optional in ICH Q8/11), which is "the multidimensional combination and interaction of input variables and process parameters that have been demonstrated to provide assurance of quality." The control strategy is a planned set of controls, which can be process parameters, process attributes, design space, facility and equipment operating conditions, and process testing that ensures process performance and product quality.

The regulatory QbD landscape continues to evolve, and AIChE conferences and sessions will continue to provide a platform to discuss and debate the latest QbD concepts and implementations.

1.4 Chapter Summaries

The contributions in this book can be divided into three sets: Chapters 2–6 address the role of key technologies, process models, process analytical technology (PAT), automatic process control, and statistical methodology, supporting QbD and establishing associated design spaces. Chapters 7–13 present a range of thoroughly developed case studies in which tools and methodologies are used to support specific drug substance and drug product QbD-related developments. Finally, Chapter 14 discusses the needs for initial efforts toward systematic data and knowledge management to support QbD and related activities. More specifically:

Chapter 2, *An Overview of the Role of Mathematical Models in Implementation of Quality by Design Paradigm for Drug Development and Manufacture* (Chatterjee, Moore, and Nasr), reviews the categories of mathematical models that can be exploited to support QbD and presents literature examples of various types of model formulations and their use. The authors emphasize that models are valuable tools at every stage of drug development and manufacture. Examples presented span early-stage risk assessment, design space development, process monitoring and control, and continuous improvement of product quality.

Chapter 3, *Role of Automatic Process Control in Quality by Design* (Braatz and coworkers), outlines how robust automatic control is an important element in actual implementation of QbD. Using phenomenologically based or data-driven models, automatic control strategies provide the active mechanism that maintains the operation of the manufacturing process within the design space despite the disturbances that inevitably arise. The authors illustrate the use of feedback control methodology combined with online process measurement for controlling critical quality attributes such as polymorphic form and particle-size distribution in batch crystallization operations.

Chapter 4, *Predictive Distributions for Constructing the ICH Q8 Design Space* (Peterson and coworkers), reviews the reported applications of response surface methodology for constructing design spaces and identifies risks associated with the use of "overlapping mean response" constructions, which are typically used in building multivariate design spaces. The authors outline two predictive distribution approaches that overcome these risks by taking into account the uncertainty associated with the experimental results used in building the response surfaces as well as the correlation that may exist among the responses. A Bayesian and a parametric bootstrapping approach are presented and illustrated with examples.

Chapter 5, *Design of Novel Integrated Pharmaceutical Processes: A Model-Based Approach* (Roman-Martinez, Woodley, and Gani), builds on model-based development of design space advanced in Chapter 2 to present a systematic strategy for identifying the best design of a process for developing an active pharmaceutical ingredient. The strategy employs a library of unit operation models and physical/chemical property prediction tools, which are used as components within a process synthesis strategy that employs mixed integer nonlinear optimization methods. This optimization-based strategy generates the optimal selection and sequence of reaction and separation unit operations as well as the associated design space. A case study is reported involving the synthesis of neuraminic acid.

Chapter 6, *Methods and Tools for Design Space Identification in Pharmaceutical Development* (Boukouvala, Muzzio, and Ierapetritou), reviews the tools and methods that have been developed in the process systems engineering literature to address issues of process feasibility and flexibility under uncertainty. It is shown that process design under uncertainty can be posed as a stochastic optimization problem. Moreover, the review notes that in general the design of the process and its automatic control system need to be treated in an integrated fashion since the automatic control system generally can serve to increase the design space. The concepts are illustrated with two single-unit examples: a powder blender and a roller compactor.

Chapter 7, *Using Quality by Design Principles as a Guide for Designing a Process Control Strategy* (Burcham and coworkers), reports a comprehensive process engineering study involving the implementation of an impurity control strategy for a new drug substance. The study makes extensive use of predictive models to determine optimal processing conditions and to map the design space around those conditions. The mechanistic models developed include complex reaction kinetics and mass transfer processes, which were developed through intensive experimentation using appropriate PAT and traditional analytical methods. The process models were used as an integral part of an in silico approach to identify the boundaries of the design space requiring experimental confirmation.

Chapter 8, *A Strategy for Tablet Active Film Coating Formulation Development Using a Content Uniformity Model and Quality by Design Principles* (Chen and coworkers), presents in detail the development of a mechanistic model to predict the relative standard deviation of table content uniformity of an active film coating unit operation. Systematic studies to identify the most important operating variables, develop model parameters, and validate model prediction across scales are reported. The model is shown to be an effective tool for developing the design space for this unit operation, including establishing the effects of scale-up.

Chapter 9, *Quality by Design: Process Trajectory Development for a Dynamic Pharmaceutical Coprecipitation Process Based on an Integrated Real-Time Process Monitoring Strategy* (Wu and Khan), describes the features, strengths, and limitations of principal component analysis of real-time process measurements as a means for process trajectory monitoring, identification of singular points of the trajectory, and development of understanding of important phenomena occurring during the dynamic process. A case study of a coprecipitation process monitored using near-infrared (NIR) and turbidity measurements is detailed. Implications for design space development are discussed.

Chapter 10, *Application of Advanced Simulation Tools for Establishing Process Design Spaces Within the Quality by Design Framework* (Khinast and coworkers), reports on the value added by advanced simulation tools in building fundamental process understanding, especially of the impact of critical sources of process variability. The use of discrete element modeling (DEM) is described to investigate a powder blending operation with the goal of screening and prioritizing potentially critical input variables and mapping out a blending experimental space. Similarly, computational fluid dynamics (CFD) is used to construct a detailed three-zone model of film formation on a tablet surface and then to characterize the critical process parameters of the coating operation. In each case, Design of Experiment (DOE) on the most important parameters are utilized to develop a design space.

Chapter 11, *Design Space Definition: A Case Study—Small-Molecule Lyophilized Parenteral* (Mockus and coworkers), builds on Chapter 4 and proposes a novel Bayesian treatment to enable the establishment of the reliability limit of a design space and describes the application of the approach to a lyophilized parenteral product. The proposed approach also provides quantitative estimates of the risk of failure of key product attributes.

Chapter 12, *Enhanced Process Design and Control of a Multiple-Input Multiple-Output Granulation Process* (Ramachandran), demonstrates the use of a previously validated mechanistic three-dimensional population balance model of a continuous wet granulation process to investigate controllability and identify the most effective input–output variable pairings for designing a feedback control system. The results are then used to suggest possible alternative process designs.

Chapter 13, *A Perspective on the Implementation of QbD on Manufacturing Through Control System: The Fluidized Bed Dryer Control with MPC and NIR Spectroscopy Case* (Velazquez and coworkers), builds on Chapter 3 and illustrates the use of automatic process control coupled with real-time NIR particle moisture content measurement to control a fluid bed dryer. Model-predictive controller design with design space constraints imposed on air velocity and inlet temperature is demonstrated for both batch and continuous operations of a pilot plant dryer.

Chapter 14, *Knowledge Management in Support of QbD* (Joglekar and coworkers), discusses the potential role of knowledge management systems in providing the structured framework for recording, using, and learning from the data that are generated in product development and manufacture. The relevant literature is reviewed, and the features of a specific workflow-based system are described. Applications to an active pharmaceutical ingredient (API) development and a drug product design space development are outlined.

The coeditors hope that these chapters serve to stimulate continued developments in tools, methods, and applications that will further solidify the role of QbD concepts and thought processes in pharmaceutical development and manufacture. In addition, we hope that the technology vendor community may be stimulated to develop software implementations that will make process model building, physical property estimation, process and control system design, and probabilistically based design space development efficient and reliable for the scientists and engineers of pharmaceutical industry.

References

1 ICH Q8 (R2), ICH Harmonised Tripartite Guidelines—Pharmaceutical Development, 2009.
2 http://www.jnj.com/connect/news/all/mcneil-ppc-inc-announces-signing-of-consent-decree-covering-manufacturing-facilities-in-las-piedras-pr-fort-washington-pa-and-lancaster-pa (accessed April 6, 2017).

3 http://www.fda.gov/AboutFDA/WhatWeDo/History/default.htm (accessed April 6, 2017).

4 Seymour, C., An Introspective of the Regulatory Influence on Drug Development, AIChE Annual Meeting, Presentation 94c, 2009.

5 ICH Q9, ICH Harmonised Tripartite Guidelines—Quality Risk Management, 2005.

6 ICH Q10, ICH Harmonised Tripartite Guidelines—Pharmaceutical Quality System, 2008.

7 ICH Q11, ICH Harmonised Tripartite Guidelines—Development and Manufacture of Drug Substances, 2012.

2

An Overview of the Role of Mathematical Models in Implementation of Quality by Design Paradigm for Drug Development and Manufacture

Sharmista Chatterjee[1], Christine M. V. Moore[2], and Moheb M. Nasr[3]

[1] *Office of New Drugs Quality Assessment, Food and Drug Administration, Silver Spring, MD, USA*
[2] *Global CMC Policy, Merck, Inc., Philadelphia, PA, USA*
[3] *GlaxoSmithKline Pharmaceuticals, Brentford, UK*

2.1   Introduction

A model is a representation of an underlying physical–chemical phenomenon. In the pharmaceutical industry, mathematical-based models can be applied at all stages of development, starting with formulation design, continuing through process development and scale-up, and extending into process monitoring and control of the commercial process. Implementation of models offers many benefits. These include, but are not limited to, (i) enhanced process understanding, (ii) reduction of experimentation cost, and (iii) improvement of productivity and product quality.

2.2   Overview of Models

Models can be broadly categorized as either qualitative or quantitative. The focus of this chapter is quantitative models. These can be classified into three broad areas: mechanistic, empirical, and hybrid. As illustrated in the knowledge pyramid in Figure 2.1, overall understanding and the information needed to derive from these models increases from empirical to mechanistic models.

Mechanistic models are based on first principles, capture the underlying physical/chemical phenomena through sets of equations, and can be time independent (i.e., steady-state) or dynamic. As indicated by Singh *et al.* [1], mechanistic models can be an excellent way to represent process knowledge. In such models, the input–output dynamics in a unit operation can be represented by a set of differential equations. Model building necessitates the

Comprehensive Quality by Design for Pharmaceutical Product Development and Manufacture,
First Edition. Edited by Gintaras V. Reklaitis, Christine Seymour, and Salvador García-Munoz.
© 2017 American Institute of Chemical Engineers, Inc. Published 2017 by John Wiley & Sons, Inc.

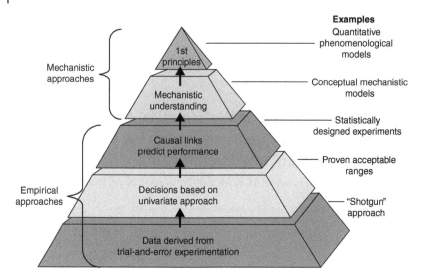

Figure 2.1 Knowledge pyramid for developing mathematical models.

availability of balance equations (e.g., mass and energy balance equations), constitutive equations, and an understanding of the constraints. Since mechanistic models are a true representation of the underlying phenomenon, predictions from these models can sometimes be extrapolated beyond the range covered by input data, depending on the validity of the underlying assumptions. Typically, the bottleneck in developing mechanistic models is coming up with equations as well as associated parameters that accurately represent the system.

Empirical modeling approaches also can be used to represent input–output dynamics. These models are particularly useful for complex systems where it is not feasible to develop mechanistic models. Empirical models treat a system as a "black box" and do not typically describe the underlying physical–chemical phenomena. These models represent input–output dynamics of a system solely in terms of observational data. One of the limitations of empirical models is that the range of applicability of these models is limited to the variation represented in the data that was used to derive the model. Hence, predictions from these models cannot be reliably extrapolated beyond the range covered by the input data. On the other hand, the advantage of empirical models is that they can be relatively easy to put together and solve, as compared with mechanistic models.

In the pharmaceutical world, empirical models are typically used for process understanding and control, such as to program software sensors associated with process analytical technology (PAT)-based tools. While mechanistic models have a distinct advantage of a wide range of predictive potential, not all processes associated with the pharmaceutical industry are understood well enough to allow them to be modeled using first principles.

Philosophically, however, there are few true mechanistic or empirical models. All mechanistic models have a degree of empiricism in them (e.g., modeling assumptions), while all empirical models have a mechanistic element (e.g., rationale for selection of input parameters that are used to derive the models). In general, models are classified into either category depending on the preponderance of mechanistic or empirical components in the model. Following this philosophy, models can be classified as semiempirical or hybrid if they have relatively equal proportion of mechanistic and empirical elements.

Hybrid models are a combination of mechanistic and empirical models. As elucidated by Gernaey *et al.* [2], the approach is to include all available process knowledge in a first-principles-based model, where the gaps in process knowledge are then represented on the basis of empirical (i.e., data-driven) approaches utilizing available experimental data. Examples of hybrid models are scale-up correlations, where the form of the equation is derived from fundamental relations, while the constants are fit from experimental data.

As shown in Figure 2.2, each model category has several potential approaches and mathematical techniques.

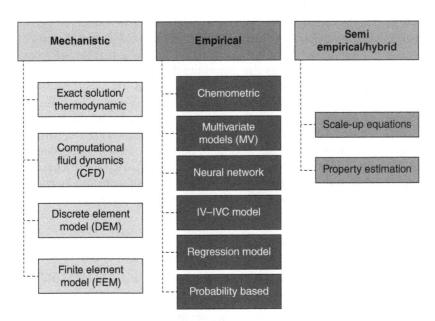

Figure 2.2 Schematic of types of models.

Mechanistic models can include, but are not limited to, (i) models that involve exact solution of equations representing the underlying physical–chemical phenomena while treating the system as one entity; (ii) computational fluid dynamics (CFD) approach, where intensive computational techniques are used to simulate fluid movements while dividing the volume occupied by the fluid into discrete cells (or the mesh); (iii) discrete element modeling (DEM) approach, which involves rigorous computation to simulate the motion of a collection of discrete particles of micrometer-scale size and above; and (iv) finite element model (FEM) that involves solving constitutive equations for a domain by discretizing the domain into small elements or nodes.

Empirical models can include, but are not limited to, (i) regression correlations (linear or nonlinear) derived between a dependent variable and one or more independent variables, (ii) statistically based latent variable (LV) models that relate a set of manifest variables to a set of LVs (multivariate models and chemometric models belong to this category), (iii) neural network models that utilize nonlinear statistical modeling tools to represent complex relations between inputs and outputs, (iv) probability-based models in which the relationship between inputs and outputs is expressed in terms of probability theory, and (v) *in vitro–in vivo* correlation (IV–IVC) models that describe the relationship between an *in vitro* property of an extended release dosage form and a relevant *in vivo* response, for example, plasma concentration. IV–IVC models include regression correlation approach as well as principles of statistical moment analysis.

Scale-up correlations based on dimensional analysis can be considered hybrid models. Dimensional analysis is based on characterization of a process in terms of dimensionless numbers. Dimensionless numbers involve a mechanistic component in identifying the factors that constitute them; however, the method of derivation of these numbers may be regarded as empirical. The objective during scale-up is to keep the dimensionless numbers constant at various scales to ensure consistency of product quality at all scales. An example of a hybrid modeling approach is the model described by Chen *et al.* [3]. This model is used for predicting active pharmaceutical ingredient content uniformity for a drug product in which the active is coated onto a core tablet. The model is based on a mechanistic description of the spray coating process in a perforated coating pan and included a number of parameters that were measured from experimental runs.

2.3 Role of Models in QbD

An example of a QbD implementation approach as outlined in ICH Q8 (R2) [4] involves the following steps:

1) Identification of quality target product profile (QTPP), which ensures the finished product's quality, safety, and efficacy

2) Identification of critical to quality attributes (CQAs)
3) Risk assessment
4) Determination of design space
5) Implementation of the control strategy
6) Continual improvement

The following paragraphs describe how models can be used at every stage of the QbD implementation approach by citing examples from published literature. This compilation is not exhaustive of various types of models that can be implemented to support QbD-based development. Instead, a few examples were selected from the literature to exemplify potential applicability for each step.

2.3.1 CQA

A CQA is a physical, chemical, biological, or microbiological property or characteristic that should be within an appropriate limit, range, or distribution to ensure the desired drug product quality. CQAs are generally associated with drug substance, excipients, intermediates, and drug product. CQAs of typical drug substance include particle size, residual solvent level, impurity levels, crystal form, and so on. In general, CQAs of a drug product are similar to the attributes that are part of the specifications, such as assay, content uniformity, dissolution, and impurity level.

One example using these models to better understand a CQA is the use of an IV–IVC that can be used to establish the link between desired clinical performances, that is, bioavailability and dissolution. IV–IVC has been defined as a predictive mathematical model describing the relationship between an *in vitro* property of a dosage form and its *in vivo* response [5]. In Rossi *et al.* [6], it is shown how IV–IVC data is used to develop and validate a dissolution test for immediate release ritonavir soft gel capsules (Norvir®). With ritonavir being a poorly soluble drug, dissolution is regarded as a predictor of *in vivo* performance, hence may be classified as a CQA. As shown in this chapter, a meaningful dissolution test (that includes test conditions as well as specification) for Norvir soft gelatin capsules was developed using *in vivo* data. A significant linear level A correlation between *in vitro* and *in vivo* parameters was established.

2.3.2 Risk Assessment

As outlined in ICH Q9 [7], risk assessment consists of the identification of hazards and the analysis and evaluation of risks associated with exposure to those hazards. Quality risk management is a systematic process for the assessment, control, communication, and review of risks to the quality of the drug (medicinal) product across the product life cycle. During development, risk assessment can be carried out to identify unit operations or drug substance synthetic steps as well as material/process parameters that have an impact on the finished product attributes. The identified parameters can then be further evaluated via either experiments or mathematical models or a combination of both.

Various quantitative or semiquantitative approaches can be used for risk assessment. ICH Q9 lists several tools that can be used for risk assessment, such as failure mode effects analysis (FMEA), failure mode effects and criticality analysis (FMECA), fault tree analysis (FTA), hazard analysis and critical control points (HACCP), hazard operability analysis (HAZOP), preliminary hazard analysis (PHA), and risk ranking and filtering.

Out of the listed tools, FMEA and FMECA can be regarded as semiquantitative approaches. These tools are commonly used for quality analysis of processes, such as those covered in six-sigma approaches. In FMEA, the risk of failure of each parameter is evaluated on the basis of frequency of occurrence (O), probability that the failure would remain undetected (D), and its severity (S). Each mode is then ranked by a group of cross-functional experts on a linear scale (e.g., a scale of 1–10), with a higher number representing a higher risk. Once the occurrence, detection, and severity are determined, the net risk is then estimated by calculating the risk priority number (RPN), which is a product of the scores for O, D, and S. A high RPN implies a greater risk.

A review of the literature showed an example where FMEA technique was implemented to improve the efficiency of a near-infrared (NIR)-based analytical procedure [8]. In this chapter, an NIR analytical procedure that was used for screening drugs for authenticity was subjected to an FMEA analysis. Each failure mode was ranked on estimated frequency of occurrence (O), probability that the failure would remain undetected later in the process (D), and severity (S), each on a scale of 1–10. Failure risks were calculated by $RPNs = O \times D \times S$. Failure modes with the highest RPN scores were subjected to corrective actions and the FMEA was repeated. Human errors turned out to be the most common cause of failure modes. Based on their findings, the authors recommended that for analytical method validation, risk analysis, for example, by FMEA, be carried out in addition to the usual analytical validation, to help in detecting previously unidentified risks. In another case, FMEA was used to identify critical formulation and process variables for a roller compaction process, and the information from FMEA was then used to build a design space by Design of Experiment (DOE) [9].

2.3.3 Design Space

As defined in ICH Q8 (R2) [4], a design space is a multidimensional combination and interaction of input variables (e.g., material attributes) and process parameters that have been demonstrated to provide assurance of quality. Often, risk assessment techniques are used to identify parameters that define a design space by identifying parameters that have a potential to impact the CQA of a drug quality. A design space can be determined via experiments and modeling at laboratory, pilot, and/or commercial scale. Design spaces can be defined for both drug substances and drug products. Modeling approaches such as mechanistic, empirical, or hybrid can be used for design space development wherein as discussed in the following examples, models can be used to support its various facets, including defining a design space at pilot scale based on DOE data, scaling up pilot scale design space to commercial scale, and understanding the limitations of the proposed design space.

a) *Design space based on DOE data*: As presented by Verma *et al.* [10], the effects of key formulation process variables for a microfluidization unit operation was investigated via fractional factorial statistically based DOE. Microfluidization was used in the preparation of nanosuspensions for poorly water-soluble drugs. Multiple linear regression and ANOVA techniques were employed to analyze the data from the DOE, in order to identify and estimate the effect of important factors, to establish their relationship with CQAs, and to create a design space and a predictive model of the microfluidization unit operation. Interactions between the variables were also depicted using contour plots.

Figure 2.3 shows a general approach for defining a design space based on DOE data. A DOE is initially carried out in terms of multiple independent input variables (i.e., inputs variables are all orthogonal to each other). Input variables are selected on the basis of the magnitude of their potential impact to product quality. Various options are available for DOE design, for example, full factorial and d-optimal. Experiments are carried out in a random fashion and response(s) is measured. Typically identified CQAs are measured as responses in a DOE. Data from the DOE is analyzed using statistical approaches such as Pareto charts to identify the statistical significance of input variables and their interactions to product quality. A regression correlation is then derived from the DOE data in terms of the significant input variables. Design space can then be represented mathematically in terms of the regression correlation or graphically, for example, as a contour surface.

b) *Design space based on hybrid model*: In this example, two mechanistic tablet film coating models were used for scale-up of tablet film coating of an established commercial immediate release product [11]. The models were the following: (a) a thermodynamic film coating model based on the first laws of thermodynamics and mass and energy balance principles that predicted exhaust air temperature and relative humidity on the basis of input conditions and (b) a physics-based film coating atomization model that described the performance of atomizers utilized in the tablet coating process. The models were used to establish an acceptable range of process parameters in a new film coater to match the proven acceptable range of operating conditions in the existing pan coaters. These are considered as hybrid models, since each model included some empirical parameters that were fitted using experimental data, to minimize the residual sum of

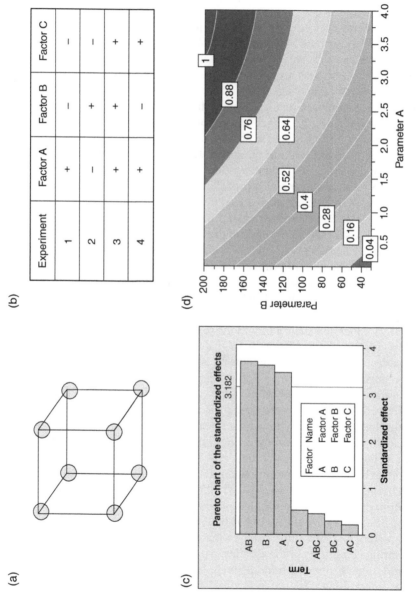

Figure 2.3 Approach for developing a design space based on DOE data. (a) Choose experimental design (e.g., full factorial, d-optimal), (b) conduct randomized experiments, (c) analyze data (determine significant factors), and (d) define design space (e.g., contour surface).

squared error between experimental data and model prediction. The established process parameters were then used to prioritize the experimental design to minimize the number of required trial runs and to support optimization. The recommendations were then provided to the commercial site to guide the design of scale-up trials.

c) *Design space based on integrated multivariate approach*: In another case, a design space was defined on the basis of DOE, optimization, and multivariate analysis (MVA) [12]. Initially, a screening DOE was carried out to identify the parameters that have an impact on the finished product's CQA. Following the screening DOE, an optimization DOE was carried out to evaluate the effects of the design factors on manufacturability and final product's CQA such as tablet blend flow and tablet dissolution and to establish a design space to ensure CQAs. Figure 2.4 is a schematic rendition of the methodology for design space development and implementation, as presented by Huang *et al.* [12].

As illustrated in Figure 2.4, design space was established as a response surface model based on DOE data. In addition, an MVA using principal component analysis (PCA) and partial least squares (PLS) was also carried out using all the variables from the DOE campaigns, to study multivariate relationships between *all* variables that include raw materials, intermediates, various unit operations, and final product. The multivariate techniques were complementary to DOE analysis and provided a representation of all multivariate interactions in the process, based on the combinations of all raw materials and process parameters. Findings from both DOE and MVA were then used to define a control strategy for the product. As elucidated by Huang *et al.*, the combined use of DOE and MVA offers a robust mechanism to explain complex multivariate relationships. Since DOEs in general deal with a limited number of experiments (due to practical limitation in the number of experiments), MVA can be considered as complementary to DOE, providing additional information about the product and processes.

d) *Mechanistic model for scale-up*: In an example by Pandey *et al.* [13], a DEM was developed to study the particle motion in pan coating. DEM simulates the prediction of individual trajectories of particles using constitutive equations. By this approach, movement due to the contact forces from neighboring particles is accounted for. An advantage of this approach is that it allows to study the changes in particle motion due to changes in operating conditions (e.g., pan speed and pan load) as well as particle properties such as tablet size, shape, and density. On the basis of DEM analysis, a modified scale-up relationship for the pan coater was proposed.

e) *Monte Carlo-based models for understanding uncertainty in design space*: A Monte Carlo-based method was applied to simulate the propagation of uncertainty in predictions performed with DOE-based design space models

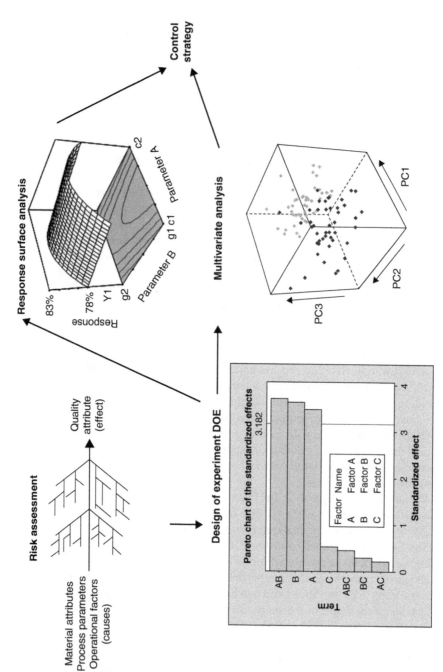

Figure 2.4 Approach for defining a design space based on DOE and MVA.

by Kauffman *et al.* [14]. In this study, the design space was represented by a polynomial model. The results of the simulations presented in this work highlighted two major benefits from the application of Monte Carlo simulation for the propagation of uncertainty in design space models. First, the simulations provided estimates of both the means and standard deviations for the predicted values of CQA. With these quantities in hand, design space was then specified on the basis of model predictions and product quality specifications with statistically meaningful confidence levels. Secondly, the simulations identified the process variable variances that have the greatest influence on the product quality variance, which can be used to prioritize control strategy and process improvement plans.

2.3.4 Control Strategy

In ICH Q10 [15], a control strategy is defined as a set of controls, derived from current product and process understanding, that assure process performance and product quality. ICH Q10 is a model for the pharmaceutical quality system that can be implemented throughout the life cycle of the product. The objective of the control strategy is to ensure that desired quality product will be manufactured. In general, a control strategy includes the following components: specifications for incoming materials and critical intermediates, ranges for process parameters, in-process monitoring and control, end-product testing at release, and other elements as described in Q10 such as change management. Furthermore, a control strategy can evolve/change during the life cycle of a product. Management of these changes is typically handled by the firm's change management procedures.

Models can be used in the implementation of a robust and efficient control strategy. In general, such models have to be updated throughout the life cycle of the product, and procedures for maintenance of these models are typically captured in the firm's change management system. Some examples of these models to support control strategy are presented as follows:

a) *LV-based models for process control*: In the example by Kourti [16], an LV approach is used to support a feed-forward control strategy. Using this approach, when a deviation is detected in the measured quality of an intermediate that could affect finished product quality, a feed-forward control strategy could be used to adjust manufacturing parameters to produce the desired quality of the finished product. An example of this approach is adjustment of tablet compression parameters based on granule bulk density. An LV model is built from multiple batch data to relate the finished product quality in terms of compression parameter settings and intermediate material attributes, for example, granule density. This LV model, as illustrated in Figure 2.5, shows the interaction between compression process parameters and intermediate material attributes [17].

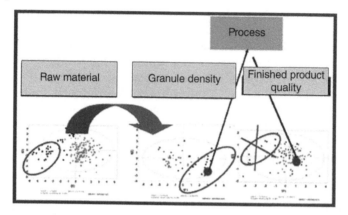

Figure 2.5 Approach for developing LV models for feed-forward control. *Source*: Kourti [17]. Reproduced with permission of John Wiley & Sons.

In another example by García-Munoz *et al.* [18], LV models were used to set quality-driven specifications for incoming raw materials. Such specifications accounted for the inherent variability in the process and the combined effect of materials with process conditions onto product quality. Additionally, Rathore *et al.* [19] demonstrated the usefulness of multivariate data analysis techniques for optimizing biopharmaceutical manufacturing, process scale-up, process comparability, and process optimization.

b) *Multivariate statistical process control (MSPC) model to support real-time release testing (RTRT)*: In ICH Q8 (R2), RTRT is defined as the ability to evaluate and ensure the quality of in-process and/or final product based on process data, which typically include a valid combination of measured material attributes and process controls. Skibsted *et al.* [20] have demonstrated how two MSPC-based models derived from data measured by two NIR instruments were used to provide an early warning during granulation and to separate good batches from potentially bad batches.

2.4 General Scientific Considerations for Model Development

Model building typically includes the following steps [21]. These steps are usually executed in a sequential manner, but many times it may be necessary to return to an earlier step, thus imparting an iterative nature of this process. The overall steps are as follows:

- Defining the purpose/objective of the model.
- Deciding on the type of modeling approach (e.g., mechanistic, empirical, or hybrid) and the experimental methodology that would be used to support

the model development. Since any model is based on a number of assumptions, it is important to understand at this stage the limitations of these assumptions in order to correctly design the experiments and to interpret the model results.

- Collecting experimental data to support model development.
- Developing model relationships, based on the scientific understanding of the process and the collected experimental data.
- Assessing the validity of the model prior to implementation, by both internal metrics and external validation.
  - o Internal validation involves comparing model prediction with the actual values, using the same data set that was used to build the model. Various techniques such as cross-validation, random (Monte Carlo) resampling, and boot strapping can be used for internal validation [20].
  - o External validation involves verification of model results with independent data set(s), that is, data that was not used to build the model. Verification of the model with an appropriate data set is especially important for empirical models to demonstrate the robustness of such models. Model validity is typically measured in terms of goodness of fit.
- Documenting model results including initial assumptions and developing plans for maintaining and updating the model throughout the life cycle of the product.

Additionally, some specific considerations are warranted when considering the implementation of models for specific purposes, as discussed in the following text.

2.4.1 Models for Process Characterization

Process characterization models can include models for process optimization (e.g., reaction kinetics model), design space determination, and scale-up. Since the term design space in general refers to a multidimensional hyperspace, it is important that models defining a design space consider multivariate interactions. In addition, for both mechanistic and empirical models, significant uncertainty can exist in the model predictions, due to the underlying assumptions and simplifications used in model derivation, variabilities in measurements in the supportive data, and error in the model fit. Evaluation of uncertainty in a design space model can lead to a more robust design space and can help identify appropriate risk mitigation steps when moving to areas of uncertainty.

Typically, if a model to define design space is developed based on laboratory or pilot scale data, it is then verified at commercial scale. Verification approaches in general consider the scale dependencies of the model parameters, the modeling approach (i.e., mechanistic or empirical), and the control strategy.

2.4.2 Models for Supporting Analytical Procedures

This category includes models used to support various PAT-based methods. In general, these are data-driven chemometric models such as PCA or PLS. These models often have the flexibility to handle noisy measurements, missing data sets, and highly correlated variables. A primary consideration when developing such models is the quality of data used to derive and to validate the model. To make a robust chemometric model, the data set should include an appropriate range of variability. For example, chemometric models meant to span a design space should contain data representative of variations within the design space.

2.4.3 Models for Process Monitoring and Control

When developing models for process monitoring and control, it is important to consider all pertinent quality attributes and in-process measurements using techniques such as risk assessment. For example, if prediction from an empirical model (e.g., LV model) is used to ensure that the process is manufacturing desired quality product when operating within the design space, it is desired to include all expected sources of variability during the model-defining stage. Including variations helps ensure that the model would be applicable to all regions within the entire design space for occurrences of material and process parameter variability. Alternatively, if the objective is to control the process in a narrow range near the target operating condition using an LV model, the model could be constructed using batches manufactured only near the target condition.

2.5 Scientific Considerations for Maintenance of Models

Typically, models may need to be updated due to an instrument or process drift. Additionally, unaccounted for variability (e.g., changes in raw material) could result in out-of-spec predictions from the model. Consequently, it can be valuable to monitor the performance of the model over the life cycle of the product. An approach for monitoring model performance could include periodic comparison of model prediction with a reference method. This approach would allow making adjustments to the model (e.g., recalibration) before failures occur.

The approach of model maintenance and update is relative to the model implementation strategy (i.e., importance of the model in the control strategy and its potential to affect product quality). Clear metrics for model update can be established depending on the level of risk of the model.

2.6 Conclusion

In the QbD paradigm, mathematical models can be an important tool for leveraging pharmaceutical process understanding and can be applicable throughout development and manufacturing including process development, scale-up, process monitoring, and continual improvement. Use of models can support efficient development and implementation of a robust process that ensures consistent manufacture of desired quality product. Although many such models have been implemented in pharmaceutical process development and manufacture, by and large these modeling approaches are still evolving and more understanding is expected to be garnered in the coming years.

References

1 Singh, R., Gernaey, K.V., Gani, R.. "Model-based computer aided framework for design of process monitoring and analysis systems," Computers and Chemical Engineering, 2009, 33, 22–42.

2 Gernaey, K.V., Gani, R. "A model based systems approach to pharmaceutical product-process design and analysis," Chemical Engineering Science, 2010, 65(21), 5757–5769.

3 Chen, W., Chang, S., Kiang, S., Marchut, A., Lyngberg, O., Wang, J., Rao, V., Desai, D., Stamato, H., and Early, W., "Modeling of pan coating processes: Prediction of tablet content uniformity and determination of critical process parameters," Journal of Pharmaceutical Sciences, 2010, 99(7), 3213–3225.

4 ICH Q8 (R2), Pharmaceutical Development, August 2009, http://www.ich.org/fileadmin/Public_Web_Site/ICH_Products/Guidelines/Quality/Q8_R1/Step4/Q8_R2_Guideline.pdf (accessed April 7, 2017).

5 FDA, "Guidance for Industry, Extended Release Oral Dosage Forms: Development, Evaluation, and Application of In Vitro/In Vivo Correlations," U.S. Department of Health and Human Services, Food and Drug Administration (FDA), Center for Drug Evaluation and Research (CDER), Rockville, 1997, pp. 1–27.

6 Rossi, R.C., Dias, C.L., Donato, E.M., Martins, L.A., Bergold, A.M., Froehlich, P.E. "Development and validation of dissolution test for ritonavir soft gelatin capsules based on in vivo data," International Journal of Pharmaceutics, 2007, 338, 119–124.

7 ICH Q9, Quality Risk Management, November 2005, http://www.ich.org/fileadmin/Public_Web_Site/ICH_Products/Guidelines/Quality/Q9/Step4/Q9_Guideline.pdf (accessed April 7, 2017).

8 Van Leeuwen, J.F., Nauta, M.J., De Kaste, D., Oderkerken-Rombouts, Y.M.C.F., Oldenhof, M.T., Vredenbregt, M.J., Barends, D.M. "Risk analysis by FMEA as an element of analytical validation," Journal of Pharmaceutical and Biomedical Analysis, 2009, 50, 1085–1087.

9 Teng, Y., Qiu, X., Wen, H. "Systematical approach of formulation and process development using roller compaction," European Journal of Pharmaceutics and Biopharmaceutics, 2009, 73, 219–229.

10 Verma, S., Lan, Y., Gokhale, R., and Burgess, D.J. "Quality by design approach to understand the process of nano suspension preparation," Internal Journal of Pharmaceutics, 2009, 377, 185–198.

11 Prpich, A., Am Ende, M.T., Katzschner, T., Lubcyzk, V., Weyhers, H., Bernhard, G. "Drug product modeling predictions for scale-up of tablet film coating- a quality by design approach," Computers and Chemical Engineering, 2010, 34, 1092–1097.

12 Huang, J., Kaul, G., Cai, C., Chatlapalli, R., Hernandez-Abad, P., Ghosh, K., Nagi, A. "Quality by design case study: An integrated multivariate approach to drug product and process development," International Journal of Pharmaceutics, 2009, 382, 23–32.

13 Pandey, P., Song, Y., Kayihan, F., Turton, R. "Simulation of particle movement in a pan coating device using discrete element modeling and its comparison with video imaging," Powder Technology, 2006, 161, 79–88.

14 Kauffman, J.F., Geoffroy, J. "Propagation of uncertainty in process model predictions by Monte Carlo simulation," American Pharmaceutical Review, 2008, 11(July/August), 75–79.

15 ICH Q10, Pharmaceutical Quality System, June 2008, http://www.ich.org/fileadmin/Public_Web_Site/ICH_Products/Guidelines/Quality/Q10/Step4/Q10_Guideline.pdf (accessed April 7, 2017).

16 Kourti, T., "Quality by design in the pharmaceutical industry: The role of multivariate analysis," American Pharmaceutical Review, June 2009, 9, 118–122.

17 Kourti, T. "Pharmaceutical manufacturing: The role of multivariate analysis in design space, control strategy, process understanding, troubleshooting, and optimization," Chemical Engineering in the Pharmaceutical Industry: R&D to Manufacturing, Am Ende, D.J. (editor), John Wiley & Sons, Inc., Hoboken, 2011.

18 García-Munoz, S., Dolph, S., and Ward, II, H. "Handling uncertainty in the establishment of a design space for the manufacture of a pharmaceutical product," Computers and Chemical Engineering, 2010, 34, 1098–1107.

19 Rathore, A., Johnson, R., Yu, O., Ozlem Kirdar, A., Annamalai, A., Ahuja, S., and Ram, K. "Applications of multivariate data analysis in biotech processing," BioPharm International.com, October 2007, 20(10).

20 Skibsted, E.T.S., Westerhuis, J.A., Smilde, A.K., Witte, D.T. "Examples of NIR based real time release in tablet manufacturing," Journal of Pharmaceutical and Biomedical Analysis, 2007, 43, 1297–1305.

21 StatSoft, Inc. "Electronic Statistics Textbook," 2011, http://www.statsoft.com/textbook/ (accessed April 7, 2017).

3

Role of Automatic Process Control in Quality by Design

Mo Jiang[1], Nicholas C. S. Kee[2,3,4], Xing Yi Woo[5], Li May Goh[2], Joshua D. Tice[2], Lifang Zhou[1], Reginald B. H. Tan[3,4], Charles F. Zukoski[2,6], Mitsuko Fujiwara[2], Zoltan K. Nagy[7], Paul J. A. Kenis[2], and Richard D. Braatz[1,2]

[1] Department of Chemical Engineering, Massachusetts Institute of Technology, Cambridge, MA, USA
[2] Department of Chemical and Biomolecular Engineering, University of Illinois at Urbana-Champaign, Urbana, IL, USA
[3] National University of Singapore, Block E5, Singapore
[4] Institute of Chemical and Engineering Sciences, Jurong Island, Singapore
[5] The Jackson Laboratory, Bar Harbor, ME, USA
[6] Department of Chemical and Biological Engineering, University at Buffalo, Buffalo, NY, USA
[7] School of Chemical Engineering, Purdue University, Forney Hall of Chemical Engineering, West Lafayette, IN, USA

3.1 Introduction

The US Food and Drug Administration defines Quality by Design (QbD) as "a systematic approach to pharmaceutical development that begins with predefined objectives and emphasizes product and process understanding and process control, based on sound science and quality risk management" [1]. Pharmaceutical development includes the following elements [2]:

- Defining the target product profile while taking into account quality, safety, and efficacy of the drug compound. This profile considers the method of drug delivery, the form of the drug compound when delivered, the bioavailability of the drug once delivered, the dosage of the drug compound to maximize therapeutic benefits while minimizing side effects, and the stability of the product.
- Identifying potential critical quality attributes of the drug product. All of the drug product characteristics that have an impact on product quality must be determined so these attributes can be studied and controlled.

Comprehensive Quality by Design for Pharmaceutical Product Development and Manufacture,
First Edition. Edited by Gintaras V. Reklaitis, Christine Seymour, and Salvador García-Munoz.

Figure 3.1 A photograph of a bench-scale crystallizer with a jacket for temperature control. Supersaturation in pharmaceutical crystallizers is usually induced by cooling and/or antisolvent addition.

- Determining the critical quality attributes of the components (e.g., drug compound, excipients) needed for the drug product to be of desired quality.
- Selecting a process capable of manufacturing the drug product and its components.
- Identifying a strategy for controlling the process to reliably achieve the target product profile.

Most pharmaceutical manufacturing processes include at least one crystallizer for purification of intermediates or the final active pharmaceutical ingredient (Figure 3.1), and usually for producing crystals to be compacted with excipients to form a tablet. The bioavailability and tablet stability depend on the crystal structure, size, and shape distribution (Figure 3.2). Typical objectives in the operation of intermediate and final crystallizations are to maximize chemical purity and yield and to produce crystals that do not cause operations problems in downstream processing. An important consideration in many crystallization concerns polymorphism, which occurs when a chemical compound can adopt different crystal structures. *Polymorphs* are the same molecular species but have different packing of molecules within the crystal structure (see Figure 3.3 for an example) [3, 4]. Another common occurrence is *solvatomorphism* or *pseudopolymorphism*, which is when varying amounts of solvent molecules can be incorporated within the crystal structure, to form

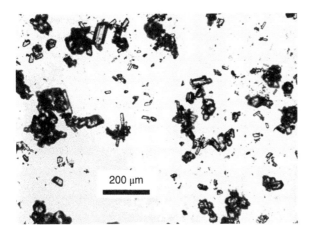

Figure 3.2 A photograph of some pharmaceutical crystals produced from a crystallizer that was inadequately controlled, showing potential variations in size and shape.

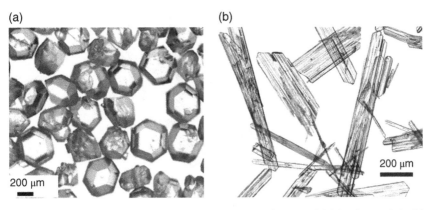

Figure 3.3 (a) α-Polymorph (metastable form, prismatic shape) and (b) β-polymorph (stable form, needlelike platelet shape) crystals of L-glutamic acid. The two polymorphs show different crystal morphologies.

crystal solvates of different composition. Different polymorphs or solvato-morphs can have very different physical properties (e.g., solubility, bioavaila-bility, and shelf-life), and hence very different efficacy. The control of crystal polymorphism is required for the manufacturing process to achieve the QbD objectives and for regulatory compliance. The most stable polymorph is typi-cally achievable with sufficient process time and suitable crystallization con-ditions. Metastable polymorphs are more challenging to manufacture reliably with a key challenge being to prevent transformation to more stable forms.

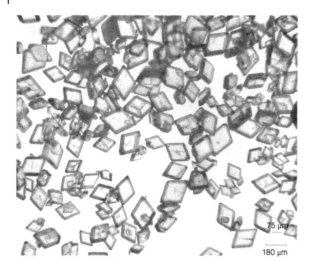

Figure 3.4 An example of a potential target CSD and polymorphic form.

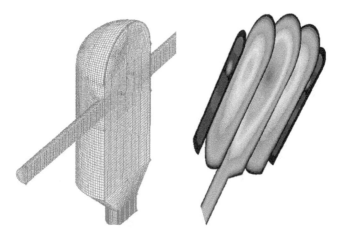

Figure 3.5 Example simulation results of the spatial distribution of turbulent Reynolds number in a confined DIJ mixer. Such mixers can be used to generate sub 10-μm crystals for many solute/solvent systems.

A QbD approach for the development of pharmaceutical crystallization processes involves two consecutive steps:

1) Design a target crystal size distribution (CSD) and polymorphic form based on bioavailability and delivery needs (Figure 3.4).
2) Design a manufacturing-scale crystallizer to produce the target CSD based on simulation models that account for all potential scale-up issues such as nonideal mixing [5] (Figure 3.5).

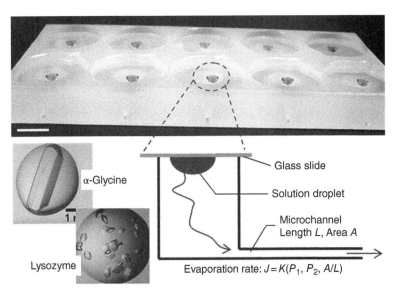

Figure 3.6 A microfluidic platform that uses evaporation to generate supersaturation in 2–10 µl-sized droplets in microwells. The rate of evaporation (*J*), and thereby the supersaturation profile, is set by the dimensions of the channel (cross-sectional area *A* and length *L*) that connects the well with the ambient environment.

Many crystallizer designs for step 2 are based on the creation of a highly intense mixing zone operating at high supersaturation to produce crystal seeds for subsequent growth [6–13]. Step 1 requires first finding all of the polymorphs by crystallizing the pharmaceutical compounds in many different solvents with many different supersaturation methods and trajectories and identifying crystallization kinetics from the small quantity of pharmaceutical compounds available for process development. Such investigations can be facilitated by high-throughput microfluidics platforms that can provide information on polymorphism, solubility, and kinetics even at high supersaturation, which is difficult to achieve in other systems (Figures 3.6 and 3.7) [14–28]. A more advanced version of the high-throughput platform in Figures 3.6 and 3.7 allows information to be collected in parallel using chips comprised of 144 wells (each 30 nl in volume) with crystal detection performed using optical spectroscopy (Figure 3.8). Further developed versions of these microfluidic array chips allow for *in situ* identification of polymorph identity via Raman spectroscopy or X-ray diffraction.

After the target product profile is defined and critical quality attributes identified, a control strategy can be identified. The elements of a control strategy include the following [2]:

- control of the input material attributes, such as the drug substance and excipients, based on an understanding of their impact on the processing or product quality

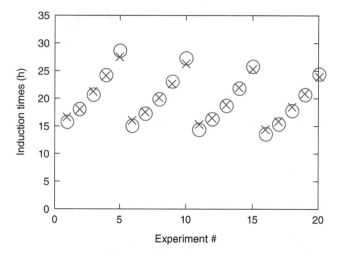

Figure 3.7 A comparison of induction times measured in wells in a microfluidic platform (crosses) with predictions of a first-principles model (circles) with nucleation kinetics estimated from the microfluidic data at high supersaturation. A detailed study demonstrated that these nucleation kinetic parameters resulted in accurate predictions of induction times in subsequent experiments.

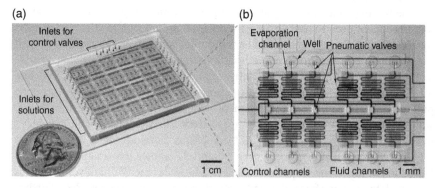

Figure 3.8 A high-throughput microfluidic chip for protein and pharmaceutical crystallization that enables 144 induction time experiments to be conducted in parallel with different precipitants and different solute concentrations: (a) a micrograph of the microfluidic chip, which contains 8 inlets for pressurized control lines, 48 inlets for different precipitants, and one inlet for the target protein or pharmaceutical, and (b) a micrograph of the microfluidic components used for metering solutions and controlling evaporation of solvent [29]. The target molecule and precipitant are metered into serpentine channels, and the ratio of each component is controlled by the placement of a pneumatic valve. After filling, the solutions are dispensed into a well that is connected to an evaporation channel. The dimensions of the evaporation channel control the rate of evaporation.

- product specifications, which define the objective of the control strategy
- controls for unit operations that have an impact on the downstream processing or end-product quality, such as crystallizers in which poor control of the CSD output can result in problems in downstream operations such as washing, filtering, drying, and/or tableting
- a monitoring program such as regular testing of intermediates and product for verifying drug product quality and accuracy of multivariate prediction models

The controls for unit operations include not only the design of feedback controllers but also other considerations associated with the integration of sensors, process design, and control required to move from input materials to optimized process design, such as [30]

- sensors, experimental design, and data analysis
- process automation
- first-principles modeling and simulation
- design of operations to form a consistent product

The remainder of this chapter provides a more detailed description of the QbD approach for the development of pharmaceutical crystallization processes, namely:

- the design of robust control strategies
- some example applications of automatic feedback control
- the role of kinetics modeling
- ideas for a deeper QbD approach

3.2 Design of Robust Control Strategies

A well-studied control objective in crystallization is to minimize the secondary nucleation of crystals that may cause potential problems in downstream filtering operations by minimizing the ratio of nucleation mass to seed mass (n/s) over the temperature profile [31–34],

$$\min_{T(t)} \left\{ \frac{\text{nucleation mass}}{\text{seed mass}} \right\} \tag{3.1}$$

The control objective is related to the temperature profile through a population balance model [35, 36] for the crystals in the slurry:

$$\frac{\partial f}{\partial t} + G(C,T)\frac{\partial f}{\partial L} = B(C,T)\delta(L) \tag{3.2}$$

where $f(L,t)$ is the CSD, which is a function of the crystal size L and time t; G is the growth rate and B is the nucleation rate, which are assumed to be functions of the solute concentration C and the temperature T; and δ is the Dirac delta function. The solute concentration is described by its mass balance

$$\frac{dC}{dt} = -3\alpha G(C,T)\int_0^\infty f(L,t)\,dL \tag{3.3}$$

where α is a conversion factor. The temperature profile must satisfy constraints due to equipment limitations

$$T_{\min} \le T(t) \le T_{\max}$$
$$R_{\min} \le \frac{dT(t)}{dt} \le R_{\max} \tag{3.4}$$

where $T_{\min}$, $T_{\max}$, $R_{\min}$, and $R_{\max}$ are user-specified scalars. The crystallization must also satisfy a constraint to ensure a high enough batch yield:

$$C(t_{\text{final}}) \le C_{\text{final,max}} \tag{3.5}$$

where t_{final} is the time in which the batch ends and $C_{\text{final,max}}$ is the limit on the concentration at t_{final}. A common practice in the crystallization control literature is to solve optimizations (3.1)–(3.5) while ignoring uncertainties in the nucleation and growth kinetics [33, 37, 38].

This example shows that the robustness of the control strategy can be strongly affected by uncertainties. A linear cooling profile results in $n/s = 10.7$, while numerical solution of the optimization (3.1)–(3.5) improves the n/s to 8.5, which is a 21% reduction in the nucleation mass to seed mass (Figure 3.9).

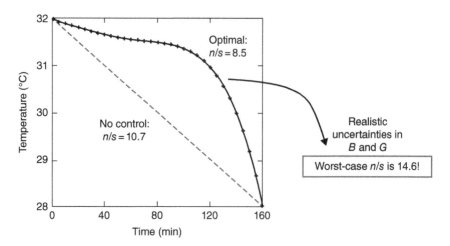

Figure 3.9 Temperature profiles for seeded cooling crystallizations. Optimal control that ignores uncertainties predicts a 21% improvement in n/s (from 10.7 to 8.5) that can be completely lost due to uncertainties (from 10.7 to 14.6).

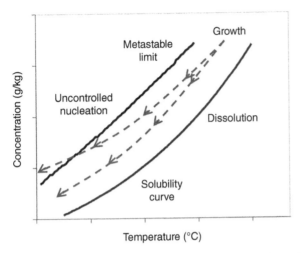

Figure 3.10 Phase diagram for crystal dissolution, growth, and uncontrolled nucleation showing robust and nonrobust operational trajectories. A control strategy that is insensitive to model uncertainties must keep the operational trajectory between the solubility curve and metastable limit.

All of the improvements due to optimization can be lost due to uncertainties in crystallization kinetics. In particular, n/s can be as large as 14.6 for realistic uncertainties in the nucleation and growth kinetics [39], which is much worse than performing no optimization at all. The poor robustness can be understood in terms of the crystallization phase diagram. When only the temperature is measured in the control strategy, uncertainties in the crystallization kinetics can cause the pathway in the phase diagram to drift across the metastable limit, which results in uncontrolled crystallization (upper dashed pathway in Figure 3.10). While not all crystallizations are as sensitive to model uncertainties, this example as well as many others [40–42] illustrates that uncertainties cannot be safely ignored, and it is important to design feedback controllers that are insensitive to these model uncertainties.

State-of-the-art *in situ* sensor technologies [43–71] (Figure 3.11) provide additional information that can be employed in real time to produce more robust control strategies. For example, a strategy that is much less sensitive to uncertainties in the kinetics employs feedback control based on *in situ* solute concentration measurement obtained by ATR-FTIR spectroscopy to follow the desired pathway in the crystallization phase diagram (lower dashed pathway in Figure 3.10). Detailed theoretical analyses indicate that this feedback control strategy more robustly operates the crystallizer at constant growth and low nucleation rates [72, 73], which has been demonstrated in numerous experimental implementations at universities and pharmaceutical companies [29, 72–77].

(a)

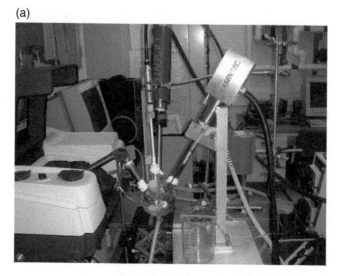

(b)

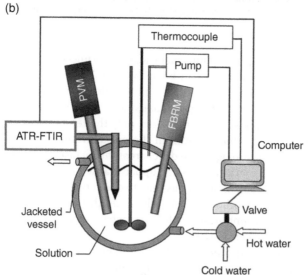

Figure 3.11 (a) A photograph and (b) a schematic of crystallization laboratory setup with state-of-the-art *in situ* sensor technologies that enable the development of robust control strategies. FBRM, focused beam reflectance measurement (laser backscattering), and PVM, particle vision and measurement (*in situ* video microscopy).

3.3   Some Example Applications of Automatic Feedback Control

Figure 3.12 shows some sample results in which automatic feedback control was applied to a Merck pharmaceutical with very high secondary nucleation kinetics [77]. The feedback control system tracks any desired trajectory in the

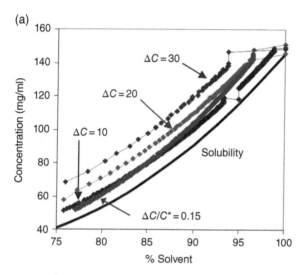

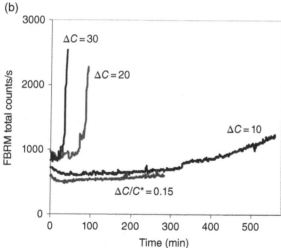

Figure 3.12 Automatic feedback control of the batch crystallization of a Merck compound: (a) concentration control at constant absolute and relative supersaturation; (b) secondary nucleation monitored by FBRM total counts per second.

crystallization phase diagram, including constant absolute or relative supersaturation (Figure 3.12a). The extent of nucleation was simultaneously monitored by tracking the total counts/second measured in real time using *in situ* laser backscattering (Figure 3.12b). The input to the feedback control system was the solubility curve and the metastable limit obtained by an automated system with software implemented in VisualBasic and MATLAB, using a similar crystallizer and sensor technology as shown in Figure 3.11 (without the PVM).

The initial three runs at constant supersaturation were run automatically, followed by inspection of the total counts/second that indicated different secondary nucleation levels at different absolute supersaturation. An absolute supersaturation $\Delta C = 30$ mg/ml resulted in excessive nucleation within an hour, whereas $\Delta C = 20$ mg/ml resulted in excessive nucleation after 70 min. An absolute supersaturation $\Delta C = 10$ mg/ml resulted in a long batch time with a slow increase in nucleation near the end of the batch. These observations motivated the selection of a pathway in the crystallization phase diagram with a constant relative supersaturation of 0.15, which has a similar supersaturation as $\Delta C = 20$ mg/ml in the early portion of the batch in which no nucleation was observed, and a somewhat lower supersaturation at 10 mg/ml near the end of the batch to suppress nucleation then. The constant relative supersaturation pathway resulted in about half of the batch time as the $\Delta C = 10$ mg/ml pathway while also having less nucleation. The automated system enabled a rapid convergence to controlled batch crystallization operations with a minimum user input. The temperature time profile resulting from the feedback control of concentration can be implemented subsequently using simple temperature controllers.

Another demonstration of automatic feedback control is for the crystallization of polymorphs of L-glutamic acid (Figures 3.3 and 3.13). The α-polymorph is easier to process in downstream operations, whereas manufacture of the stable β-polymorph crystals is more reliably produced by operating between the solubility curves of the two polymorphs (Figure 3.14). Operating an unseeded crystallizer at various cooling rates produced crystals of both polymorphs, with widely varying proportions (Figure 3.15). The objective of this experimental study was to produce crystals of pure single polymorphic form.

Pure stable β-polymorph crystals in an unseeded crystallizer can be achieved at any cooling rate, as long as the operations in the last portion of the batch remain for a long enough time between the two solubility curves (Figure 3.16). The batch time can be reduced by maximizing the supersaturation while operating between the two solubility curves, by increasing the temperature after the initial mixture of α- and β-polymorph crystals is produced by rapid cooling. The products were large β-polymorph crystals (Figure 3.16).

The manufacture of the metastable α-polymorph crystals requires operation above the solubility curves for both forms (Figure 3.14), but not at such a high supersaturation that nucleation of the stable β-polymorph crystals occurs.

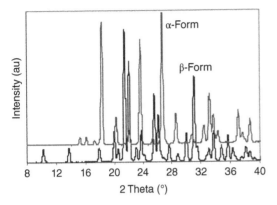

Figure 3.13 Powder X-ray diffraction patterns for the metastable α (upper curve) and stable β (lower curve) polymorphic forms of L-glutamic acid.

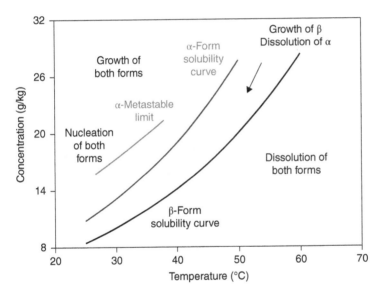

Figure 3.14 Crystallization phase diagram for L-glutamic acid. For such a monotropic system, β-form crystals grow and α-form crystals dissolve in the region between the solubility curves of the α- and β-forms.

The most reliable operation is to seed with the metastable crystals while cooling to operations above the solubility of the α-polymorph while remaining below the metastable limit (Figure 3.17). Seeding with α-polymorph crystals is required because the initial unseeded operation produced a mixture of α- and β-polymorph crystals for the full range of cooling rates (Figure 3.15), which would always result in a large quantity of β-polymorph crystals at the end of the batch.

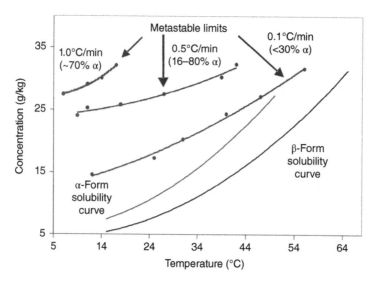

Figure 3.15 Crystallization phase diagram for L-glutamic acid, showing metastable limits at various cooling rates (0.1, 0.5, and 1.0°C/min) with corresponding polymorphic portions of nuclei.

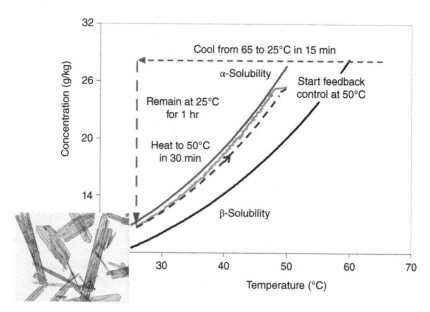

Figure 3.16 Controlled operations to manufacture the L-glutamic acid stable β-polymorph. The operation begins with a quick cooling from 65 to 25°C to generate nuclei, followed by holding the temperature at 25°C for 1 h to deplete supersaturation to maximize surface area of crystals, then a quick heating to 50°C (while operating between the two solubility curves) to reach a much higher temperature to increase the growth rate of the β-form crystals. The last step is to apply concentration feedback control (solid lines) to grow larger β-form product crystals.

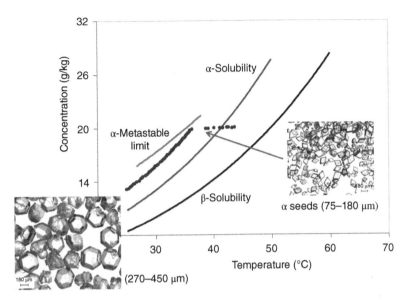

Figure 3.17 Controlled operations to manufacture a pure batch of metastable α-form crystals of ʟ-glutamic acid. α-Form seeds (75–180 μm in size) are grown to large sizes (270–450 μm in size) by following an operational trajectory between the α-form solubility curve and the α-form metastable limit.

Automatic feedback control can be employed to force the crystallizer to follow a desired pathway within the desired operating window of the crystallization phase diagram, to produce large metastable crystals without the addition of any additives (Figure 3.17) [78]. The same approach has been used to robustly manufacture stable and metastable solvatomorphs [79–81]. These experimental implements also demonstrate how *in situ* sensor technologies can be used to design and implement a robust control strategy.

The feedback control of concentration to follow a target pathway in the crystallization phase diagram (Figure 3.18b) has several advantages over the application of a predetermined temperature or antisolvent addition profile (Figure 3.18a) [72, 73, 82]:

- lower costs and process development times
- lower trial-and-error experimentation
- insensitivity to most variations in kinetics and most disturbances

Concentration feedback control can result in high variability in the batch time, which is indirectly used as an extra degree of freedom to produce the reduced sensitivity. A drawback of both control approaches in Figure 3.18, as well as nearly all other crystallization control approaches described in the literature, is sensitivity to variations in the solubility due to impurities. The simplest way to

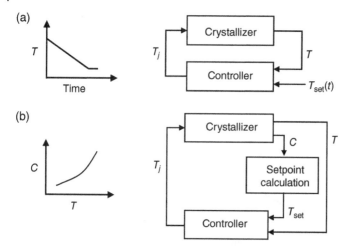

Figure 3.18 Implementations of (a) temperature control and (b) concentration feedback control for cooling crystallization. The implementations for antisolvent crystallization are similar, with T replaced by % solvent.

deal with impurities is to have fine control of the input material attributes, which is one of the tenets of the QbD approach [2]. Unfortunately, often the chemical feed streams for pharmaceutical production are outsourced, and suppliers are switched or remain the same but modify their operations, which results in a change in the impurity profile for the input materials to the crystallizer. For pharmaceutical–solvent combinations in which only one form occurs, alternative feedback control strategies have been developed that employ *in situ* laser backscattering measurements so as to be insensitive to variations in the solubility as well as other disturbances [83–85]. It will be interesting to see whether such approaches can be modified to reduce their weaknesses, such as reduced molecular purity of the product crystals, and how well these approaches can be extended to various types of polymorphic crystallizations.

3.4   The Role of Kinetics Modeling

The aforementioned robust control strategies are insensitive to variations in crystallization kinetics and actually did not require a determination of explicit expressions for the crystallization kinetics. Instead, information on nucleation and growth kinetics were incorporated implicitly through the position of the metastable limit within the crystallization phase diagram (Figure 3.10). Explicit models for the crystallization kinetics can still be useful, in particular, in improving the characteristics of the seed crystals and in reducing risk during scale-up from bench to manufacturing scale. The quantity and size distribution

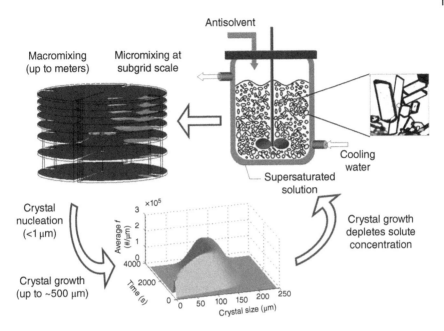

Figure 3.19 A manufacturing-scale crystallizer couples phenomena (micromixing, macromixing, nucleation, and growth) over a wide range of length scales (submicron to meter scale).

of seed crystals can be optimized from first-principles population balance models that incorporate the crystallization kinetics [32]. These same models can be coupled with computational fluid dynamics (CFD) simulations to identify potential scale-up problems, revise batch operations (such as antisolvent addition rates) to reduce effects of nonideal mixing at the manufacturing scale, and optimize process equipment such as the geometries and physical locations of baffles and feed streams [5, 86].

Nonideal mixing is a common scale-up problem in batch crystallizers. Many industrial researchers apply vanilla CFD simulation to try to describe some of the effects of imperfect mixing, but simulation codes that do not include a population balance model for the crystals do not adequately describe all of the crystallization phenomena important in manufacturing operations. A typical manufacturing-scale crystallizer as shown in Figure 3.19 involves processes that encompass a wide range of length scales, which include crystal nucleation (<1 μm), crystal growth (from submicron to hundreds of microns), micromixing (less than the size of a grid cell in a CFD simulation), and macromixing (up to meters)—such systems are *multiscale* [87]. High-resolution simulations [88, 89] of pharmaceutical crystallizers that couple micromixing, macromixing, and population balance models can provide insights that would be difficult to observe in experiments. For example, simulation of an antisolvent crystallization showed increased nucleation near the mixing blade at intermediate times, which was due to increased mixing intensity [86]. Observing the source of nuclei is rarely possible in a large-scale crystallizer.

3.5 Ideas for a Deeper QbD Approach

The previous discussions were mostly focused on the design of a crystallizer that manufactures large uniform-sized crystals while minimizing nucleation, which is motivated by the fact that large crystals are easier to wash, filter, and dry. Most crystallizers in the pharmaceutical industry add seeds at the beginning of batch, with the remainder of the control strategy specified by optimization of cooling and/or antisolvent addition.

An idea explored has been the continuous addition of seed crystals during a batch, to provide extra degrees of freedom to enable the manufacture of crystals with a target size distribution that is not necessarily large uniform-size crystals [6]. For many pharmaceutical–solvent combinations, the application of high supersaturation enables the manufacture of very small seed crystals with a narrow size distribution, which can be coupled with a jacketed stirred-tank crystallizer to subsequently grow these crystals to a desired size (Figure 3.20). For some systems, high-quality seed crystals can be manufactured by performing cooling crystallization within a DIJ mixer, in which the fluids in the two jets are saturated solutions of the same solvent at two different temperatures.

For either mode of supersaturation generation, many target CSDs can be achieved by manipulating seed generation continuously using a DIJ mixer [6]. The slurry exiting the DIJ mixer is directed into the aging vessel with crystal growth rate optimally controlled to get as close as possible to the target size distribution. A unimodal CSD is easily obtained by continuously manipulating the inlet jet

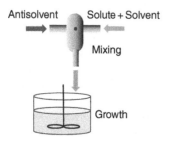

Figure 3.20 A schematic of antisolvent crystallization occurring within a confined DIJ mixer operating at high supersaturation that continuously produces seed crystals that enter a stirred tank. The high supersaturation in the DIJ mixer is generated in a mixing zone where opposing streams from two jets (containing saturated solution and antisolvent, respectively) collide. A feedback control system operates the stirred-tank crystallizer at lower supersaturation so that only crystal growth (no nucleation) occurs.

(a)

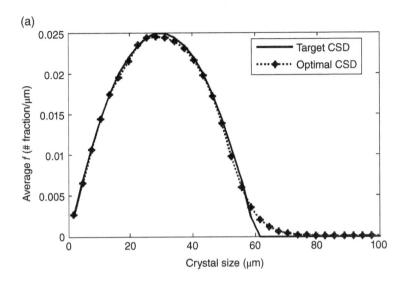

(b)

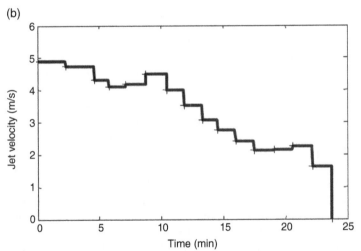

Figure 3.21 (a) Comparison of a unimodal target CSD (line) with the CSD (dash star) achieved by optimization of continuously varying inlet velocities to a DIJ mixer coupled to an stirred tank (simulation results). (b) The time profile of inlet jet velocities for the simulated achievable CSD, with the jets constrained to have the same inlet velocity.

velocities of the DIJ mixer throughout the batch, to continuously vary the seeding (Figure 3.21). The inlet velocities were computed by solution of the optimization:

$$\min_{v_{jet}(t)} \left\| f_{target}\left(L\right) - f\left(L, t_{final}\right) \right\| \tag{3.6}$$

where f_{target} is the target CSD, v_{jet} is the inlet velocities for the two jets, and $\| \; \|$ is the integral of the square of its argument over the crystal dimension L. This optimization was numerically solved subject to the constraints

$$\frac{\partial f(L,t)}{\partial t} + G\frac{\partial f(L,t)}{\partial L} = f_{jet}\left(L,t; v_{jet}(t)\right) \tag{3.7}$$

$$v_{jet,min} \le v_{jet}(t) \le v_{jet,max} \tag{3.8}$$

where f_{jet} is the CSD of the crystals existing in the DIJ mixer, which was computed using a coupled model for micromixing, macromixing, and population balance within the DIJ mixer [6],[1] and $v_{jet,min}$ and $v_{jet,max}$ are user-specified scalars set to satisfy constraints on the DIJ mixer operation. Equation (3.7) is the population balance model for the stirred tank operating under a constant growth rate by using feedback control of the concentration.

The largest deviation between the target CSD and the nearest achievable CSD in Figure 3.21 occurs at the largest crystal size where the CSD drops to zero. Simulation results for other target CSDs also show the largest deviation occurring wherever the target CSD is sharp. The most challenging unimodal target CSD is a uniform distribution, with the optimal inlet jet velocities being close to a uniform distribution—a constant value of particle numbers per unit length between some minimum and maximum size (Figure 3.22). Theoretical analysis [6] indicates that the proposed control strategy is reasonably robust to variations in the inlet jet velocity profiles, which was tested experimentally by attempting to achieve a nearly uniform CSD by using a constant inlet jet velocity followed by a sharp drop to zero, to approximate the profile in Figure 3.22b. The experimental implementation resulted in a CSD very close to the CSD determined by simulations to most closely match a uniform distribution (see Figure 3.23) [90]. Similar to the simulated achievable CSD, the experimentally achieved CSD also has the largest deviation with the target uniform distribution at the largest crystal size (sharp side). In addition to the inlet jet velocity (Figures 3.21b and 3.22b), the supersaturation trajectory can be optimized over time, to take advantage of the ability to robustly control the supersaturation (Figures 3.12a and 3.17) [91].

3.6 Summary

This chapter discussed the role and application of automatic control in QbD with examples taken from pharmaceutical crystallization. Microfluidic platforms were discussed as a technology for early-stage pharmaceutical process

---

1 The same optimization formulation applies whether f_{jet} is determined experimentally or by simulation.

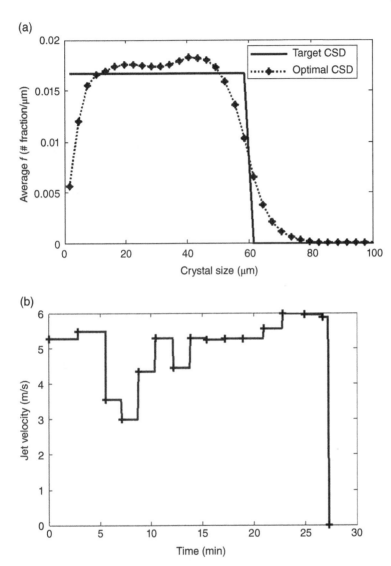

Figure 3.22 (a) Comparison of a uniform target CSD (line) with the CSD (dash star) achieved by optimization of continuously varying inlet velocities to a DIJ mixer coupled to an aging tank (simulation results). (b) The time profile of inlet jet velocities for the simulated achievable CSD, with the jets constrained to have the same inlet velocity.

development to screen for polymorphs, identify desirable process parameters such as choice of solvents, and rapidly determine key properties such as solubility and crystallization kinetics. Throughout the drug development and manufacturing processes, automatic feedback control can be employed together with microfluidic

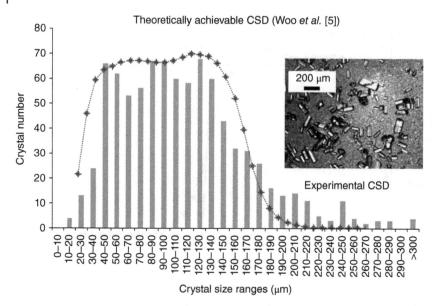

Figure 3.23 A nearly uniform crystal size distribution (CSD) for L-asparagine monohydrate obtained by coupling a cooling DIJ mixer with a stirred-tank crystallizer controlled to follow a preset constant supersaturation profile. The product crystals are shown in the photograph (scale bar = 200 μm). The experimentally measured CSD (histogram) is very close to the CSD predicted by Woo *et al.* [5] (line star) as being the closest that is achievable to a uniform distribution. (The width of the CSD depends on the crystallization system (e.g., compound/solvent combination and growth rate), whereas the overall shape does not.)

platforms and/or process sensor technology to reduce process/product sensitivity to uncertainties and disturbances for control of properties such as CSD and polymorphism. The key points were illustrated by simulation results and experimental implementations. Kinetics modeling was described as a step toward improving process designs, especially for reducing risk during scale-up by quantifying the effects of nonideal mixing on product quality. A deeper QbD approach to pharmaceutical crystallization was described that designs controlled operations to obtain a target product CSD.

Acknowledgments

Financial support is acknowledged from OSIsoft, Eli Lilly Pharmaceuticals, AbbVie Pharmaceuticals, Bristol-Myers Squibb, and the Singapore Agency for Science, Technology, and Research.

References

1 Lionberger RA, Lee SL, Lee L, Raw A, Yu LX. Quality by design: Concepts for ANDAs. AAPS J. 2008;10(2):268–276.

2 International Committee on Harmonization. *ICH Harmonized Tripartite Guideline: Pharmaceutical Development Q8(R2).* August 2009, pp. 10 and 14.

3 Brittain HG, editor. Polymorphism in Pharmaceutical Solids, 2nd edition, Informa Healthcare, London/New York, 2009.

4 Bernstein J. Polymorphism in Molecular Crystals. International Union of Crystallography, Chester, England.

5 Woo XY, Tan RBH, Braatz RD. Modeling and computational fluid dynamics-population balance equation-micromixing simulation of impinging jet crystallizers. Cryst. Growth Des. 2009;9(1):156–164.

6 Woo XY, Tan RBH, Braatz RD. Precise tailoring of the crystal size distribution by controlled growth and continuous seeding from impinging jet crystallizers. CrystEngComm. 2011;13:2006–2014.

7 Midler M, Paul EL, Whittington EF, Futran M, Liu PD, Hsu J, Pan S-H. Crystallization method to improve crystal structure and size. US Patent 5,314,506. 1994.

8 am Ende DJ, Crawford TC, Weston NP. Reactive crystallization method to improve particle size. US Patent 6,558,435. 2003.

9 Dauer R, Mokrauer JE, McKeel WJ. Dual jet crystallizer apparatus. US Patent 5,578,279. 1996.

10 Lindrud MD, Kim S, Wei C. Sonic impinging jet crystallization apparatus and process. US Patent 6,302,958. 2001.

11 Lindenberg C, Mazzotti M. Experimental characterization and multi-scale modeling of mixing in static mixers. Part 2. Effect of viscosity and scale-up. Chem. Eng. Sci. 2009;64(20):4286–4294.

12 Chiou H, Chan HK, Prud'homme RK, Raper JA. Evaluation on the use of confined liquid impinging jets for the synthesis of nanodrug particles. Drug. Dev. Ind. Pharm. 2008;34(1):59–64.

13 Mahajan AJ, Kirwan DJ. Nucleation and growth-kinetics of biochemicals measured at high supersaturations. J. Cryst. Growth. 1994;144(3–4): 281–290.

14 Talreja S, Kim DY, Mirarefi AY, Zukoski CF, Kenis PJA. Screening and optimization of protein crystallization conditions through gradual evaporation using a novel crystallization platform. J. Appl. Crystallogr. 2005;38:988–995.

15 Goh LM, Chen KJ, Bhamidi V, He G, Kee NCS, Kenis PJA, Zukoski CF, Braatz RD. A stochastic model for nucleation kinetics determination in droplet-based microfluidic systems. Cryst. Growth Des. 2010;10(6):2515–2521.

16 Anderson MJ, Hansen CL, Quake SR. Phase knowledge enables rational screens for protein crystallization. PNAS. 2006;103(45):16746–16751.

17 Hansen CL, Skordalakes E, Berger JM, Quake SR. A robust and scalable microfluidic metering method that allows protein crystal growth by free interface diffusion. PNAS. 2002;99(26):16531–16536.

18 Alvarez AJ, Singh A, Myerson AS. Polymorph screening: Comparing a semi-automated approach with a high throughput method. Cryst. Growth Des. 2009;9(9):4181–4188.

19 Kim K, Lee IS, Centrone A, Hatton AT, Myerson AS. Formation of nanosized organic molecular crystals on engineered surfaces. J. Am. Chem. Soc. 2009;131(51):18212+.

20 Li L, Ismagilov RF. Protein crystallization using microfluidic technologies based on valves, droplets and SlipChip. Annu. Rev. Biophys. 2010;39:139–158.

21 He G, Tan RBH, Kenis PJA, Zukoski CF. Generalized phase behavior of small molecules and nanoparticles. J. Phys. Chem. B. 2007;111(43):12494–12499.

22 He G, Tan RBH, Kenis PJA, Zukoski CF. Metastable states of small-molecule solutions. J. Phys. Chem. B. 2007;111(51):14121–14129.

23 Talreja S, Kenis PJA, Zukoski CF. A kinetic model to simulate protein crystal growth in an evaporation-based crystallization platform. Langmuir. 2007;23(8):4516–4522.

24 Talreja S, Perry SL, Guha S, Bhamidi V, Zukoski CF, Kenis PJA. Determination of the phase diagram for soluble and membrane proteins. J. Phys. Chem. B. 2010;114(13):4432–4441.

25 Kee NCS, Woo XY, Goh LM, Rusli E, He G, Bhamidi V, Tan RBH, Kenis PJA, Zukoski CF, Braatz RD. Design of crystallization processes from laboratory research and development to the manufacturing scale: Part I. Am. Pharm. Rev. 2008;11(6):110–115.

26 Kee NCS, Woo XY, Goh LM, Rusli E, He G, Bhamidi V, Tan RBH, Kenis PJA, Zukoski CF, Braatz RD. Design of crystallization processes from laboratory research and development to the manufacturing scale. Part II. Am. Pharm. Rev. 2008;11(7):66–74.

27 Kenis PJA, Tice JD, Perry SL, Roberts GW, Talreja S. Microfluidic chips for membrane protein crystallization. *Eleventh International Conference on Miniaturized Systems for Chemistry and Life Sciences (μTAS 2007)*, Paris, France, October 7–11, 2007.

28 Chen K, Goh LM, He GW, Bhamidi V, Kenis PJA, Zukoski CF, Braatz RD. Identification of nucleation rates in droplet-based microfluidic systems. Chem. Eng. Sci. 2012;77:235–241.

29 Cote A, Zhou G, Stanik M. A novel crystallization methodology to ensure isolation of the most stable crystal form. Org. Proc. Res. Dev. 2009;13(6):1276–1283.

30 Yu LX, Lionberger RA, Raw AS, D'Costa R, Wu H, Hussain AS. Applications of process analytical technology to crystallization processes. Adv. Drug. Deliv. Rev. 2004;56(3):349–369.

31 Matthews HB, Rawlings JB. Batch crystallization of a photochemical: Modeling, control, and filtration. AIChE J. 1998;44(5):1119–1127.

32 Chung SH, Ma DL, Braatz RD. Optimal seeding in batch crystallization. Can. J. Chem. Eng. 1999;77(3):590–596.

33 Corriou JP, Rohani S. A new look at optimal control of a batch crystallizer. AIChE J. 2008;54(12):3188–3206.

34 Sarkar D, Rohani S, Jutan A. Multi-objective optimization of seeded batch crystallization processes. Chem. Eng. Sci. 2006;61(16):5282–5295.

35 Hulbert HM, Katz S. Some problems in particle technology: A statistical mechanical formulation. Chem. Eng. Sci. 1964;19:555–574.

36 Randolph AD, Larson MA. Theory of Particulate Processes: Analysis and Techniques of Continuous Crystallization. Academic Press, New York, 1971.

37 Rawlings JB, Miller SM, Witkowski WR. Model identification and control of solution crystallization processes—A review. Ind. Eng. Chem. Res. 1993;32(7):1275–1296.

38 Braatz RD, Hasebe S. Particle size and shape control in crystallization processes. In: Rawlings JB, Ogunnaike BA, and Eaton JW, eds. Sixth International Conference on Chemical Process Control, AIChE Symposium Series. AIChE Press, New York, 2002;vol. 98, issue 326, pp 307–327.

39 Nagy ZK, Braatz RD. Distributional uncertainty analysis using power series and polynomial chaos expansions. J. Process Control. 2007;17(3):229–240.

40 Hermanto MW, Chiu MS, Woo XY, Braatz RD. Robust optimal control of polymorphic transformation in batch crystallization. AIChE J. 2007;53(10):2643–2650.

41 Nagy ZK, Braatz RD. Worst-case and distributional robustness analysis of finite-time control trajectories for nonlinear distributed parameter systems. IEEE Trans. Control Syst. Technol. 2003;11(5):694–704.

42 Nagy ZK, Braatz RD. Open-loop and closed-loop robust optimal control of batch processes using distributional and worst-case analysis. J. Process Control. 2004;14(4):411–422.

43 Fevotte, G. *In situ* Raman spectroscopy for in-line control of pharmaceutical crystallization and solids elaboration processes: A review. Chem. Eng. Res. Des. 2007;85(A7):906–920.

44 Thompson DR, Kougoulos E, Jones AG, Wood-Kaczmar MW. Solute concentration measurement of an important organic compound using ATR-UV spectroscopy. J. Cryst. Growth. 2005;276(1–2):230–236.

45 Dunuwila DD, Carroll LB, Berglund KA. An investigation of the applicability of attenuated total-reflection infrared-spectroscopy for measurement of solubility and supersaturation of aqueous citric-acid solutions. J. Cryst. Growth. 1994;137(3–4):561–568.

46 Feng LL, Berglund KA. ATR-FTIR for determining optimal cooling curves for batch crystallization of succinic acid. Cryst. Growth Des. 2002;2(5):449–452.

47 Billot P, Couty M, Hosek P. Application of ATR-UV spectroscopy for monitoring the crystallisation of UV absorbing and nonabsorbing molecules. Org. Proc. Res. Dev. 2010;14(3):511–523.

48 Fevotte, G. New perspectives for the on-line monitoring of pharmaceutical crystallization processes using *in situ* infrared spectroscopy. Int. J. Pharm. 2002;241(2):263–278.

49 Grön H, Mougin P, Thomas A, White G, Wilkinson D, Hammond RB, Lai XJ, Roberts KJ. Dynamic in-process examination of particle size and crystallographic form under defined conditions of reactant supersaturation as associated with the batch crystallization of monosodium glutamate from aqueous solution. Ind. Eng. Chem. Res. 2003;42(20):4888–4898.

50 Togkalidou T, Fujiwara M, Patel S, Braatz RD. Solute concentration prediction using chemometrics and ATR-FTIR spectroscopy. J. Cryst. Growth. 2001;231(4):534–543.

51 De Anda JC, Wang XZ, Lai X, Roberts KJ. Classifying organic crystals via in-process image analysis and the use of monitoring charts to follow polymorphic and morphological changes. J. Process Control. 2005;15(7):785–797.

52 Wang XZ, De Anda JC, Roberts KJ. Real-time measurement of the growth rates of individual crystal facets using imaging and image analysis—A feasibility study on needle-shaped crystals of L-glutamic acid. Chem. Eng. Res. Des. 2007;85(A7):921–927.

53 Parsons AR, Black SN, Colling R. Automated measurement of metastable zones for pharmaceutical compounds. Chem. Eng. Res. Des. 2003;81(A6):700–704.

54 Gunawan R, Ma DL, Fujiwara M, Braatz RD. Identification of kinetic parameters in multidimensional crystallization processes. Int. J. Modern Physics B. 2002;16(1–2):367–374.

55 Wong SW, Georgakis C, Botsaris GD, Saranteas K, Bakale R. Online estimation and monitoring of diastereomeric resolution using FBRM, ATR-FTIR, and Raman spectroscopy. Ind. Eng. Chem. Res. 2008;47(15): 5576–5584.

56 Pollanen K, Hakkinen AW, Reinikainen SP, Louhi-Kultanen A, Nystrom L. A study on batch cooling crystallization of sulphathiazole—Process monitoring using ATR-FTIR and product characterization by automated image analysis. Chem. Eng. Res. Des. 2006;84(A1):47–59.

57 Togkalidou T, Braatz RD, Johnson BK, Davidson O, Andrews A. Experimental design and inferential modeling in pharmaceutical crystallization. AIChE J. 2001;47(1):160–168.

58 Worlitschek J, Mazzotti M. Model-based optimization of particle size distribution in batch-cooling crystallization of paracetamol. Cryst. Growth. Des. 2004;4(5):891–903.

59 Larsen PA, Patience DB, Rawlings JB. Industrial crystallization process control. IEEE Control Syst. Mag. 2006;26(4):70–80.

60 Togkalidou T, Tung HH, Sun Y, Andrews AT, Braatz RD. Parameter estimation and optimization of a loosely bound aggregating pharmaceutical crystallization using *in situ* infrared and laser backscattering measurements. Ind. Eng. Chem. Res. 2004;43(19):6168–6181.

61 Lewiner F, Fevotte G, Klein JP, Puel F. Improving batch cooling seeded crystallization of an organic weed-killer using on-line ATR FTIR measurement of supersaturation. J. Cryst. Growth. 2001;226(2–3):348–362.

62 Barrett P, Smith B, Worlitschek J, Bracken V, O'Sullivan B, O'Grady D. A review of the use of process analytical technology for the understanding and optimization of production batch crystallization processes. Org. Proc. Res. Dev. 2005;9(3):348–355.

63 Kempkes M, Eggers J, Mazzotti M. Measurement of particle size and shape by FBRM and *in situ* microscopy. Chem. Eng. Sci. 2008;63(19):4656–4675.

64 Alatalo HM, Hatakka H, Louhi-Kultanen M, Kohonen J, Reinikainen SP. Closed-loop control of reactive crystallization. Part I: Supersaturation-controlled crystallization of L-glutamic acid. Chem. Eng. Technol. 2010;33(5):743–750.

65 Chen ZP, Morris J, Borissova A, Khan S, Mahmud T, Penchev R, Roberts KJ. On-line monitoring of batch cooling crystallization of organic compounds using ATR-FTIR spectroscopy coupled with an advanced calibration method. Chemom. Intell. Lab. Syst. 2009;96(1):49–58.

66 Borissova A, Khan S, Mahmud T, Roberts KJ, Andrews J, Dallin P, Chen ZP, Morris J. *In situ* measurement of solution concentration during the batch cooling crystallization of L-glutamic acid using ATR-FTIR spectroscopy coupled with chemometrics. Cryst. Growth Des. 2009;9(2):692–706.

67 Simon LL, Nagy ZK, Hungerbuhler K. Endoscopy-based *in situ* bulk video imaging of batch crystallization processes. Org. Proc. Res. Dev. 2009;13(6):1254–1261.

68 Simon LL, Nagy ZK, Hungerbuhler K. Comparison of external bulk video imaging with focused beam reflectance measurement and ultra-violet visible spectroscopy for metastable zone identification in food and pharmaceutical crystallization processes. Chem. Eng. Sci. 2009;64(14):3344–3351.

69 Simon LL, Oucherif KA, Nagy ZK, Hungerbuhler K. Bulk video imaging based multivariate image analysis, process control chart and acoustic signal assisted nucleation detection. Chem. Eng. Sci. 2010;65(17):4983–4995.

70 Chow K, Tong HHY, Lum S, Chow AHL. Engineering of pharmaceutical materials: An industrial perspective. J. Pharm. Sci. 2008;97(8):2855–2877.

71 Hukkanen EJ, Braatz RD. Measurement of particle size distribution in suspension polymerization using in situ laser backscattering. Sens. Actuators B. 2003;96:451–459.

72 Nagy ZK, Fujiwara M, Chew JW, Braatz RD. Comparative performance of concentration and temperature controlled batch crystallizations. J. Process Control. 2008;18(3–4):399–407.

73 Fujiwara M, Nagy ZK, Chew JW, Braatz RD. First-principles and direct design approaches for the control of pharmaceutical crystallization. J. Process Control. 2006;15(5):493–504.

74 Fujiwara M, Chow PS, Ma DL, Braatz RD. Paracetamol crystallization using laser backscattering and ATR-FTIR spectroscopy: Metastability, agglomeration, and control. Cryst. Growth Des. 2002;2(5):363–370.

75 Grön H, Borissova A, Roberts KJ. In-process ATR-FTIR spectroscopy for closed-loop supersaturation control of a batch crystallizer producing monosodium glutamate crystals of defined size. Ind. Eng. Chem. Res. 2003;42(1):198–206.

76 Liotta V, Sabesan V. Monitoring and feedback control of supersaturation using ATR-FTIR to produce an active pharmaceutical ingredient of a desired crystal size. Org. Proc. Res. Des. 2004;8(3):488–494.

77 Zhou GX, Fujiwara M, Woo XY, Rusli E, Tung H-H, Starbuck C, Davidson O, Ge Z, Braatz RD. Direct design of pharmaceutical antisolvent crystallization through concentration control. Cryst. Growth Des. 2006;6(4):892–898.

78 Kee NCS, Tan RBH, Braatz RD. Selective crystallization of the metastable alpha-form of L-glutamic acid using concentration feedback control. Cryst. Growth Des. 2009;9(7):3044–3051.

79 Kee NCS, Arendt PD, Tan RBH, Braatz RD. Selective crystallization of the metastable anhydrate form in the enantiotropic pseudo-dimorph system of L-phenylalanine using concentration feedback control. Cryst. Growth Des. 2009;9(7):3052–3061.

80 Kee NCS, Arendt PD, Goh LM, Tan RBH, Braatz RD. Nucleation and growth kinetics estimation for L-phenylalanine hydrate and anhydrate crystallization. CrystEngComm. 2011;13:1197–1209.

81 Kee NCS, Tan RBH, Braatz RD. Semiautomated identification of the phase diagram for enantiotropic crystallizations using ATR-FTIR spectroscopy and laser backscattering. Ind. Eng. Chem. Res. 2011;50:1488–1495.

82 Nagy ZK, Braatz RD. Advances and new directions in crystallization control. Annu. Rev. Chem. Biomol. Eng. 2012;3:55–75.

83 Doki N, Seki H, Takano K, Asatani H, Yokota M, Kubota N. Process control of seeded batch cooling crystallization of the metastable alpha-form glycine using an in-situ ATR-FTIR spectrometer and an in-situ FBRM particle counter. Cryst. Growth Des. 2004;4(5):949–953.

84 Woo XY, Nagy ZK, Tan RBH, Braatz RD. Adaptive concentration control of cooling and antisolvent crystallization with laser backscattering measurement. Cryst. Growth Des. 2009;9(1):182–191.

85 Abu Bakar MR, Nagy ZK, Saleemi AN, Rielly CD. The impact of direct nucleation control on crystal size distribution in pharmaceutical crystallization processes. Cryst. Growth Des. 2009;9(3):1378–1384.

86 Woo XY, Tan RBH, Chow PS, Braatz RD. Simulation of mixing effects in antisolvent crystallization using a coupled CFD-PDF-PBE approach. Cryst. Growth Des. 2006;6(6):1291–1303.

87 Braatz RD, Alkire RC, Seebauer E, Rusli E, Gunawan R, Drews TO, Li X, He Y. Perspectives on the design and control of multiscale systems. J. Process Control. 2006;16(3):193–204, and citations therein.

88 Ma DL, Tafti DK, Braatz RD. High-resolution simulation of multidimensional crystal growth. Ind. Eng. Chem. Res. 2002;41(25):6217–6223.

89 Gunawan R, Fusman I, Braatz RD. High resolution algorithms for multidimensional population balance equations. AIChE J. 2004;50(11):2738–2749.

90 Jiang M, Wong M, Zhu Z, Zhang J, Zhou L, Wang K, Ford AN, Si T, Hasenberg L, Li Y, Braatz RD. Towards achieving a flattop crystal size distribution by continuous seeding and controlled growth. Chem. Eng. Sci. 2012;77:2–9.

91 Zhou L. PhD thesis. Massachusetts Institute of Technology, Cambridge, MA, 2013.

4

Predictive Distributions for Constructing the ICH Q8 Design Space

John J. Peterson, Mohammad Yahyah, Kevin Lief, and Neil Hodnett

GlaxoSmithKline Pharmaceuticals, Brentford, UK

4.1   Introduction

Currently, regulators are encouraging the pharmaceutical industry to acquire improved understanding of how their manufacturing processes work and to be able to assess the reliability and robustness of their manufacturing systems. Recent guidances for industry such as the Critical Path Initiative, Food and Drug Administration (FDA)'s current good manufacturing practice (cGMP) for the twenty-first century, the Process Analytical Technology Initiative, and Quality by Design have provided some broad strategies for the pharmaceutical industry to improve the robustness and reliability of their manufacturing processes. In particular, the International Congress on Harmonisation (ICH) Q8 guidance has outlined quality by design principles for pharmaceutical development. An important concept for Quality by Design in the pharmaceutical industry is design space (DS), which was introduced in the ICH Q8 guidance.

The ICH Q8 defines DS as "The multidimensional combination and interaction of input variables (e.g., material attributes) and process parameters that have been demonstrated to provide assurance of quality." It states further that "Working within the design space is not considered as a change. Movement out of the design space is considered to be a change and would normally initiate a regulatory post approval change process. Design space is proposed by the applicant and is subject regulatory assessment and approval." The creation of a DS for a manufacturing process not only offers an opportunity to make changes within the DS without regulatory approval but also provides opportunities for further experimentation, after regulatory approval of the manufacturing process, which can help further process improvement. The phrase "demonstrated to provide assurance of quality" from the ICH Q8

Comprehensive Quality by Design for Pharmaceutical Product Development and Manufacture, First Edition. Edited by Gintaras V. Reklaitis, Christine Seymour, and Salvador García-Munoz. © 2017 American Institute of Chemical Engineers, Inc. Published 2017 by John Wiley & Sons, Inc.

definition of DS begs the question, "How much assurance?" Use of an appropriate predictive distribution for quality responses associated with a DS can help to answer such a question.

However, with regard to practical implementation, there are three issues that are working against the construction of a DS based on a reliability measure that can address the question of "How much assurance?" One such issue is the nature of classical response surface methodology that primarily focuses on inference about mean response surfaces. This may lead practitioners to naïvely follow examples in classical texts that illustrate the approach of overlapping mean responses (OMRs) even though the problem for DS requires a reliability-based analysis to quantify the assurance of meeting process specifications.

The second issue synergizing the temptation to use an OMR approach is the availability of point-and-click-oriented statistical packages such as Design Expert^{TM} or JMP$^®$, which make the construction of an OMR plot very easy. However, as shown by Peterson [1], the OMR approach has two serious flaws associated with multiple-response optimization in general. More specifically for DSs, the OMR approach does not provide a way to quantify the level of assurance of meeting product specifications.

Thirdly, it appears that the opportunity for easily constructing DSs based on OMR software is more than just an academic concern. A few recent (and two possibly very influential) papers have been published that appear to propose the use of OMR as a way to construct DSs. It is hoped that this chapter can provide insight into the risk of using the OMR approach to DS construction and, in addition, underscore two alternative approaches based on *predictive distributions* for a process that will allow pharmaceutical scientists to ascertain the reliability of a DS to provide assurance of meeting product specifications.

In the next section, we discuss the background of the OMR approach in more detail and then describe its flaws for process optimization in general and for DS in particular. In Section 4.3, we describe two approaches to multiple-response surface optimization and show how they can be naturally extended to construct a DS that provides an answer to the question of assurance for meeting product specifications. Finally, we provide a summary and a discussion section.

4.2 Overlapping Means Approach

The earliest and simplest approach to multiple-response surface optimization is the OMR method. Apparently, this dates back to Lind *et al.* [2]. This method involves simply looking at OMR surfaces (e.g., by way of contour plots) to ascertain where two or more mean response surfaces possess a region of overlap with a desirable multiple-response configuration. Mathematically, this can be expressed as

$$\left\{x \in \chi : \hat{E}(Y \mid x) \in S\right\},$$

where Y is an $r \times 1$ vector of response types, x is a vector of predictive factors, χ is the experimental region or region of interest, and S is an r-dimensional quality-related specification region. Here, $\hat{E}$ represents the estimated expectation for Y given x. Further discussion of the history of this approach can be found in Montgomery and Bettencourt [3].

The OMR plotting approach is mentioned in several popular applied statistics books on experimental design and response surface optimization, such as Montgomery [4] (pp. 425 and 453), Khuri and Cornell [5] (p. 282), Myers and Montgomery [6] (p. 249), Cornell [7] (p. 509), and Box, Hunter, and Hunter [8] (p. 447). Some of these texts point out that such an approach has graphical challenges when there are more than two predictive variables. However, progress in computer graphics has made it easier to view the effect of several variables on the mean response surfaces. Design Expert has a slider tab whereby the user can change the values of the third, fourth, and so forth predictive factors to see the change in the OMR surface plot. Statistical applications such as S-plus®, R, and SAS® allow the user to create response surface plots in a matrix (trellis-like) layout to see how an OMR surface plot of two variables changes as other predictive factors are changed.

In addition to OMR plots, popular point-and-click statistical packages such as Design Expert and JMP also permit the construction of *desirability functions* [9] for multiple-response process optimization. These functions map the mean responses onto a scalar desirability function, which is typically the geometric mean of individual desirability functions. However, as shown by Peterson [1], both the OMR and the desirability function approaches have two serious flaws associated with multiple-response optimization in general. More specifically for DSs, neither the OMR nor the desirability function approach provides a way to quantify the level of assurance of meeting product specifications.

For basic multiple-response surface optimization, the use of OMR plots (or desirability functions) has two serious flaws. First, they do not account for the model parameter uncertainty, which can be substantial [10]. Second, they ignore the correlation structure of the regression model residuals, which can have serious consequences [11]. For the construction of a DS, it is further required to have a method for quantification of assurance of meeting process specifications. The OMR plots and desirability functions available in commercial software do not provide for such a method.

The OMR region has been called the *sweet spot* of a process [12]. However, as shown by Peterson [1] and del Castillo [13], even the best point within this region may possess a disappointingly small reliability with regard to meeting process specifications. Considering the following simple example makes this clear. Suppose that $Y_1, \ldots, Y_r$ represent r different quality response types and that each Y_i has a normal distribution with mean, μ_i, and variance, σ_i^2. For simplicity, assume that we have only upper bound product specifications, $u_1, \ldots, u_r$. If $\mu_1 < u_1$, then we know that $\Pr(Y_1 < u_1) < 0.5$. If $Y_1, \ldots, Y_r$ are independent and $\mu_i < u_i(i = 1, \ldots, r)$, then we know that $\Pr(Y_1 < u_1, \ldots, Y_r < u_r) < 0.5^r$. If $Y_1, \ldots, Y_r$ are positively correlated, then $\Pr(Y_1 < u_1, \ldots, Y_r < u_r)$ will be larger; likewise if $Y_1, \ldots, Y_r$ are negatively correlated, then the joint probability of meeting specifications will be smaller.

Of course, for specifications having both lower and upper bounds, the probability of a future response meeting specifications can be arbitrarily small even if the response means are dead center in the specification interval. This is because for such responses, the probability of meeting specifications can go down toward zero as the variability increases.

Peterson [11] also provides an example that shows that the correlation structure among the Y_is can have a noticeable (and sometimes negative) influence on a joint probability such as $\Pr(Y_1 < u_1, \ldots, Y_r < u_r)$. Therefore, any attempt to use an OMR region (which ignores the variance and covariance structure of the Y_is) "to provide assurance of quality" with regard to meeting process specifications is likely to result in a DS that is too large and possesses process conditions with a poor probability of meeting specifications for future process runs.

It may be that classical response surface methodology, in concert with available software for easily constructing OMR plots, has facilitated the apparent proposal of using the OMR approach to construct a DS that has appeared in some recent publications. As far as we know, the first published example of using the OMR approach to create an example of a DS can be found in the ICH Q8 guidance document, Pharmaceutical Development, annex to Q8 (2007). Section 2.4.2 states that "Examples of different potential approaches to presentation of a design space are presented in Appendix 2." Figure 2c in Appendix 2 of the ICH Q8 Annex shows an overlay plot for a granulation operation with two quality responses, friability and dissolution. The figure title taken from the ICH Q8 Annex states: "Potential process design space, comprised of the overlap region of design ranges for friability and/or dissolution." The ICH Q8 Annex does not state statistically how the response surfaces are formed (e.g., mean or median), but many would naturally assume a mean response surface (see Figure 4.1).

A recent position paper [14] summarizes progress made by the Design Space Task Team within the International Society for Pharmaceutical Engineering's Product Quality Lifecycle Implementation initiative. Figure 11 of that paper appears to suggest that an OMR plot could be used to represent a DS.

It is unfortunate that both of these two papers discussed previously could have substantial influence on the willingness of their readers to use OMR plots to construct a DS. A few other recent articles also appear to suggest the use of OMR plots for the construction of a DS [15–17]. It is perhaps a curious blind spot that none of these publications mentioned the statistical nature of their fitted response surfaces (e.g., mean or median).

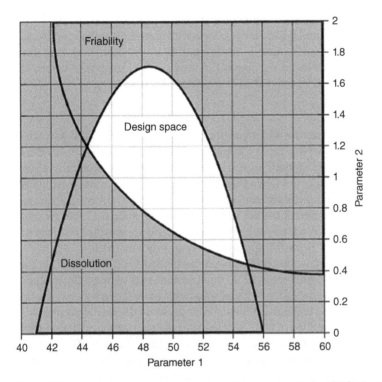

Figure 4.1 The overlapping response surface plot appearing in the ICH Q8 Annex.

While it is commendable that these articles recommend the utilization of statistically designed experiments (i.e., design of experiments) for building prediction models to be used in the construction of a DS, one further step needs to be taken. In addition to predicting means, we need to use a predictive *distribution* that will allow the quantification of process reliabilities needed "to provide assurance of quality" as required by the ICH Q8 definition of DS.

4.3 Predictive Distribution Approach

Every process produces a different (to a lesser or greater degree) set of responses each time it is run. Over time, a process will produce a distribution of responses. This distribution will have a location of central tendency (e.g., the mean) and some degree of dispersion. Ideally, one would engineer the process to have a mean close to a desired target with a small degree of dispersion about its mean. The process response distribution can also be used to compute a probability measure of the reliability of the process for meeting specification limits. Such a reliability measure is helpful for making risk assessments about the process.

A predictive distribution for a process can be obtained by combining distributions relating to natural noise in the process with uncertainty about unknown model parameters, which must be estimated from experimental data (see Peterson [18], for a detailed discussion). A process with response vector, Y, may be related to a model (empirical or mechanistic) as follows:

$$Y = f(x \mid \theta) + e, \tag{4.1}$$

where $f(x \mid \theta)$ represents the underlying functional form of the process as a function of a vector of controllable factors, x, and vector of unknown model parameters, θ. Here, e, is a vector of random error (i.e., noise) terms, whose distribution may also depend on a matrix of unknown parameters (e.g., Σ), which must be estimated from experimental data. Typically, e may be modeled to have a multivariate normal distribution with mean vector $\mathbf{0}$ and variance–covariance matrix, Σ. This can also be expressed as $e = \Sigma^{\frac{1}{2}'} \varepsilon$, where ε is a vector of independent standard normal random variables and $\Sigma^{\frac{1}{2}'}$ is the transpose of the square-root matrix for Σ.

As the values that make up (θ, Σ) must be estimated from the data, one can model the uncertainty inherent in (θ, Σ) and combine that with ε to form a predictive distribution for the process. Using standard Monte Carlo procedures, the distribution for ε is easy to simulate, but in order to obtain the predictive distribution for the model described in (4.1), we must also obtain a statistical distribution for (θ, Σ). There are two basic procedures for doing this. One is to use Bayesian statistical methods to obtain what is called a posterior probability distribution expressing uncertainty about (θ, Σ). Such a distribution uses information from experimental data, and possibly prior knowledge as well. In the 1990s, Markov Chain Monte Carlo (MCMC) techniques [19] became popular as a general way to sample from a posterior distribution for unknown model parameters.

Using the posterior probability distribution for (θ, Σ) and the probability distribution for ε, one can sample from the posterior predictive distribution for Y, the vector of process responses. This is more easily seen if we rewrite the model in (4.1) as

$$Y = f(x \mid \theta) + \Sigma^{\frac{1}{2}'} \varepsilon \tag{4.2}$$

For each set of process conditions, x, we can use the model in (4.2), along with the distributions for (θ, Σ) and ε to simulate the posterior predictive distribution for Y. We can then estimate process conformance probabilities of the sort, $p(x) = \Pr(Y \in S \mid x, \text{data})$, where each $p(x)$ is estimated by

$$\frac{1}{N} \sum_{i=1}^{N} I(Y_i \in S).$$

Here, N is the number of Monte Carlo simulations, $I(\cdot)$ is an indicator function, $(Y_1, \ldots, Y_N)$ are the simulated process responses from (4.2) at process condition, x, and S is a specification region.

An ICH Q8 DS can then be quantitatively defined as

$$\{x \in \chi : p(x) \geq R\}, \tag{4.3}$$

for some inferential region, χ, and some reliability level, R [11]. The set in (4.3) can be used to create a Bayesian DS that takes into account all of the model parameter uncertainty and the correlation structure among the quality responses (at fixed x points) by way of the posterior predictive distribution. The use of a probability measure for assessing process reliability and risk has been espoused by some FDA regulators [20, 21].

However, Bayesian statistical methods still remain as a somewhat specialist area of statistical inference. An alternative approach is to use *parametric bootstrapping* [22] as a more transparent way to get a distribution expressing the uncertainty of the unknown model parameters. A simple form of the parametric bootstrap approach can be described as follows. First, fit your process model to experimental data to estimate the unknown model parameters. Using computer simulation, simulate a new set of (artificial) responses so that you will have new responses similar to the ones obtained from your real experiment. Using the artificial responses, estimate a new set of model parameters. Do this many times (say, 1000–2000) to obtain more sets of model parameter estimates. Each of these model parameters (obtained from each artificial data set) will be somewhat different, due to the variability in the (simulated) data. The collection of all these sets of model parameter estimates forms a bootstrap distribution of the model parameters that expresses their uncertainty. The parametric bootstrap distribution can then be used in place of the Bayesian posterior distribution to enable one to obtain a predictive distribution for the quality responses. This simple parametric bootstrap approach will approximate the Bayesian posterior distribution of model parameters, but it will tend to produce a parameter distribution that is slightly tighter than the one obtained using the Bayesian approach (unless the Bayesian approach also brings substantial prior information to bear). In many cases, the parametric bootstrap approach should provide a good approximation to the Bayesian posterior predictive distribution when the data sample size is large. More sophisticated parametric bootstrap approaches are possible [22], which would result in more accurate approximations to the Bayesian posterior distribution of model parameters. These approaches should be considered if there is not much information with which to estimate the model parameters.

4.4 Examples

In this section, we consider two different DS examples. One involves a mechanistic model for a particular synthesis stage of an active pharmaceutical ingredient (API). The other example illustrates the use of an empirical model for a pharmaceutical tableting process. While these examples represent only prototype experiments, rather than those performed in actual manufacturing plants for regulatory submission, they nonetheless provide useful illustrations, based on real data, for comparison purposes. For both examples, we compare the OMR DS with DSs created using predictive distributions from Bayesian and parametric bootstrap analyses.

4.4.1 A Mechanistic Model Example

A kinetic study was performed to enhance process understanding and support optimization and scale-up of a telescoped process for the manufacture of an API. The final chemical transformation of the telescoped sequence comprises the addition of cyclopropylamine (CPA) to a solution of aromatic chloride (ArCl) at 85–90°C, followed by a 2-h stir to yield the desired CPA adduct (Product). A stoichiometric excess of CPA is required since the HCl liberated will neutralize one molar equivalent of the amine to afford CPA hydrochloride (CPA HCl) (Figure 4.2). In order to ensure product quality specifications are met, the reaction is progressed until <0.1% ArCl remains.

The reaction displays simple second-order kinetics with two equivalents of CPA being consumed per mole of product formed. The proposed reaction mechanism is

$$\text{ArCl} + \text{CPA} \xrightarrow{\text{rate determining}} \text{product} + \text{HCl}$$

$$\text{CPA} + \text{HCl} \xrightarrow{\text{fast}} \text{CPA} \cdot \text{HCl}$$

Typically acid–base neutralizations occur rapidly, in solution; therefore, the aromatic substitution step is assumed to be rate limiting and the overall rate equation can be simplified to

$$\frac{d\text{product}}{dt} = 60k\left[\text{ACl}\right]_t \left[\text{CPA}\right]_t,$$

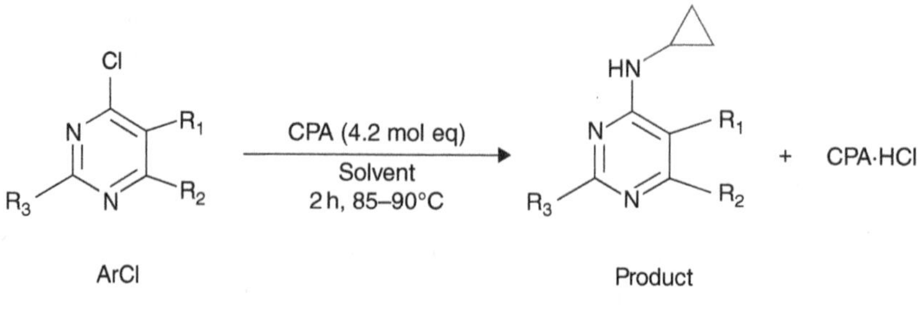

ArCl

Product

Figure 4.2 Nucleophilic aromatic substitution of aromatic chloride ArCl with CPA.

where $k = z\exp\left(-\dfrac{E_a}{R}\left[\dfrac{1}{(temp + T_r)}\right]\right)$, $T_r = 273$ (to convert to Kelvin scale) and $R = 8.31\,J/kmol$ (universal gas constant).

Solving the equations analytically gives the functional form of the rate equation, which describes the concentration of Product at time t. The resulting model (with random noise term) is then

$$Y = \frac{x_2(x_3 - 2x_2)}{x_3\left[\exp\left[t(x_3 - 2x_2)60z\exp\left(-\dfrac{E_a}{R(x_1 + T_r)}\right)\right]\right] - 2x_2} + \sigma\varepsilon \tag{4.4}$$

Here, x_1 = temperature of the reaction (°C), x_2 = concentration of initial ArCl amount, x_3 = concentration of initial CPA amount, and t = sampling time (seconds) are the controllable process factors, while E_a and z are the unknown model parameters (activation energy and Arrhenius constant, respectively) that need to be estimated. Furthermore, σ is also an unknown model parameter that must be estimated. For the model in (4.4), ε has the standard normal distribution.

It is desired to have Y less than or equal to 0.1%. The model parameters are E_a, z, and σ. The controllable factor variables are $x = (x_1, x_2, x_3, x_4)$. The inferential region for the DS is $\chi = [0,100] \times [0,100] \times [0.15,0.5] \times [1.33,1.25]$.

As such, the DS has the form

$$\{x \in \chi : \Pr(Y\langle 0.1|x, data) \geq R\}. \tag{4.5}$$

Here, we take R to be 0.99.

An OMR DS was computed by simply fitting the model in (4.4) to the data and gridding over the experimental region to determine a set of factor variables such that the predicted mean response was less than or equal to 0.1. Since we have only one response type, such an OMR DS should have the form in (4.5) with $R = 0.5$.

A Bayesian DS was created by BRugs Bayesian statistical package available for use with the R statistical software package. The BRugsFit function in the BRugs package uses a sophisticated Bayesian MCMC algorithm to produce a posterior distribution for the three unknown model parameters, E_a, z, and σ. The prior distributions used were weak priors using uniform distributions over wide intervals relative to what was known about the limits for the model parameter values. The limits for the uniform distributions for E_a, z, and σ are [45 000, 55 000], [0.00045,0.0014], and [0,0.0067], respectively. The upper bound value for σ is a 99% upper confidence bound for σ. This posterior distribution, along with ε, is then used in (4.5) to produce a posterior predictive distribution for Y to compute the DS of the form in (4.5).

A parametric bootstrap DS was computed by first obtaining a parametric bootstrap distribution for the unknown model parameters, E_a, z, and σ, by using SAS statistical package to do the necessary model fitting and computer simulations. Analogous to the Bayesian approach, this bootstrap parameter distribution was then used to compute a predictive distribution for Y to compute the DS of the form in (4.5).

The OMR and parametric bootstrap DSs are shown in Figure 4.3. The contour corresponding to the 0.5 probability level represents the OMR DS, while the contour corresponding to the 0.99 probability level represents the parametric bootstrap DS. The Bayesian DS was virtually identical to the parametric bootstrap DS. For this particular example, the error standard deviation estimate was rather small, 0.004. As such, the OMR DS is not too much larger than the Bayesian or parametric bootstrap DS. However, it should be noted that the small error variation for this process does induce an *edge of risk* for this DS. If one operates this process within the OMR DS but outside of the Bayesian or parametric bootstrap DS, then minor changes in the process factors could result in a modest reliability (i.e., close to 0.5) of meeting the process specification limit of 0.1. As such, even though the OMR DS is not much bigger than the Bayesian or parametric bootstrap DS for this example, it is prudent to operate the process within the DS created by the predictive distribution rather than the one created by the predicted mean response. Furthermore, the difference would likely be much larger if the experiment had contained some true replications (at identical operating conditions), which could have been used to extract and model batch-to-batch variation (in addition to the measurement error variation).

4.4.2 An Empirical Model Example

The second example is that of a tablet formulation experiment. Here, two primary unit operations, granulation and compression, were studied with regard to their influence on three critical quality responses. The three quality responses (and their specifications) were

y_1 = tablet disintegration time (at most 15 min), y_2 = friability (at most 0.8%), and y_3 = hardness (8–14 kp).

Three granulation factors were x_1 = quantity of water added, x_2 = rate of water addition, and x_3 = wet massing time, while the three compression factors utilized were x_4 = main compression force, x_5 = main compression/precompression ratio, and x_6 = speed. The factors $x_1 - x_6$ were analyzed in a coded form $\left(\text{i.e.,} -1 \leq x_i \leq 1, i = 1, \ldots, 6 \right)$. The experimental design takes seven combinations of the three granulation factors (plus two center points) and passes the resulting granule batches onto eight combinations (plus three center points) of the three compression factors. This resulted in 99 experimental runs. The aim of the series of experiments was to investigate the physical characteristics of the tablets, given a diverse range of granulation batches being compressed under a wide range of factor settings.

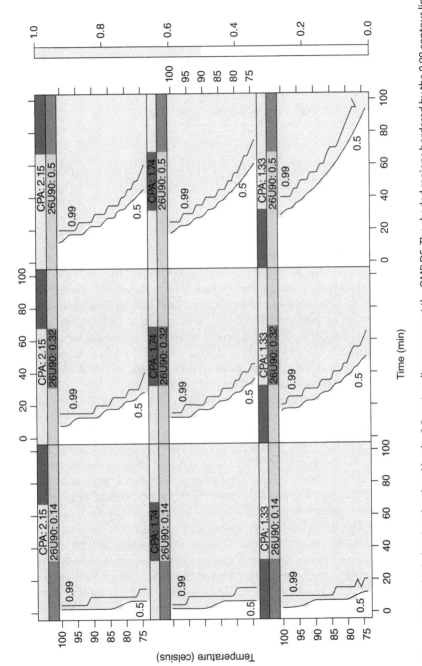

Figure 4.3 The shaded regions bordered by the 0.5 contour lines represent the OMR DS. The shaded regions bordered by the 0.99 contour lines represent the parametric bootstrap and Bayesian posterior predictive DSs.

Fitting regression models to the data produced the following three models with y_1 = tablet disintegration time, y_2 = friability, and y_3 = hardness:

$$\hat{y}_1 = 14.86 - 0.49x_1 - 0.21x_2 - 1.01x_3 + 0.78x_4 - 0.64x_1x_3 + 0.40x_2x_4 + 0.57x_3x_4$$

$$\hat{y}_2^* = 23.86 + 3.68x_1 + 0.59x_2 + 0.003x_3 + 0.24x_4 + 0.03x_5 + 2.83x_1x_3$$
$$+ 0.42x_1x_4 + 0.53x_1x_5 + 0.17x_2x_3 + 0.76x_2x_4 + 0.39x_3x_4 + 0.43x_1x_2x_4$$
$$+ 0.21x_1x_3x_4 + 0.35x_2x_3x_4 + 0.70x_1x_2x_3x_4$$

$$\hat{y}_3 = 12.94 - 7.13x_1 + 0.49x_2 - 1.97x_3 + 0.29x_1x_2 - 0.56x_2x_3.$$

$$(4.6)$$

Here, y_2 (friability) was transformed as $y_3^* = (y_3 + 0.5)^{-\frac{1}{2}}$ to improve normality. The acceptance region for the three quality responses was defined to be

$$S = \left\{ (y_1, y_2, y_3) : y_1 \leq 15, \quad y_2 \leq 0.8, \quad 8 \leq y_3 \leq 14 \right\}.$$

Using the three estimated models in (4.6) and the bounds in S, the Design Expert package was able to easily produce an OMR plot for any two of the factors, with the other factors at fixed values. The x_6 factor had no apparent effect and was not included in any of the three models, while x_5 had only a small effect for the second model. As such, x_5 was fixed at its center point for purposes of displaying the DS. The experimental region employed for creating the OMR DS was $\chi = [-1,1] \times [-1,1] \times [-1,1] \times [-1,1]$. A 3×3 matrix of OMR plots is shown in Figure 4.4. This provides an overview of the OMR DS, which is shown by the (large) light-gray regions on the plot.

The multiple-response model in (4.6) is often referred to as the seemingly unrelated regressions (SUR) model [23]. According to the SUR model assumptions, we assume that the three regression error terms (e_1, e_2, e_3) (corresponding to the three responses in (4.6)) are correlated among themselves. Using the Bayesian MCMC, R function rsurGibbs (from the bayesm R package) and posterior predictive probabilities, $p(x)$, as in (4.3), were computed across a grid of x points in the experimental region. The black region in Figure 4.4 is the Bayesian DS (with R = 0.8). The region encompassing the black and gray regions is the parametric bootstrap DS, which is only slightly larger than the Bayesian DS. The largest Bayesian and bootstrap probabilities of meeting all quality specifications turned out to be 0.84 and 0.85, respectively, while the minimum probability turned out to be only 0.15 and 0.155, respectively. As discussed in the Overlapping Means Approach section, it is not surprising that a low probability of acceptance value such as 0.15, even with only one-sided acceptance limits, can occur since $0.5^3 = 0.125$. The estimated correlation matrix of the residuals from a SUR fit of the data is

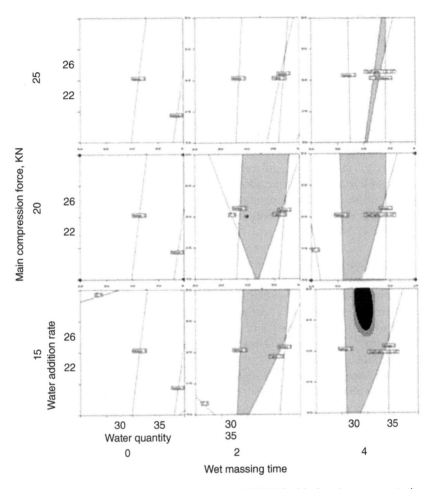

Figure 4.4 The light-gray regions represent the OMR DS. The black region represents the Bayesian DS with reliability value *R* = 0.8. The black and darker gray regions combined represent the parametric bootstrap DS with reliability value *R* = 0.8.

$$\begin{bmatrix} 1 & 0.2 & 0.13 \\ 0.2 & 1 & -0.02 \\ 0.13 & -0.02 & 1 \end{bmatrix}.$$

This is a distribution where the means are all within the acceptance limits, but only 15% of the responses are within the acceptance limits. This shows the risk of using an OMR DS. Peterson and Lief [24] use Bayesian posterior predictive distributions to assess five other DS examples based on empirical models.

They find that in each case, the OMR DSs contain minimum probabilities of conformance, $p(x) = \Pr(Y \in S \mid x, \text{data})$, between 0.11 and 0.34.

4.5 Summary and Discussion

DS is a key issue in pharmaceutical quality by design. It is seen by pharmaceutical manufacturers as a way to enhance regulatory flexibility and to be able to experiment and make changes to a manufacturing process (within the DS) without requiring formal regulatory approval. It appears that regulators see DS as an opportunity for pharmaceutical companies to improve their understanding of their manufacturing processes before and after regulatory approval. It is also an opportunity to establish a procedure for quantifying the reliability and risk associated with the process or unit operation for which the DS was created. As such, it is important that the actual DS constructed be a clear quantitative representation of the definition of DS in the ICH Q8 guidance. In particular, the DS should be constructed in such a manner that it can be "demonstrated to provide assurance of quality." We believe that a reliability-based DS of the type in (4.6) is important as it provides a probability that can be used to quantify reliability and risk. The use of probability measures and distributions to assess uncertainty and quantify risk has been espoused by some regulators [20, 21].

However, three situations currently exist that may make OMR DSs appear attractive to pharmaceutical scientists. One situation is that there are abundant examples of OMR plots in popular books on experimental design and response surface methodology. (However, such books predate the ICH Q8 DS issue and do not address themselves specifically to the DS problem.) The second situation is that OMR plots are easy to create using popular software such as Design Expert or SAS/JMP. Thirdly, there are several papers, two of which may be quite influential, that appear to suggest the use of OMR plots for constructing a DS.

It is hoped that this article can persuade pharmaceutical statisticians and scientists to instead construct their DSs based on predictive distributions. The predictive distribution approach provides a holistic and flexible procedure with which to generate predictive distributions that can be used to construct a DS. The Bayesian method can be used to construct a posterior predictive distribution. If adequate software is available, this may be the best approach, particularly if prior information is available regarding the model parameters. On the other hand, the parametric bootstrap approach is statistically more transparent and, in many cases, easier to program if software for the Bayesian approach is not available. The SAS statistical package is particularly amenable to programming for doing parametric bootstrapping. In addition, the R statistical programming language has a function called boot, which can be used in conjunction with other *R* statistical model-fitting functions to perform parametric bootstrapping.

Recently, MacGregor and Bruwer [25] have pointed out the potential for partial least squares (PLS) to be used to construct a DS. Chen *et al.* [26] provided an MCMC Bayesian approach to PLS. Furthermore, the R packages pls and chemometics both have PLS functions with which one can apply the boot function. As such, it may also be possible to produce an approximate predictive distribution for a PLS-type model to be used in the formation of a DS.

Acknowledgment

We acknowledge Mark Hughes for the part he played in carrying out the original kinetics experiments and reaction modeling for the first case study.

References

1 Peterson, J. J. A posterior predictive approach to multiple response surface optimization. Journal of Quality Technology, 2004; 36:139–153.

2 Lind, E. E., Goldin, J., and Hickman, J. B. Fitting yield and cost to response surfaces. Chemical Engineering Progress, 1960; 56: 62–68.

3 Montgomery, D. C., and Bettencourt, V. M. Multiple response surface methods in computer simulation. Simulation, 1977; 29: 113–121.

4 Montgomery, D. C. Design and analysis of experiments (6th edition). John Wiley & Sons, Inc., Hoboken, 2005.

5 Khuri, A. and Cornell, J. A. Response surfaces: designs and analyses (2nd edition), Marcel-Dekker, New York, 1996.

6 Myers, R. and Montgomery, D. C. Response surface methodology, John Wiley & Sons, Inc., New York, 1995.

7 Cornell, J. A. Experiments with mixtures: designs, models, and the analysis of mixture data (3rd edition), John Wiley & Sons, Inc., New York, 2002.

8 Box, G. E. P., Hunter, J. S., and Hunter, W. G. Statistics for experimenters (2nd edition), John Wiley & Sons, Inc., New York, 2005.

9 Derringer, G. and Suich, R. Simultaneous optimization of several response variables. Journal of Quality Technology, 1980; 12: 214–219.

10 Hunter, J. S. Discussion of response surface methodology—current status and future directions. Journal of Quality Technology, 1999; 31: 54–57.

11 Peterson, J. J. A Bayesian approach to the ICH Q8 definition of design space. Journal of Biopharmaceutical Statistics, 2008; 18: 959–975.

12 Anderson, M. J. and Whitcomb, P. J. Find the most favorable formulations. Chemical Engineering Progress, 1998; April, 63–67.

13 del Castillo, E. Process optimization: a statistical approach, Springer, New York, 2007.

14 LePore, J. and Spavins, J. PQLI design space. Journal of Pharmaceutical Innovation, 2008; 3: 79–87.

15 Kamm, J. Can you win the space race? Pharmaceutical Manufacturing, 2007; 6, issue 5: at http://www.pharmamanufacturing.com/articles/2007/091.html (accessed April 9, 2017).

16 am Ende, D., Bronk, K. S., Mustakis, J., O'Connor, G., Santa Maria, C. L., Nosal, R., and Watson, T. J. N. API quality by design example form the Torcetrapib manufacturing process. Journal of Pharmaceutical Innovation, 2007; 2: 71–86.

17 Harms, J., Wang, X., Kim, T., Yang, X., and Rathore, A. S. Defining product design space for biotech products: case study of *Pichia pastoris* fermentation. Biotechnology Progress, 2008; 24: 655–662.

18 Peterson, J. J. What your ICH Q8 design space needs: a multivariate predictive distribution. Pharmaceutical Manufacturing, 2009; 8(10): 23–28.

19 Gamerman, D. Markov chain Monte Carlo: stochastic simulation for Bayesian inference, Chapman & Hall/CRC Press, Suffolk, 2006.

20 Claycamp, H. G. Room for probability in ICH Q9: quality risk management, presented at the *Pharmaceutical Statistics 2008: Confronting Controversy conference*, March 2008, Arlington, VA (sponsored by the Institute of Validation Technology).

21 Kauffman, J. and Geoffroy, J.-M. Uncertainty in design space—the application of Monte Carlo methods to process modeling, *22nd International Forum: Process Analytical Technology—IFPAC*, 2008, January 27–30, Baltimore, MD.

22 Davison, C. and Hinkley, D. V. Bootstrap methods and their application, Cambridge University Press, Cambridge, 1997.

23 Johnston, J. Econometric methods (2nd edition), McGraw-Hill Book Company, New York, 1963.

24 Peterson, J. J. and Lief, K. The ICH Q8 definition of design space: a comparison of the overlapping means and the Bayesian predictive approaches. Statistics in Biopharmaceutical Research, 2010; 2: 249–259.

25 MacGregor, J. F. and Bruwer, M.-J. A framework for the development of design and control spaces. Journal of Pharmaceutical Innovation, 2008; 3: 15–22.

26 Chen, H., Bakshi, B. R., and Goel, P. K. Bayesian latent variable regression via Gibbs sampling: methodology and practical aspects. Journal of Chemometrics, 2007; 21: 578–591.

5

Design of Novel Integrated Pharmaceutical Processes: A Model-Based Approach

Alicia Román-Martínez[1,2], John M. Woodley[1], and Rafiqul Gani[1]

[1] *Department of Chemical and Biochemical Engineering, Technical University of Denmark, Lyngby, Denmark*
[2] *Facultad de Ciencias Químicas, Universidad Autónoma de San Luis Potosí, San Luis Potosí, Mexico*

5.1 Introduction

Pharmaceutical industry is constantly facing challenges to meet demands from the regulatory bodies as well as consumers for higher and reliable product qualities, while at the same time having to find more sustainable manufacturing options. These challenges can be efficiently tackled through pharmaceutical Quality by Design (QbD), which is a means to achieve a desired (target) product quality through better understanding and design of the product and its corresponding manufacturing process. Therefore, an important element of the product–process design control problem is to identify the design space within which the desired product quality can be reliably achieved. This design space, which is a multidimensional combination and interaction of input variables (e.g., raw material attributes), process parameters, and process operating conditions, has been reported to provide assurance of product quality [1]. Different approaches to define a process design space have been proposed [2–7]. These approaches have a common characteristic, namely, they all employ what could be termed as "trial and error" based experimental efforts that lack a systematic search within the domain space. From a practical point of view, even with a very good design of experiment, it would not be feasible to experimentally verify all potentially interesting candidates. Consequently, it cannot be guaranteed that a more sustainable alternative design cannot be found to achieve the target product quality. A model-based methodology, however, can overcome this problem because validated process models can be systematically used to define (identify) the process design search space within which the optimal

Comprehensive Quality by Design for Pharmaceutical Product Development and Manufacture,
First Edition. Edited by Gintaras V. Reklaitis, Christine Seymour, and Salvador García-Munoz.
© 2017 American Institute of Chemical Engineers, Inc. Published 2017 by John Wiley & Sons, Inc.

solution can be found. This model-based methodology could save development time and resources as virtual experiments could be performed to increase the process understanding and its impact on product quality.

Integrated enzymatic and chemo-enzymatic processes are involved in the production of active pharmaceutical ingredients and/or intermediates. The integration of chemo-enzymatic and enzymatic processes involves combinations of at least two unit operations. For example, membrane bioreactors (where enzymatic biotransformations takes place simultaneously with separation) could be coupled with another chemical or enzymatic reaction (one-pot synthesis) forming integrated simultaneous product recovery (ISPR) [8–10]. This approach to design of biochemical processes offers opportunities for improvements in terms of process performance, flexibility, reduction of processing steps, and better utilization of resources. Also, limitations causing low productivity, such as unfavorable equilibrium, very dilute concentrations of the product, enzyme inhibition by substrates and/or products, and difficult downstream processing, could be overcome. Incorporating opportunities for the generation of alternatives such as the one-pot synthesis and/or ISPR and evaluating their feasibility in a systematic and efficient manner would enhance the application of QbD.

In this chapter, a model-based framework that helps to define the design space within which feasible and integrated process options can be screened and evaluated in a systematic manner is presented. In particular, the integration issues related to reaction and separation steps of processes involving enzymatic and chemo-enzymatic synthesis for the production of pharmaceuticals and/or intermediate pharmaceuticals are highlighted. The model-based framework has the following main characteristics:

- A generic model generation toolbox
- A superstructure-based generation and screening of all potential synthesis routes
- A decomposition-based approach that manages the combinatorial explosion by effectively reducing the search space and limiting the number of feasible alternatives
- Evaluation and detailed analysis of a small number of alternatives to locate the best available option

Special modeling toolbox containing information on a large range of chemo-enzymatic processes has been developed and implemented within the framework. Other tools needed by the method, such as databases, property prediction routines, solvers, and so on, are also available through the framework.

The chapter is organized as follows: The introduction is followed by the general problem description together with its mathematical formulation and solution strategy. Next, the model-based methodology implemented within the framework for design and analysis of the network (flowsheet) options is presented. Next, application of the framework to the study of the production of neuraminic acid is highlighted, followed by the concluding remarks.

5.2 Problem Description

The design problem is defined as follows:

> *Given* the details of the raw materials (substrates) and the product(s) in terms of critical quality attributes, the reactions taking place and database of properties, *determine* the optimal enzymatic or chemo-enzymatic process route (network) giving the configuration of the reactors, separators, and their integration in the network that matches the target specifications and constraints.

All possible options are to be considered and analyzed based on the knowledge of the available technologies.

The following assumptions are necessary:

1) Only the reaction(s) and/or separation(s) steps of the process are considered.
2) The system may contain a minimum of one enzymatic reaction and a maximum of two reactions where at least one is enzymatic and the products are of low molecular weight.
3) The integration options related to one-pot synthesis and ISPR are considered.
4) The reactions occur in the liquid phase.
5) Each processing unit has a maximum of two phases.
6) All alternatives can be generated from the superstructure with maximum four processing steps.
7) The individual process models corresponding to each alternative are derived from a generic model representing the superstructure.

Based on these assumptions and information, a superstructure containing all possible processing routes is created (see Figure 5.1).

5.2.1 Mathematical Formulation

The design problem defined earlier is expressed mathematically in Equations 5.1–5.5 corresponding to the superstructure shown in Figure 5.1.

$$\text{Minimize/maximize } F_{\text{OBJ}} = \sum f_j \left(\underline{Y}, \underline{X}, \underline{Z}, \underline{d}, \underline{\theta} \right) \tag{5.1}$$

subject to $\underline{Y}, \underline{Z}$

and

logical constraints

$$g_{\text{logical,LB}} \le g_{\text{logical}} \left(\underline{Y} \right) \le g_{\text{logical,UB}}, \tag{5.2}$$

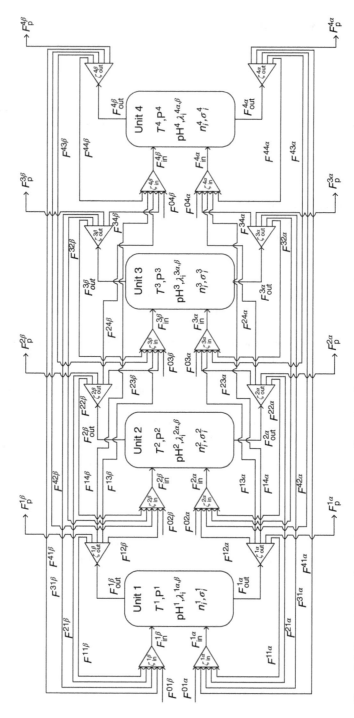

Figure 5.1 Superstructure for reaction–separation configuration of enzymatic processes (symbols: streams F (mol), temperature T (K), pressure P (bar), total number o moles n (mol), separation factors σ (–), binary existence variable ζ [0,1]; subscripts: bottom flow α, top flow α, product β).

structural constraints

$$g_{\text{structural,LB}} \leq g_{\text{structural}}\left(\underline{Y}\right) \leq g_{\text{structural,UB}}, \tag{5.3}$$

operational constraints

$$g_{\text{operational,LB}} \leq g_{\text{operational}}\left(\underline{Y},\underline{X},\underline{Z},\underline{d},\underline{\theta}\right) \leq g_{\text{operational,UB}}, \tag{5.4}$$

and process model

$$h_{\text{p}}\left(\underline{Y},\underline{X},\frac{\partial \underline{X}}{\partial \underline{Z}},\underline{d},\underline{\theta}\right) = \frac{\partial \underline{X}}{\partial t}. \tag{5.5}$$

In Equations 5.1–5.5, F_{OBJ} is the objective function representing a set of predefined performance criteria that may be minimized or maximized according to the design problem being solved; $\underline{Z}$ are the vector of design variables; $\underline{Y}$ are the vector of decision variables; $\underline{X}$ are the vector of process variables; $\underline{d}$ are the known variables; $\underline{\theta}$ are the model parameters; g are a set of constraint functions; and h_{p} is the process model. The decision variables $\underline{Y}$ define the existence of processing units, streams, and operations in network u. The known variables $\underline{d}$ can be, for example, equipment parameters, while $\underline{\theta}$ can be the kinetic model constants. For given values of $\underline{Y},\underline{Z},\underline{d},\underline{\theta}$, the process model equations are solved for $\underline{X}$. The constraint functions include logical, structural, and operational constraints bounded by lower bounds LB and upper bounds UB.

The number of processing routes (options) covered by the superstructure is given by

$$\text{NPO} = \sum_{u=1}^{u}\left(\prod_{u=1}^{u}j,u\right)\left(\frac{(r \cdot u)!}{u!(r \cdot u - u)!}\right) \tag{5.6}$$

In Equation 5.6, j,u is the number of identified unit operations in each network (flowsheet) u. The objective here is to find the values of the optimization (design and decision) variables that satisfy all the constraints and minimize or maximize the objective function. Depending on the design problem, this optimization problem may consist of highly nonlinear equations and continuous and discrete variables, resulting in a mixed integer nonlinear programming (MINLP) model. The size of alternatives found in the superstructure increases the size of the MINLP model in terms of variables and constraints.

5.2.2 Solution Approach

A decomposition-based solution approach [11] has been selected and implemented in the model-based framework for design of enzymatic and chemo-enzymatic processes. The framework guides the user through a stage-by-stage solution procedure (methodology), described in the next section.

5.3 Methodology

The methodology for design and analysis of integrated enzymatic and chemo-enzymatic processes consists of six stages. Its workflow is outlined in Figure 5.2. Every stage in the methodology is supported by different tools like computational software (e.g., ICAS-MoT®) and databases (e.g., BRENDA®). In Stage 1, the design problem is defined, including the determination of the objective function, the process/operation scenario, and the constraints that the options

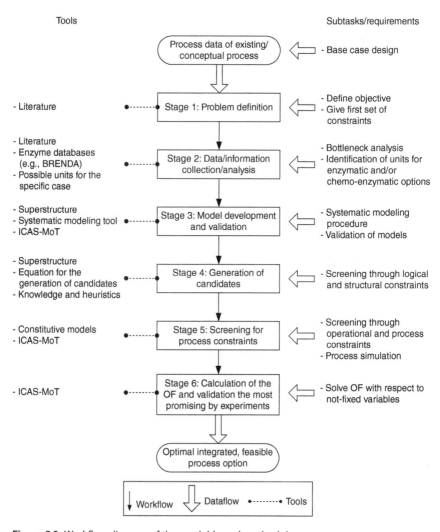

Figure 5.2 Workflow diagram of the model-based methodology.

need to match. In Stage 2, available information/data are collected for the product–process being studied. Where data such as pure compound property data are not available, they are generated with suitable property prediction models. The objective of Stage 3 is to develop the superstructure and the generic process models. First, the number of processing options present (N1) in the superstructure is counted, and then by specifying values of a subset of decision variables that can be defined based on the collected data, the superstructure is simplified and the number of processing options (N2 < N1) reduced. The model consists of mass and energy balance equations, constitutive model equations, and connection equations. Depending on batch, fed-batch, or continuous operation modes, dynamic or steady-state models are generated for each process–operation scenario. In Stage 4, the objective is to generate and evaluate all feasible integrated flowsheet (network) options through synthesis rules (logical constraints) and structural constraints, by fixing values to specific binary decision variables in the superstructure (Equations 5.2 and 5.3). Consecutively, in this stage, all options are screened for feasibility by additional structural and operational constraints (Equation 5.4), to match the target defined in Stage 1. Identified redundant options are removed. Subsequently, all remaining options are screened using specific process models (Equation 5.5) derived from the generic model of Stage 3. At this point, the number of alternatives are N3 < N2. In Stage 5, through the screening achieved in the previous stage, only for a small number of options (N3), the objective function (Equation 5.1) defined in Stage 1 is calculated. In Stage 6, the remaining options (N4 < N3) are ordered in terms of their calculated objective function values and the best is identified. This together with another 1–2 options could now be verified through rigorous simulation models as well as properly designed experiments.

The main features of the methodology are briefly discussed in the following text. More details on the model-based methodology can be found in Roman-Martinez [12] and Lutze *et al.* [13].

5.3.1 Superstructure

The superstructure features a number of different process units and their interconnections. For the design and development of enzymatic and chemo-enzymatic processes, the superstructure (see Figure 5.1) consists of a maximum four processing steps involving reactors, separators, or reactive separators. Also, it is considered that, for two reaction systems, both reactions can take place in a single reactor (one-pot synthesis). When continuous removal or addition of a reactant or product is taking place, *in situ* substrate supply/product removal is considered. The reactors, separators, and reactive separators may have, at the most, two phases (α and β). In each processing unit, there are inlet streams and outlet streams corresponding to the phases α and β.

Bypass streams to all processing units and recycle streams are allowed. The reaction(s) can take place in aqueous and organic phases (no gaseous or solid phases are formed). The catalyst (chemical or enzymatic) can be in solution, immobilized or supported. In Figure 5.1, $T^k, P^k, \mathrm{pH}^k, n_i^k$, and σ_i^k are the temperature, pressure, pH, moles of compound i, and separation factor of the compound i in the unit k, respectively.

$F_i^k, F_i^{k\alpha}, F_i^{k\beta}, F_i^{k\alpha p}, F_i^{k\beta p}, F_i^{k\alpha R}, F_i^{k\beta R}, F_i^{k\alpha R1}, F_i^{k\alpha R2}, F_i^{k\beta R1}, F_i^{k\beta R2}, F_i^{k\beta p1}$, and $F_i^{k\alpha p1}$ are the inlet, outlet, product, and recycle streams; $Y^{k\alpha}, Y^{k\beta}, Y^{k\alpha m}, Y^{k\beta m}, Y^{k\alpha R}$, and $Y^{k\beta R}$ are decision variables (values between 0 and 1) that connect the streams to a second unit directly, by bypass or by recycling.

5.3.2 Model Development

As explained in Stage 3 of the methodology, a generic process model for any enzymatic or chemo-enzymatic process represented by the superstructure in Figure 5.1 has been derived. The model consists of molar and energy balance equations around the superstructure represented in Figure 5.1, connection equations, and constitutive equations. Dynamic model is used always when batch or fed-batch processes are considered. If steady-state process model is selected, the left-hand side of the balance equations is set to zero.

Molar balance for compound i around each unit of the superstructure:

$$
\begin{aligned}
\frac{\partial n_i}{\partial t} = &\sum_{u=1}^{u} Y^u \left(Y_{in}^{u,\alpha} x_{i,in}^{u,\alpha} F_{in}^{u,\alpha} + Y_{in}^{u,\beta} x_{i,in}^{u,\beta} F_{in}^{u,\beta} \right) \\
&- \sum_{u=1}^{u} Y^u \left(Y_{out}^{u,\beta} x_{i,out}^{u,\alpha} F_{out}^{u,\alpha} + Y_{out}^{u,\beta} x_{i,out}^{u,\beta} F_{out}^{u,\beta} \right) + \sum_{u=1}^{u} \left(\lambda_i^{u,\alpha} + \lambda_i^{u,\beta} \right)
\end{aligned}
\tag{5.7}
$$

Energy balance around each unit of the superstructure:

$$
\begin{aligned}
\frac{\partial H}{\partial t} = &\sum_{u=1}^{u} Y^u \left(Y_{in}^{u,\alpha} \sum_{i=1}^{NC} x_{i,in}^{u,\alpha} F_{in}^{u,\alpha} h_{i,in}^{u,\alpha} + Y_{in}^{u,\beta} \sum_{i=1}^{NC} x_{i,in}^{u,\beta} F_{in}^{u,\beta} h_{i,in}^{u,\beta} \right) \\
&- \sum_{u=1}^{u} Y^u \left(Y_{out}^{u,\alpha} \sum_{i=1}^{NC} x_{i,out}^{u,\alpha} F_{out}^{u,\alpha} h_{i,out}^{u,\alpha} + Y_{out}^{u,\beta} \sum_{i=1}^{NC} x_{i,out}^{u,\beta} F_{out}^{u,\beta} h_{i,out}^{u,\beta} \right) \\
&+ \sum_{u=1}^{u} \left(Q^u \right)
\end{aligned}
\tag{5.8}
$$

There are $k = 4$ connection streams to the inlet streams for each processing unit u, therefore,

$$
Y^{u,\alpha} F_{in}^{u,\alpha} = \sum_{k=0}^{4} Y^{u,k,\alpha} F^{u,k,\alpha}
\tag{5.9}
$$

$$
Y^{u,\beta} F_{in}^{u,\beta} = \sum_{k=0}^{4} Y^{u,k,\beta} F^{u,k,\beta}
\tag{5.10}
$$

The outlet streams of each unit are given by

$$Y_{\text{out}}^{u,\alpha} F_{i,\text{out}}^{u,\alpha} = f\left(F_{\text{in}}^{u,\alpha}, F_{\text{in}}^{u,\beta}, x_i^{u,\alpha}, x_i^{u,\beta}, \sigma_i^{u}\right) \tag{5.11}$$

$$Y_{\text{out}}^{u,\beta} F_{i,\text{out}}^{u,\beta} = f\left(F_{\text{in}}^{u,\alpha}, F_{\text{in}}^{u,\beta}, x_i^{u,\alpha}, x_i^{u,\beta}, \sigma_i^{u}\right) \tag{5.12}$$

The outlet streams to the environment for each phase are given by

$$Y_{\text{p}}^{u,\alpha} F_{\text{p}}^{u,\alpha} = Y_{\text{out}}^{u,\alpha} F_{\text{out}}^{u,\alpha} \zeta_u^{u,\alpha} \tag{5.13}$$

$$Y_{\text{p}}^{u,\beta} F_{\text{p}}^{u,\beta} = Y_{\text{out}}^{u,\beta} F_{\text{out}}^{u,\beta} \zeta_u^{u,\beta} \tag{5.14}$$

The connection streams k to each unit are given by

$$Y_{\text{out}}^{u,\alpha} F_{\text{out}}^{u,\alpha} = Y^{u,k,\alpha} F^{u,k,\alpha} \zeta_k^{u,\alpha} \tag{5.15}$$

$$Y_{\text{out}}^{u,\beta} F_{\text{out}}^{u,\beta} = Y^{u,k,\beta} F^{u,k,\beta} \zeta_k^{u,\beta} \tag{5.16}$$

And the conversion rate is given by

$$\lambda = f\left(r_i, n_{\text{catalyst/enzyme}}, x_i, T, P\right) \tag{5.17}$$

The superstructure and the process model are generic, that is, applicable for a wide range of enzymatic and chemo-enzymatic processes and would only need to be developed once. From this generic model, different problem-specific models are generated and tested. Process options based on unreliable models are removed if model predictions are found to be unreliable.

5.3.3 Decomposition Strategy

Due to the complexity of the problem, solution of the model equations can be time-consuming and in some cases even unsolvable. Therefore, a decomposition-based solution strategy is selected here and it can help to reduce the computational time and effort. The decomposition approach divides the mathematical problem into smaller solvable subproblems. The decomposition approach (Stages 4 and 5 of the methodology) is divided into five steps (at the start, there are N1 options):

- Step I: Generation of options and screening by logical constraints in Equation 5.2 to fix a subset of $\underline{Y}$.
- Step II: Screening by structural constraints in Equation 5.3 to fix the remaining $\underline{Y}$. At the end of this step, N2 feasible options are found.
- Step III: Screening by operational constraints to fix some process conditions (Equation 5.4).
- Step IV: Screening by process constraints in Equation 5.5 (the generic process model: Equations 5.7–5.17) to fix the remaining $\underline{X}$ for N3 options.
- Step V: Evaluation of the feasible options—the objective function in Equation 5.1 is calculated for all N3 feasible solutions.

After obtaining the optimal solution(s), they are selected (Stage 6 of the methodology) as the best one(s) for experimental validation, experimentation, and rigorous simulation.

5.4 Application: Case Study

The framework for design of integrated enzymatic and chemo-enzymatic processes is highlighted through the neuraminic acid (Neu5Ac) synthesis, which is an important intermediate pharmaceutical due to its antiviral, anticancer, and anti-inflammatory effect [14]. The reactions taking place during the synthesis are shown in Figure 5.3.

Both reactions suffer from low yields due to the thermodynamic equilibrium and a very complicated or rather expensive downstream processing. The equilibrium in the first reaction favors the substrate, so the production of high amounts of ManNAc is not recommended. To overcome this, usually an excess of pyruvate (Pyr) is added, which complicates downstream processing because the product (Neu5Ac) is difficult to separate from pyruvate as the compounds have similar pK_a values (2.5 and 2.6, respectively) [15].

As shown in Figure 5.4, the two-step synthesis of Neu5Ac (product) may be conceptually divided into an upstream process to produce the ManNAc (intermediate), which acts as the substrate for the aldolase reaction (a biotransformation step converting ManNAc to Neu5Ac), and a downstream process to recover the desired product (Neu5Ac).

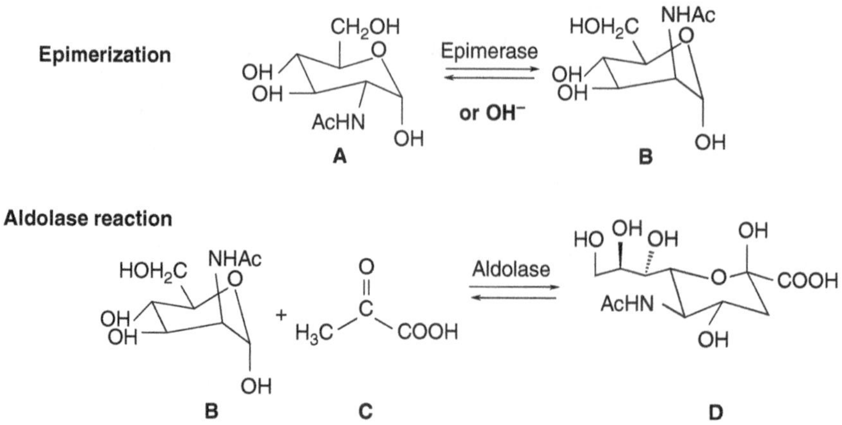

Figure 5.3 Synthesis of *N*-acetyl-neuraminic acid from GlcNAc in two steps (**A**, *N*-acetyl-ᴅ-glucosamine (GlcNAc); **B**, *N*-acetyl-ᴅ-manosamine (ManNAc); **C**, pyruvic acid (Pyr); **D**, *N*-acetylneuraminic acid (NeuAc); epimerase (EPi), *N*-acylglucosamine-2-epimerase (E.C. 5.1.3.8); aldolase (Ald), *N*-acetylneuraminic acid aldolase (E.C. 4.1.3.3)).

Figure 5.4 Conventional process scheme for chemo-enzymatic synthesis of Neu5Ac.

Most of the known production systems for Neu5Ac employ old and conventional methods, that is, a classical batch process with a considerable excess of pyruvate and precipitation of the product with glacial acetic acid. Therefore, the objective is to propose an improved and integrated alternative for the production of Neu5Ac, which achieves improvements in terms of product yield and productivity. To reach this objective, the model-based framework for design and analysis of integrated enzymatic and chemo-enzymatic processes is used.

5.4.1 Stage 1: Problem Definition

For this system, the problem definition statement is having GlcNAc (**A**) and Pyr (**B**) as substrates; the objective is to make the desired product, Neu5Ac (**D**), through two enzyme-based reactions. At the same time, find also the best processing route (network) with respect to a specified objective function. All possible combinations should be considered and analyzed, including the integration of reaction/reaction (e.g., one-pot synthesis), reaction/separation (e.g., *in situ* product removal), and separation/separation. Sequential search based on exhaustive enumeration and testing of alternatives will be done to identify the best.

The criterion for optimal selection as given by the objective function is defined for this case study, as maximizing the percentage yield of the product.

$$\text{Maximize,} \, F_{\text{OBJ}} = \frac{n_{\text{Neu5Ac}}}{n_{\text{GlcNAc},0}} \times 100 \qquad (5.18)$$

5.4.2 Stage 2: Data/Information Collection/Analysis

Process data (components properties, reactions) have been collected [12, 15–29]. The process bottlenecks/limitations are identified as follows:

- *Difficult downstream separation.* The pK_a values for Neu5Ac (2.5) and pyruvate (2.6) are similar, which may lead to a potential difficulty in downstream separation unless the pyruvate concentration is kept low. In the current process, a large amount of pyruvate is used (up to 10-fold [15]) to shift the equilibrium toward the product side, which increases the complexity of downstream processing.
- *Unfavorable equilibrium.* While neither of the reactions gives byproducts, for both reactions the equilibrium is not favorable: for the epimerization

reaction it is 0.24 l/mol, while for the aldolase reaction it is 28.7 l/mol with respect to Neu5Ac synthesis.

- *Dilute concentrations.* Since the equilibrium conversion is proportional to the initial reactant concentrations, for both reactions there should be a compromise in operating with the highest possible substrate molarities but with minimum product inhibition [22].
- *Substrate and product inhibition.* The enzyme for the aldolase reaction is inhibited by both Neu5Ac and GlcNAc, while the enzyme for the epimerization reaction is inhibited by both pyruvate and Neu5Ac.

5.4.3 Stage 3: Superstructure, Model Development, and Decomposition Strategy

After collecting the data concerning the reactions and the methods of separation, the following problem-specific superstructure (Figure 5.5) is created from the original superstructure (Figure 5.1). The maximum number of options that could be generated from the new superstructure, which represents the different possible operations for each processing unit, is now computed. The number of total combinations of flowsheets (processing options) with $u = 4$ processing units interconnected by $r = 16$ streams is given by Equation 5.6:

$$
\mathrm{NPO} = 3\left(\frac{(16 \cdot 1)!}{1!(16 \cdot 1 - 1)!}\right) + 3 \cdot 5\left(\frac{(16 \cdot 2)!}{2!(16 \cdot 2 - 2)!}\right) + 3 \cdot 5 \cdot 5\left(\frac{(16 \cdot 3)!}{3!(16 \cdot 3 - 3)!}\right)
$$
$$
+ 3 \cdot 5 \cdot 5 \cdot 4\left(\frac{(16 \cdot 4)!}{4!(16 \cdot 4 - 4)!}\right)
$$

Total number of options at this stage is N1 = *191,917,488*.

5.4.4 Stage 4: Generation of Feasible Candidates and Screening

Since it is possible to represent every process configuration derived from the superstructure with the generic model, it is therefore also possible to evaluate them. However, before doing this, first the decomposition strategy (explained in Section 5.3) is employed to reduce the number of processing options. All the specified reactions and the identified feasible separation techniques (from Stage 2) are included in the generation of feasible candidates. First the logical constraints, then the structural constraints, and then the operational constraints and finally process models are applied.

According to the superstructure in Figures 5.1 and 5.5, integer variables and equations for the selection of the logical sequence of processing steps and option j in each processing unit are introduced. Therefore, the logical and structural constraints (Equations 5.2 and 5.3) are as follows.

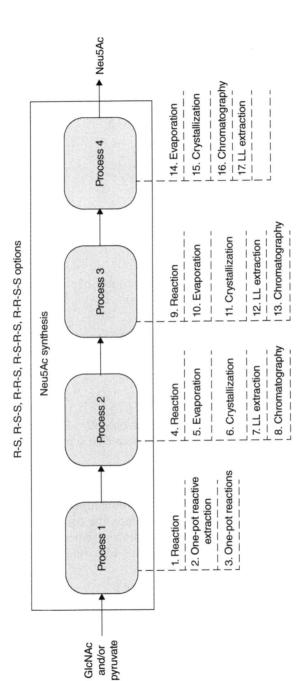

Figure 5.5 Reaction–separation options for the chemo-enzymatic synthesis of Neu5Ac.

For network 1:

$$Y_j^1 = 0 \quad \text{or} \quad 1 \quad j = 1,2,3 \quad \sum_{j=1}^{3} Y_j^1 = 1 \tag{5.19}$$

For network 2:

$$Y_j^2 = 0 \quad \text{or} \quad 1 \quad j = 4,5,6,7,8 \quad \sum_{j=4}^{8} Y_j^2 = 1 \tag{5.20}$$

For network 3:

$$Y_j^3 = 0 \quad \text{or} \quad 1 \quad j = 9,10,11,12,13 \quad \sum_{j=9}^{13} Y_j^3 = 1 \tag{5.21}$$

For network 4:

$$Y_j^4 = 0 \quad \text{or} \quad 1 \quad j = 14,15,16,17 \quad \sum_{j=14}^{17} Y_j^4 = 1 \tag{5.22}$$

And the logical constraints that limit the number of processing units:

$$Y_j^1 + Y_j^2 + Y_j^3 + Y_j^4 \leq 4 \tag{5.23}$$

With the purpose to exclude scenarios that are not feasible, the following structural constraints summarized in Table 5.1 are employed. They also include matches that are not feasible and/or allowed.

5.4.5 Stage 5: Screening by Process Model

After screening in terms of structural constraints, only N2 = 27 options (summarized in Table 5.2) are remaining, which are subjected to screening by operational and process constraints.

A sample calculation is illustrated for processing option 3, which is one-pot reaction followed by crystallization (OPR-CRYST, represented in Figure 5.6). The specific process model for this process configuration is derived from the generic model developed in Stage 3. Therefore, the equations of the specific process model are taken from the generic molar balance Equation 5.9 and its corresponding constitutive equations together with other process constraints.

$$\frac{\partial n_i}{\partial t} = \sum_{u=1}^{u} Y^u \left(Y_{in}^{u,\alpha} x_{i,in}^{u,\alpha} F_{in}^{u,\alpha} + Y_{in}^{u,\beta} x_{i,in}^{u,\beta} F_{in}^{u,\beta} \right)$$

$$- \sum_{u=1}^{u} Y^u \left(Y_{out}^{u,\beta} x_{i,out}^{u,\alpha} F_{out}^{u,\alpha} + Y_{out}^{u,\beta} x_{i,out}^{u,\beta} F_{out}^{u,\beta} \right)$$

$$+ \sum_{u=1}^{u} \left(\lambda_i^{u,\alpha} + \lambda_i^{u,\beta} \right) \tag{5.24}$$

Table 5.1 Structural constraints.

	Integer variables values	Matches not allowed	Rationale
R-S options $Y_j^1 + Y_j^2 = 2$	$Y_1^1 = 0$	Y_2^1 with Y_5^2, Y_7^2	Redundancy of liquid–liquid extraction after a reactive extraction
	$Y_4^2 = 0$	Y_3^1 with Y_5^2, Y_7^2	No availability of compounds of LL extraction
R-S-S options $Y_j^1 + Y_j^2 + Y_j^3 = 3$	$Y_1^1 = 0,\ Y_4^2 = 0,\ Y_9^3 = 0$	Y_2^1 with Y_5^2, Y_7^2	Evaporation may cause a problem in solubility since it is needed to operate in diluted concentrations
		Y_3^1 with Y_7^2	Redundancy of operations
		Y_5^2 with Y_{10}^3, Y_{12}^3	No availability of compounds for LL extraction
		Y_6^2 with Y_{12}^3, Y_{10}^3	
		Y_8^2 with Y_{10}^3, Y_{12}^3	
R-R-S options $Y_j^1 + Y_j^2 + Y_j^3 = 3$	$Y_2^1 = 0, Y_3^1 = 0, Y_5^2 = 0, Y_6^2 = 0,$ $Y_7^2 = 0, Y_8^2 = 0, Y_9^3 = 0$	Y_4^2 with Y_{10}^3, Y_{12}^3	No available compounds for LL extraction Evaporation may cause a problem in solubility since it is needed to operate in diluted concentrations
R-S-R-S options $Y_j^1 + Y_j^2 + Y_j^3 + Y_j^4 = 4$	$Y_1^1 = 0, Y_3^1 = 0, Y_4^2 = 0, Y_{10}^3 = 0,$ $Y_{11}^3 = 0, Y_{12}^3 = 0, Y_{13}^4 = 0, Y_{14}^4 = 0$	Y_1^1 with Y_7^2, Y_8^2 Y_9^3 with Y_{14}^4, Y_{17}^4	No available compounds for LL extraction Evaporation may cause a problem in solubility since it is needed to operate in diluted concentrations
R-R-S-S options $Y_j^1 + Y_j^2 + Y_j^3 + Y_j^4 = 4$	$Y_2^1 = 0, Y_3^1 = 0, Y_5^2 = 0, Y_6^2 = 0,$ $Y_7^2 = 0, Y_8^2 = 0, Y_9^3 = 0, Y_{14}^4 = 0$	Y_4^2 with Y_{12}^3	Redundancy of operations
		Y_{10}^3 with Y_{14}^4, Y_{17}^4	No available compounds for LL extraction
		Y_{11}^3 with Y_{16}^4, Y_{17}^4	Evaporation may cause a problem in solubility since it is needed to operate in diluted concentrations
		Y_{13}^3 with Y_{17}^4	

Table 5.2 Options after screening by structural constraints.

Option no.	Process	Option no.	Process
1	OPRS-CRYST	15	R-R-CRYST
2	OPRS-CHRO	16	R-R-CHRO
3	OPR-CRYST	17	R-EVAP-R-CRYST
4	OPR-CHRO	18	R-EVAP-R-CHRO
5	OPR-CRYST-EVAP	19	R-CRYST-R-CRYST
6	OPR-CRYST-CRYST	20	R-CRYST-R-CHROM
7	OPR-CHRO-CRYST	21	R-R-EVAP-CRYST
8	OPR-CHRO-CHRO	22	R-R-EVAP-CHRO
9	OPR-EVAP-CRYST	23	R-R-CRYST-EVAP
10	OPR-EVAP-CHRO	24	R-R-CRYST-CRYST
11	OPR-CRYST-EVAP	25	R-R-CHRO-EVAP
12	OPR-CRYST-CRYST	26	R-R-CHRO-CRYST
13	OPR-CHRO-CRYST	27	R-R-CHRO-CHRO
14	OPR-CHRO-CHRO		

CHRO, chromatography; CRYST, crystallization; EVAP, evaporation; OPR, one-pot reactions; OPRS, one-pot reactive extraction; R, reaction.

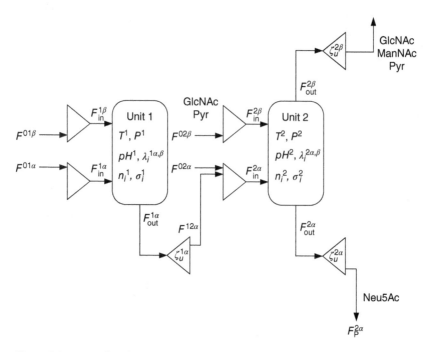

Figure 5.6 Process flowsheet (network) for option 3 (OPR-CRYST).

For this case, the following binary variables exist, while all other binary variables in the superstructure are zero:

$$Y^{01\alpha} = Y_{in}^{1\alpha} = Y_{out}^{1\alpha} = Y^{12\alpha} = 1 \tag{5.25}$$

$$Y^{02\alpha} = Y_{in}^{2\alpha} = Y_{out}^{2\alpha} = Y_{out}^{2\beta} = Y_p^{2\alpha} = Y_p^{2\beta} = 1 \tag{5.26}$$

As shown in Figure 5.6, there is no splitting of streams; therefore, the connection scheme of the existing flow streams F is given by

$$\zeta_u^{1\alpha} = \zeta_u^{2\alpha} = \zeta_u^{2\beta} = 1 \tag{5.27}$$

$$F^{01\alpha} = F_{in}^{1\alpha} \tag{5.28}$$

$$F_{out}^{1\alpha} = F^{12\alpha} \tag{5.29}$$

$$F_{in}^{2\alpha} = F^{12\alpha} + F^{02\alpha} \tag{5.30}$$

$$F_{out}^{2\beta} = F_p^{2\beta} \tag{5.31}$$

$$F_{out}^{2\alpha} = F_p^{2\alpha} \tag{5.32}$$

Since the defined scenario is run in batch, the generic mass balance Equation 5.7 is split into two time domains, one for the reaction and another for the separation afterward:

$$\frac{dn_i}{dt} = \left[\frac{dn_i}{dt}\right]_{t=0}^{t=t_{reaction}} + \left[\frac{dn_i}{dt}\right]_{t=t_{reaction}}^{t=t_{end}} \tag{5.33}$$

During the reaction time, the initial concentration (Equation 5.34) is changing due to two reactions. The conversion rates are functions of component and enzyme concentrations determined from the literature:

$$\left[\frac{dn_i}{dt}\right]_{t=0} = x_{i,in}^{1\alpha} F_{in}^{1\alpha} \tag{5.34}$$

$$\frac{dn_i}{dt_{reaction}} = \lambda_i^{reaction1} + \lambda_i^{reaction2} \tag{5.35}$$

The conversion rates are replaced by derived kinetic reaction expressions [27, 28, 30].

When the reaction is in equilibrium or complete, the contents of the reactor are emptied and sent to the crystallization step where the feed is initially mixed with glacial acetic acid.

$$x_{i,\text{out}}^{1\alpha} F_{\text{out}}^{1\alpha} = \left[\frac{dn_i}{dt} \right]^{t=t_{\text{reaction}}} \tag{5.36}$$

$$\left[\frac{dn_i}{dt} \right]_{t=t_{\text{reaction}}} = x_i F_{\text{in}}^{2\alpha} \tag{5.37}$$

The product stream $F_p^{2\beta}$ can be calculated with Equation 5.38 in which the crystallization time t_{cryst} is a function of the volume of the solution and the concentrations:

$$x_D F_p^{2\beta} = \sigma^\beta \cdot \frac{t}{t_{\text{cryst}}} \cdot x_D F_{\text{in}}^{2\alpha} \tag{5.38}$$

All the other components are found in the product stream $F_p^{2\alpha}$.

The following operational constraints have been identified, that is, the ratio of A over D [20] and C over D [31]:

$$\frac{n_A}{n_D} \leq 0.3 \tag{5.39}$$

$$\frac{n_C}{n_D} \leq 2.2 \tag{5.40}$$

The results of the process simulations for all the 27 remaining options, using the tool ICAS-MoT [32] of ICAS12®, are listed in Table 5.3. In this table, the following N3 = 10 options: 1, 2, 3, 5, 6, 7, 8, 10, 17, and 18 have been found to satisfy all process constraints and these are further investigated in the next stage.

5.4.6 Stage 6: Evaluation of the Feasible Options: Calculation of the Objective Function

In this stage, the most promising options are further investigated, first in terms of the objective function (Equation 5.18) and then in terms of Equation 5.40. The ten most promising candidates are listed in Table 5.4, and they are the ones that satisfy all the constraints including the additional constraint (Equation 5.40). From the results in Table 5.4, it can be noted that option 2 (OPRS-CHRO: One-pot reactive extraction followed by a chromatographic step) is the best in terms of product yield percentage (53.85%). This option (see Figure 5.7) consists of an integrated reaction and extraction in a batch reactor with continuous renewal of the organic phase, followed by a chromatographic step.

Table 5.3 Results of process simulation at same initial substrates amounts ($n_{GlcNAc,0} = 1.3$ mol, $n_{Pyr,0} = 2.6$ mol) and same enzyme concentrations ($C_{epi} = 1500$ U/l, $C_{ald} = 24\,000$ U/l).

Option no.	Process	Neu5Ac production after reaction(s) (g/day)	(Neu5Ac recovered) after separation (g/day)
1*	OPRS-CRYST	269	201.02
2*	OPRS-CHRO	269	216.49
3*	OPR-CRYST	213.4	160.82
4	OPR-CHRO	213.4	120.61
5*	OPR-CRYST-EVAP	269	185.56
6*	OPR-CRYST-CRYST	269	151.54
7*	OPR-CHRO-CRYST	269	160.82
8*	OPR-CHRO-CHRO	269	173.19
9	OPR-EVAP-CRYST	213.4	145.35
10*	OPR-EVAP-CHRO	213.4	157.06
11	OPR-CRYST-EVAP	213.4	147.24
12	OPR-CRYST-CRYST	213.4	120.04
13	OPR-CHRO-CRYST	213.4	128.04
14	OPR-CHRO-CHRO	213.4	128.04
15	R-R-CRYST	72.37	54.28
16	R-R-CHRO	72.37	57.90
17*	R-EVAP-R-CRYST	217.11	162.83
18*	R-EVAP-R-CHRO	217.11	173.69
19	R-CRYST-R-CRYST	180.92	135.69
20	R-CRYST-R-CHROM	180.92	144.73
21	R-R-EVAP-CRYST	72.37	49.93
22	R-R-EVAP-CHRO	72.37	53.26
23	R-R-CRYST-EVAP	72.37	46.39
24	R-R-CRYST-CRYST	72.37	40.71
25	R-R-CHRO-EVAP	72.37	49.93
26	R-R-CHRO-CRYST	72.37	43.42
27	R-R-CHRO-CHRO	72.37	43.42

Note: * in column 1 indicates the options retained after Stage 5.

Table 5.4 Calculation of the objective function of the feasible alternatives.

Option no.	Process	Product yield (%)
1	OPRS-CRYST	50.00
2*	OPRS-CHRO	53.85
3	OPR-CRYST	40.00
5	OPRS-CRYST-EVAP	46.15
6	OPRS-CRYST-CRYST	37.70
7	OPRS-CHRO-CRYST	40.00
8	OPRS-CHRO-CHRO	43.08
10	OPR-EVAP-CHRO	39.23
17	R-EVAP-R-CRYST	40.77
18	R-EVAP-R-CHRO	43.20

Note: * indicates the best processing option.

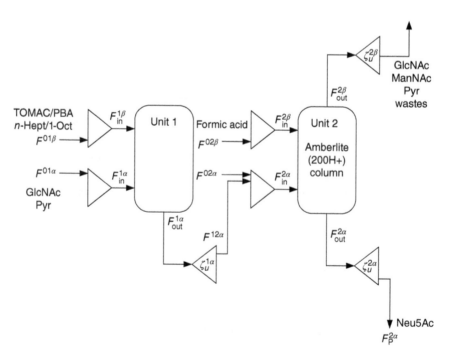

Figure 5.7 Process flowsheet for option 2 (OPRS-CHRO).

5.5 Conclusions

Integration of enzymatic and chemo-enzymatic processes together with their corresponding separation techniques can help to create novel pharmaceutical processing options that may lead to better control of product quality. Two of the most important methods of integration of this kind of processes are the one-pot synthesis and the ISPR. The introduction of these two integration approaches together with a systematic model-based methodology for process design allows one to generate and evaluate systematically numerous options to identify the most promising ones for the final step of verification by experiments. This chapter presented a model-based framework where the methodology has been implemented together with other necessary computer-aided tools. Applicability of the framework is limited by the availability of the necessary data and the models in the model library. Integration of model identification options together with process synthesis tools that also incorporate process intensification would be a natural extension of this framework. Also, systematic classification of reaction systems where the aforementioned concepts can be directly applied is under development.

References

1 Hsu S, Reklaitis GV, Venkatasubramanian V. Modeling and control of roller compaction for pharmaceutical manufacturing. Part I: Process dynamics and control framework. Journal of Pharmaceutical Innovation. 2010; 5: 14–23.

2 Harms J, Wang X, Kim T, Xiaoming Y, Rathore AS. Defining process design space for biotech products: Case study of *Pichia pastoris* fermentation. Biotechnology Progress. 2008; 24: 655–662.

3 Rathore AS. Roadmap for implementation of quality by design (QbD) for biotechnology products. Trends in Biotechnology. 2009; 27: 546–553.

4 Rathore AS, Branning R, Cecchini D. Design space for biotech products. BioPharma International. 2007; 20(4): 36–40.

5 Lepore J, Spavins J. PQLI design space. Journal of Pharmaceutical Innovation. 2008; 3: 79–87.

6 Garcia T, Cook G, Nosal R. PQLI key topics—criticality, design space and control strategy. Journal of Pharmaceutical Innovation. 2008; 3: 60–68.

7 Nosal R, Schultz T. PQLI definition of criticality. Journal of Pharmaceutical Innovation. 2008; 3: 69–78.

8 Hailes HC, Dalby PA, Woodley JM. Perspective. Integration of biocatalytic conversion into chemical synthesis. Journal of Chemical Technology and Biotechnology. 2007; 82: 1063–1066.

9 Kim P, Pollard DJ, Woodley JM. Substrate supply for effective biocatalysis. Biotechnology Progress. 2007; 23: 74–82.

10 Woodley JM, Bisschops M, Straathof AJJ, Ottens M. Perspective. Future directions for *in-situ* product removal (ISPR). Journal of Chemical Technology and Biotechnology. 2008; 83: 121–123.

11 Karunanithi AT, Achenie LEK, Gani R. A new decomposition-based computer-aided molecular/mixture design methodology for the design of optimal solvents and solvent mixtures. Industrial and Engineering Chemistry Research. 2005; 44: 4785–4797.

12 Roman-Martinez A. A model-based framework for design of intensified enzyme-based processes. PhD thesis. Technical University of Denmark: Lyngby, 2011.

13 Lutze P, Roman-Martinez A, Woodley JW, Gani R. A systematic synthesis and design methodology to achieve intensification in (bio) chemical processes. Computers and Chemical Engineering. 2012; 36: 189–207.

14 Tao F, Zhang Y, Ma C, Xu P. Biotechnological production and applications of N-acetyl-D-nueraminic acid: current state and perspectives. Applied Microbial Biotechnology 2010; 87: 1281–1289.

15 Dawson MJ, Noble D, Mahmoudian M. Process for the preparation of N. acetylneuraminic acid. 2000. US Patent 6156544.

16 Augé C, David S, Gautheron C. Synthesis with immobilized enzyme of the most important acid. Tetrahedron Letters. 1984; 25: 4663–4664.

17 Juneja LR, Koketsu M, Nishimoto K, Kim M, Takehiko Y, Itoh T. Large-scale of sialic acid from chalaza and egg-yolk membrane. Carbohydrate Research. 1991; 214: 179–186.

18 Kragl U, Gygax D, Ghisalba O, Wandrey C. Enzymatic two-step synthesis of N-acetyl-neuraminic acid in the enzyme membrane reactor. Angewandte Chemie International Edition English. 1991; 30: 827–828.

19 Ohta Y. Method for preparing N-acetylneuraminic acid by N-acetylneuraminic acid lyase at a pH of 10-12. 1995. US Patent 5472860.

20 Mahmoudian M, Noble D, Drake CS, Middleton RF, Montgomery DS, Piercey JE, Ramlakhan D, Todd M, Dawson MJ. An efficient process for production of N-acetylneuramic acid using n-acetylneuraminic acid aldolase. Enzyme and Microbial Technology. 1997; 20: 393–400.

21 Maru I, Ohnishi J, Ohta Y, Tsukada Y. Simple and large-scale production of N-acetylneuraminic acid from N-acetyl-D-glucosamine and pyruvate using N-acyl-D-glucosamine 2-epimerase and N-acetylneuraminate lyase. Carbohydrate Research. 1998; 306: 575–578.

22 Blayer S, Woodley JM, Dawson MJ, Lilly MD. Alkaline biocatalysis for the direct synthesis of N-acetyl-D-neuraminic acid (Neu5Ac) from N-acetyl-D-glucosamine (GlcNAc). Biotechnology and Bioengineering. 1999; 66: 131–136.

23 Tabata K, Koizumi S, Endo T, Ozaki A. Production N-acetyl-D-neuraminic acid by coupling bacteria expressing N-acetyl-D-glucosamine 2-epimerase and N-acetyl-D-neuraminic acid synthetase. Enzyme and Microbial Technology. 2002; 30: 327–333.

24 Lee J, Yi J, Lee S, Takahashi S, Kim B. Production of N-acetylneuraminic acid from N-acetylglucosamine and pyruvate using recombinant human rennin binding protein and sialic acid aldolase in one pot. Enzyme and Microbial Technology. 2004; 35: 121–125.

25 Xu P, Qiu JH, Zhang YN, Chen J, Wang PG, Yan B, Son J, Xi RM, Deng ZX, Ma CQ. Efficient whole-cell biocatalytic synthesis of N-acetyl-D-neuraminic acid. Advanced Synthesis and Catalysis. 2007; 349: 1614–1618.

26 Lee Y, Chien HR, Hsu W. Production of N-acetyl-D-neuraminic acid by recombinant whole cells expressing *Anabaena* sp. CH1 N-acetyl-D-glucosamine 2-epimerase and *Escherichia coli* N-acetyl-D-neuraminic acid lyase. Journal of Biotechnology. 2007; 129: 453–460.

27 Zimmermann V, Hennemann HG, Daußmann T, Kragl U. Modelling the reaction course of N-acetylneuraminic acid synthesis from N-acetyl-D-glucosamine—new strategies for the optimization of neuraminic acid synthesis. Applied Microbiology and Biotechnology. 2007; 76: 597–605.

28 Zimmermann V, Masuck I, Kragl U. Reactive extraction of N-acetylneuraminic acid—a new method to recover neuraminic acid from reaction solutions. Separation and Purification Technology. 2008; 61: 60–67.

29 Wang T, Chen Y, Pan H, Wank F, Cheng C, Lee W. Production of N-acetyl-D-neuraminic acid using two sequential enzymes over expressed as double-tagged fusion proteins. BMC Biotechnology. 2009; 9: 63.

30 Zimmermann V, Masuck I, Kragl U. Reactive extraction of N-acetylneuraminic acid-kinetic model and simulation of integrated product removal. Separation and Purification Technology. 2008b; 63: 129–137.

31 Yamaguchi S, Ohnishi J, Maru I, Ohta Y. Simple and large scale production of N-acetylneuraminic acid and N-acetyl-D-mannosamine. Trends in Glycoscience and Glycotechnology. 2006; 18: 245–252.

32 Gani R, Hytoft G, Jaksland C, Jensen AK. An integrated computer aided system for integrated design of chemical processes. Computers and Chemical Engineering. 1997; 21: 1135–1146.

6

Methods and Tools for Design Space Identification in Pharmaceutical Development

Fani Boukouvala, Fernando J. Muzzio, and Marianthi G. Ierapetritou

Department of Chemical and Biochemical Engineering, Rutgers University, Piscataway, NJ, USA

6.1 Introduction

The need for a more structured approach to process development has been recently identified in the pharmaceutical industry in order to consistently guarantee quality and value to processes and products [1–4]. The concept of design space (DS) is a key aspect of the pharmaceutical quality by design (QbD) initiative and was formally introduced in the International Conference on Harmonisation (ICH) Q8 guideline for pharmaceutical development [5]. According to the formal definition, it is "the multidimensional combination and interaction of input variables (e.g., material attributes) and process parameters that have been demonstrated to provide assurance of quality. Working within the design space is not considered as a change. Movement out of the design space is considered to be a change and would normally initiate a regulatory post approval change process." The importance of the design space definition lies firstly in the "assurance of quality," which is the key goal of QbD, and secondly in the broadening from an acceptable operating set point, to a collection of tolerable operating regions, which make up the DS. Thus, if a process DS is accurately identified, one can have the freedom to operate within this entire region without additional regulatory approval and simultaneously have higher confidence about the final product quality. For all the aforementioned reasons, the DS is a critical component of the recent effort of introducing QbD in pharmaceutical development and has attracted a lot of attention in the literature.

In ICH Q8 two critical aspects of the DS are defined, namely, the selection of variables and the final presentation of the DS. The first step to identifying a DS is the recognition of the critical variables that should be included. The presentation of a DS should provide the ranges of these critical variables that

Comprehensive Quality by Design for Pharmaceutical Product Development and Manufacture, First Edition. Edited by Gintaras V. Reklaitis, Christine Seymour, and Salvador García-Munoz.

result in a product meeting a desired quality, but should not be confused with a simple combination of proven acceptable ranges. As a result of this introductory report, two more elaborate guidance papers [6, 7] were published from specialized teams, which aim to provide a more detailed description of the DS. Specifically, the Design Space Task Team of the International Society for Pharmaceutical Engineering (ISPE) outlines the components and steps of a DS through the Product Quality Lifecycle Implementation (PQLI) initiative.

In Ref. [7] it is emphasized that the sources for accurately defining the DS of a process may include the available literature, experience and knowledge of the process, first principles, experimental data, empirical models, or—most often—some combination of all these methods. However, the choice of the tools used to characterize the DS depends on the availability of these resources. For example, when the process is well characterized and the underlying principles are known, the DS should be obtained by predictive first-principles models. The main advantage of the existence of a first-principles model is the ability to simulate the process under any given conditions for predicting the outputs of interest. In fact, it is only recently that computer-aided process design and simulation has become an essential tool for pharmaceutical process development and optimization of manufacturing [8]. On the other hand, if a process is new and its scientific principles are not well understood, it can be treated as a black-box process. In the later case, the DS dynamically evolves, since as additional knowledge and information about a process is obtained or as new raw materials, evolving specifications, and new technology become available, the DS of the process is better characterized. In addition, the DS should be insensitive to the scale of a process even if this requires the performance of additional scale-up experiments.

Even though recent discussion has focused on general guidelines [7] and the types of variables that a DS should include [8], relatively little has been put forth in terms of defining methods to construct a DS of a process. Clearly, there is no unique approach in defining a DS, but once the methodology and tools to determine the DS are chosen, the DS has to be presented efficiently, in a way that can be easily interpreted. Figure 6.1 is a very common representation of the DS, used to explain graphically its relationship to the knowledge space and the normal operating ranges. The larger knowledge space contains all the information about all regions of the process that have been investigated. Subsequently, the difference between the knowledge and design space can be defined as the region of the knowledge space that generates unacceptable product.

The DS is interconnected with the two remaining PQLI key topics: criticality and control strategy [6, 9, 10]. Criticality refers to the determination of which quality attributes and operating conditions of a process are necessary in order to predict a desired output so that they will only be included in the DS. A risk assessment procedure is defined that filters the variables through a series of

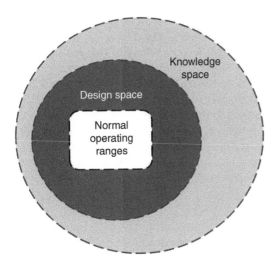

Figure 6.1 Link between knowledge space, design space, and normal operating ranges.

questions in order to identify their effects on safety, quality, and efficacy of a process output. As a result of this procedure, the variables are classified according to their level of criticality. The necessary set of controls for assurance of acceptable process performance and product quality is the final PQLI key topic, that is, the required control strategy. Clearly, the three topics are highly connected since it is necessary to first identify the critical variables (criticality) that define a DS and these critical variables should be manipulated based on an efficient control strategy in order for a process to remain within the DS at all times.

A framework for the development of the design and control space was presented in Ref. [11] where the importance of simultaneously identifying the material properties, process operating parameters, and control space is discussed. The significance of raw material properties in the DS characterization is important since these are properties that cannot be easily manipulated or controlled, as opposed to operating conditions. In addition, the need to take account of control strategies in the DS can be illustrated by identifying the—possibly significant—changes in a DS if the applied control strategy is modified [11, 12]. However, even by implementing a very efficient control strategy, specific raw materials cannot meet the acceptable final product specifications. Recently, the work of Peterson *et al.* [13, 14] has focused on showing the limitations of a simplistic overlapping means approach for mapping a DS that fails to answer the question about how much assurance of quality such an approach can provide. Specifically a Bayesian approach for the identification of the ICH Q8 DS is adopted in order to calculate the probability of meeting specifications. Finally, an introduction to probabilistic risk assessment through the introduction of the Taguchi quality loss concept and its connection to pharmaceutical quality, risk, and DS can be found in Ref. [15].

Several examples of specific experimental applications to define the DS of pharmaceutical processes can be found in the literature. For example, in Ref. [16], the operating window for the process of a fluidized bed granulation has been identified by assessing the impact of critical parameters, such as inlet air humidity, on fluidization behavior and granule size. A similar procedure was followed in Ref. [17] in order to define the confidence zone of chromatographic analytical methods. In Ref. [18], three different data-based methodologies are used to generate the graphical representation of the region in which acceptable performance of unit black-box processes can be ensured. The efficiency of these approaches is compared through two pharmaceutical case studies: predicting the DS of a continuous powder mixer and a loss-in-weight feeder of a continuous tablet manufacturing process. Integration of feed-forward control in the DS of a high-shear wet granulation process was studied by Garcia-Munoz *et al.* [12] where it is shown that a DS can be significantly enlarged if a correct control strategy is employed. This work signifies the advantage of a successful control sequence, which increases the flexibility of a process toward uncertainty and inevitable variations.

This work aims to summarize all the developed methods and modeling tools used for mapping the DS of pharmaceutical processes, as well as present new perspectives, by introducing into the DS discussion concepts that are well established in the process systems engineering community and have been successfully used in several other industrial fields. The remaining of this chapter is organized as follows. Section 6.2 describes the DS as a multidisciplinary concept since for its definition and identification, different concepts from fields such as statistics, optimization, and process systems design are used. The importance of integrating process DS with control strategies is outlined in Section 6.3. Finally, two case studies are described, which aim to exemplify the majority of the described tools and methods in Section 6.4.

6.2 Design Space: A Multidisciplinary Concept

By definition, design space is not an unknown concept. The use of risk-based analyses to identify design constraints and the necessary control scheme for a process are concepts and ideas that have been extensively studied in other industries and literatures [19]. Through the work of Peterson *et al.* [20], the incessant increase in the role of statistics for the identification of DS is identified. In addition, the work introduced by the authors [18] that makes use of concepts introduced in the process systems design literature, such as process feasibility and flexibility, is also proven to be useful in both DS identification and its final presentation. This section aims to connect all the developed methods and tools for the DS identification in order to identify similarities and differences as well as to introduce new concepts that can be constructive for the same purpose.

The role of statistics in pharmaceutical development and manufacturing is becoming more and more significant as the industry is undergoing through rapid changes in order to incorporate QbD and PAT principles [21]. Through a Bayesian framework the quantification of the amount of reliability that a DS provides can be achieved through a clearly defined figure of merit (Equation 6.1) [13, 14, 21, 22]:

$$\Pr\left(Y \in A \mid x, \text{data}\right) \geq R \tag{6.1}$$

where for a given acceptance region A, the probability of a response Y belonging to A, given the input vector x and a set of data, should be greater than or equal to a predefined reliability level R. For this calculation, a posterior predictive distribution for Y must be assumed and all the uncertainty of the process model parameters should be taken into account. Several applications of Bayesian DS of different pharmaceutical and biopharmaceutical processes are presented in Refs. [14, 22]. Additional statistical tools that can be applied for the identification of critical parameters and critical uncertainty sources, which is the first step in DS identification, are sensitivity and correlation analysis [23]. Sensitivity analysis can be used for the classification of criticality of the process variables as well as their possible interactions. On the other hand, correlation analysis aims to improve the understanding of the uncertain multiple-unit integrated system by identifying the interactions of the different unit operations of the system under uncertainty. It is reasonable to assume that reducing the uncertainty in earliest stages may result in a more robust definition of the process since uncertainty propagates through the different stages. Correlation analysis, however, can help identify, for example, the most critical stage in this uncertainty propagation.

The Bayesian framework for DS identification is similar to a concept that has been introduced in the process systems design community, namely, stochastic flexibility (SF) [24–26]. SF is a probabilistic measure that was established in the 1990s to measure the system's ability to tolerate continuous uncertainty [24, 27–29]. In other words, it can be defined as the probability of a given process design to operate feasibly. Mathematically, any process can be modeled through the following set of equality and inequality constraints:

$$\begin{aligned} h\left(d, z, x, \theta\right) &= 0 \\ g\left(d, z, x, \theta\right) &\leq 0 \end{aligned} \tag{6.2}$$

where d represents the design variables, z represents the control variables, x are the state variables, and θ are the uncertain parameters of a process. Since the uncertain parameters are usually assumed to follow a distribution function, the determination of SF involves the calculation of a multiple integral of the joint probability of all the uncertain parameters over the parametric space. SF corresponds to the cumulative probability of the joint distribution that lies within the feasible region. SF, however, was an extension of the concept of process flexibility that was introduced earlier (1980s) as the identification of the boundaries of a process where feasible, profitable, and acceptable performance of the process is guaranteed [30–32]. For many years, this has been considered as a major concern in many process industries, and a substantial amount of work has been done in order to define concepts such as "operability," "feasibility," and "flexibility" of processes that involve uncertain parameters. Uncertainty can occur for a variety of reasons; most common among them is the variability of certain process parameters during plant operation. Uncertainties might have a major impact on the process performance, and it is very important to assess—and ideally quantify—whether the uncertainty in some parameters may potentially cause problems that may be interpreted in the real process.

Optimal process design under uncertainty was defined as a mathematical formulation in Ref. [31] where the effects of parameters that contain considerable uncertainty on the optimality and feasibility of a chemical plant were studied. The objective of solving such problems was to ensure optimality and feasibility of operation for a given range of uncertain parameter values by identifying a measure of the size of the feasible region of operation. In Ref. [33], the flexibility of such processes was defined and quantified by the flexibility index (FI), which represented the maximum allowed deviation of uncertain parameters from their nominal values, such that feasible operation could be guaranteed by changing the control variables. A series of papers dealing with flexibility analysis and the formulation and optimization of processes under uncertainty were published in the following years for cases where the process model is known in closed form [30, 34–36] as well as for cases where the process is treated as a black box [37–40].

A typical mathematical formulation of an optimization problem under uncertainty is the following:

$$\max/\min_{d,z,x} P\left(d,z,x,\theta\right)$$

s.t.

$$h\left(d,z,x,\theta\right)=0$$
$$g\left(d,z,x,\theta\right)\leq 0$$

(6.3)

where the notation is identical to Equation 6.2 and P is the desired objective function. It is by now very common in all areas of science and engineering to utilize mathematical programming techniques to solve such problems. In general, the objective function to be optimized is representative of the performance criteria of the process, for example, minimization of cost and output variability or maximization of profit [41]. The solution of this problem is a set of optimal values of the discrete and continuous variables that simultaneously minimize/maximize the objective(s) and satisfy the constraints. When dealing with multiple-objective problems (i.e., choosing between maximizing expected product performance and minimizing process variability), this can generate trade-off problems that will need to be solved for making a decision [28].

Another concept that can be linked to the DS and the concept of trade-off optimization problems is robustness. The definition of robustness is the "ability of a process to operate under changing conditions with a relatively stable performance" (insensitive to uncertainty) [42]. The way process variability has been handled so far by performing deterministic feasibility analysis can be interpreted as the allowance of a maximum variance (uncertain parameter varies between upper and lower bounds). Another way, however, to penalize variability is to model it in the form of a continuous loss function (Taguchi's perspective) [24]. For example, if a problem contains an uncertain variable θ with a nominal value of θ^*, it is assumed that any deviation in this variable will affect the objective (i.e., cost function), and since variability is unwanted, this can be modeled by introducing a penalty loss term in the objective function such as

$$C_q = k\left(\theta - \theta^*\right)^2 \tag{6.4}$$

where k is a penalty parameter. The expected value of this loss function is equal to

$$E\left(C_q\right) = k\left[\sigma^2 + \left(\mu - \theta^*\right)^2\right] \tag{6.5}$$

where σ, μ are the standard deviation (SD) and mean of parameter θ. Equation 6.5 can be divided into two components: (i) variability loss and (ii) loss due to deviation from the mean value. This introduces an additional choice to the user since a constraint can be modeled through setting strict upper and lower bounds or else through a continuous loss term. Employing such an approach contributes toward the same goal as the probabilistic DS approach introduced by Peterson *et al.* [13, 14], that is to say, there needs to be a distinction between product that is exactly on target (high reliability) and product that just meets specifications (low reliability).

Most of the aforementioned concepts and methods of this multidisciplinary review will be made clear through the case studies presented in Section 6.4. What is realized by the authors is that all of the aforementioned well-established process systems engineering tools can be very useful in developing methods for pharmaceutical DS identification, where critical input parameters and variables might contain some level of uncertainty, or simply may be varied during process operation. The use of the aforementioned tools is expected to be very significant in the near future due to the current development of more first-principles process models for various unit operations [3, 43–47]. In the next sections, it will be demonstrated that the purpose of a well-developed DS should convey the maximum deviations and combinations of all the critical inputs that provide assurance of quality, and this can be achieved through coupling of process flexibility with process models.

6.3   Integration of Design Space and Control Strategy

Process control in engineering is the online modification of a process based on the results of real-time process monitoring. In pharmaceutical engineering, the control space is the region within which a process is normally operated and it should be a subset of the DS to attain robustness. Otherwise, effective process control is necessary in order to assure that a process is constantly operated within the DS. Through [11, 12] it is shown that a successful control strategy can significantly affect the DS of a process in a positive manner since the acceptable ranges of material property or process input variable variations can be enlarged. Thus, it is important to account for the online manipulations of input variables based on a control scheme in order to get a correct form of the process DS.

Recently, the advantages and the need of integrating process design and control have been recognized for ensuring both optimality and controllability of a process [48, 49]. Specifically, the ability of a process to move from one steady state to another is quantified in order to avoid unacceptable disturbances and minimize unnecessary variability in the final product. First, the desired outputs and the possible expected disturbances are taken into account and then the operability of a process is defined. Operability can be assured by controlling the set of input parameters while identifying the optimum out of possibly multiple competing designs. The applications of operability analysis can be classified into linear or nonlinear and also steady-state or dynamic models [50–52]. In Ref. [53], an interested reader can find a systematic comparison of the terms of operability and flexibility and their applications in real-life steady-state case problems. In this work, the significance of a control strategy and its effects on the DS are shown through Case Study 2.

6.4   Case Studies

6.4.1   Design Space of a Continuous Mixer: Use of Data-Driven-Based Approaches

In this section a simple continuous mixer case study is chosen in order to illustrate the capabilities of coupling data-driven modeling and feasibility analysis techniques—for the mapping of the DS of a process. In this case study, the goal is to extract useful information from a data set based on a design of experiment (DOE) to characterize the robustness of a mixer as a function of its operating conditions and design configuration. The effect of material properties is not assessed in this DOE, but following the same methodology, the DS can incorporate material property effects if the necessary data set is available. The uniqueness of this case study is the existence of different possible designs, whose effects are incorporated into the DS by solving an optimization problem.

The performance of a commercial continuous powder mixer (Gericke GCM 250) is studied as a function of operating parameters and design variables. The parameters that have been identified as significant—through preliminary statistical analysis of variance (ANOVA)—involve the impeller rotation rate and the powder flow rate, and the design variable is the blade configuration ("All forward" and "Alternate Blades"). Forward direction imposes a forward flow along the axis of the mixer; backward direction imposes powder flow in the opposite direction. The first two variables are continuous, whereas the last one has to be represented as a discrete variable.

The output measured to characterize the performance of the mixer is the relative standard deviation (Equation 6.6) of the concentration of the active pharmaceutical ingredient (API), which in this case is acetaminophen. Initially pure acetaminophen (API) was pre-blended with a small amount of silicon dioxide (0.25%) in a V-blender, and finally Avicel PH-200 was blended with acetaminophen (APAP). The RSD measurements are a result of the ratio of the SD of acetaminophen concentration over the average concentration of a large number of samples. For each experimental run, 20 samples were collected from the outlet. Concentration of acetaminophen in each sample was measured using a NIR spectroscopy analytical method:

$$\mathrm{RSD} = \frac{s}{\bar{C}} = \frac{\text{standard deviation}}{\text{average concentration}} \quad s = \sqrt{\frac{\sum_1^N \left(C_i - \bar{C}\right)^2}{N-1}}. \tag{6.6}$$

where $\bar{C}$ is the average concentration of the total samples (N) collected in each mixing run and C_i is the concentration of each sample. s is the standard deviation between the sample concentrations. Even though in this example the methodology used can be considered as deterministic, the aspect of reliability of the produced DS is introduced by the measured output itself, since it is a standard deviation of a large number of measurements.

Based on the data obtained, three different data-based methodologies (response surface methodology (RSM), Kriging, and high dimensional model representation (HDMR)) are used to predict the output as a function of the two operating conditions at different design configurations. In other words, each design is treated as a separate model in the proposed methodology. In the case where RSM is used, the output is given in closed form as a second-order polynomial with interaction terms and fitted parameters based on the experimental data. If the other two interpolating methods are used, the output is not given in a closed-form equation, but in an interpolating look-up table structure. More details about the three methodologies can be found in Ref. [18]. In order to incorporate and compare the effects of different design configurations, an optimization approach is used to determine the best design. In this case there is one discrete variable—blade configuration—and the objective function formed (Equation 6.7) identifies which design results in minimum RSD over a range of rotation rates and feed speeds. The predicted optimum RSD mapping within the knowledge space is compared in Figure 6.2. It can be observed that all the methods lead to similar results:

$$
\left.
\begin{array}{l}
\min \ y_f \mathrm{RSD}_f + y_a \mathrm{RSD}_a \\
\mathrm{s.t.} \ y_f + y_a = 1 \\
\mathrm{fr}_{min} \le \mathrm{fr} \le \mathrm{fr}_{max} \\
\mathrm{rr}_{min} \le \mathrm{rr} \le \mathrm{rr}_{max}
\end{array}
\right\}. \tag{6.7}
$$

where RSD_f and RSD_a represent the predicted RSD values for all-forward design and alternate, respectively, based on one of the three aforementioned methods (RSM, Kriging, or HDMR); fr_{min}, rr_{min}, fr_{max}, and rr_{max} represent the minimum and maximum values of the input variables (flow rate and rotation rate) in their experimentally investigated range (knowledge space). The discrete variables y_f and y_a can only take values of either 0 or 1 (binary variables) and are linked to two different design configurations. Based on the formulation of the problem, a value of 1 indicates the use of the corresponding design. For example, if y_f is equal to 1, then the all-forward design configuration is the optimum design for specific conditions.

Setting the upper limit threshold of RSD_{max} of output acetaminophen concentration to be 10%, the acceptable DS calculated using all methods are mapped in Figure 6.3. In order to produce the DS of Figure 6.3, feasibility analysis of the problem defined by Equation 6.7 is performed, that is to say, for different combinations of operating conditions within the knowledge space, the threshold can be satisfied by using the appropriate design. It should be mentioned that the relatively high value of the RSD mentioned here (the usual acceptability limit for powder blends is 5%) is an artifact of the online spectroscopic technique used to quantify composition; RSDs measured by dissolving tablets generated by the process range around 2.5%.

As shown in Figure 6.3, which compares the DS, all three methods predict similar feasible operating ranges. It is clear that a feasible operation can be achieved for the entire range of flow rates, but only for a midrange of the impeller rotation rates. The outer dashed lines represent the limits of the investigated knowledge space.

The results obtained by the three data-driven methodologies agree with the expected results based on the existing knowledge of mixing process. In general, the rotation rate is an important factor in continuous mixing. Increasing the rotation rate leads to a higher degree of dispersion of powder in the mixer.

(a)

(b)

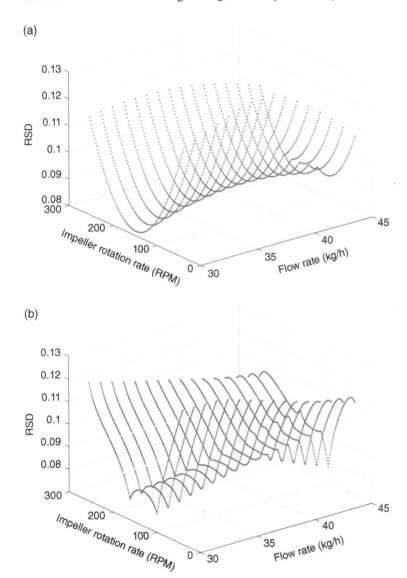

Figure 6.2 Estimated optimum RSD within the range of investigated flow rate and impeller rotation rate values predicted by (a) RSM, (b) Kriging, (c) HDMR, and (d) RSM, Kriging, and HDMR for flow rate of 30 kg/h in 2D.

(c)

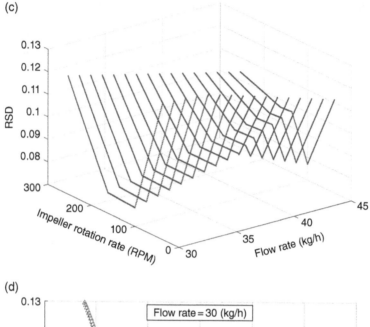

(d)

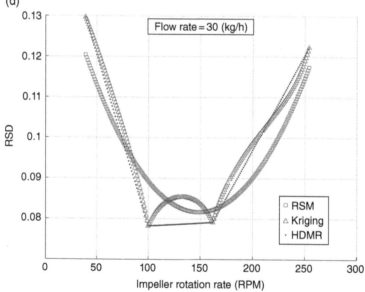

Figure 6.2 (Continued)

However, higher impeller rotation rate also decreases the time available for mixing (lower residence time). Understanding these opposing effects is the key to achieving optimal mixing and can explain the fact that the lowest *RSD* is achieved in midrange rotation rates. In contrast, flow rate is found to be the

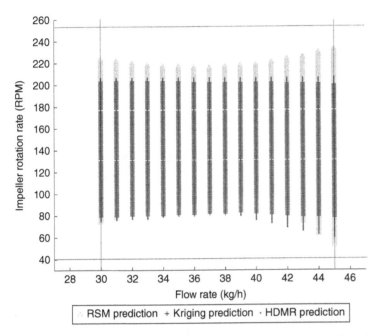

Figure 6.3 Design space (flow rate and impeller rotation rate ranges) of continuous mixer for achieving RSD$_{max}$ = 10%.

least significant factor using ANOVA, and this can explain the fact that for the investigated rotation rates, feasible operation can be achieved for all flow rates in the range of 30–45 kg/h. However, this variable must be included in the DS since it is a significant factor for feeder performance. Specifically, increasing the flow rate can decrease the output flow variability of the feeder. The fact that this variable does not affect the variability in concentration of this specific case study implies that the mixer can efficiently filter out the feeder variability, but this might not be the case for other feeding–mixing integrated systems.

For almost the entire DS, all three methods predict better performance when using the "Alternate Blades" configuration, while for high flow rates and low rotation rates, the "All Forward" design results in better mixing performance (Figure 6.4).

6.4.2 Roller Compaction Case Study: Integration of Control Strategy and Its Effects on the Design Space

In this case study, a process model based on first principles has been developed and previously validated by experimental data [54]. Roller compaction is a process that is employed to overcome problems such as poor flowability, segregation, and dust generation of powders with very small particle sizes. It is

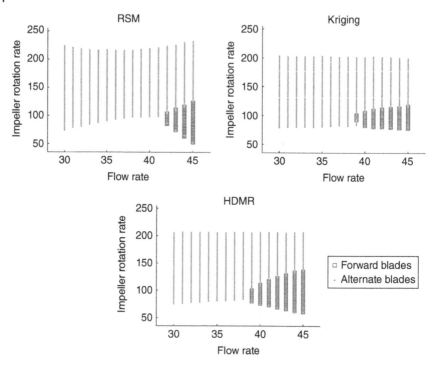

Figure 6.4 Design space produced by (a) RSM approach, (b) Kriging approach, and (c) HDMR approach showing the optimum design configurations for each point.

observed that dry granulation processes improve flow properties and prevents segregation of ingredients of a powder mixture. The process involves the continuous feeding of powder to the roller compactor by augers while the two rotating rolls compress the powder into a ribbon. The roll gaps are adjustable resulting in possible variations in the roll gap width according to the desired operating conditions.

The model used is based on Johanson's rolling theory, which predicts the steady-state stress and density profiles of the ribbon during compaction [54]. In order to incorporate the dynamic behavior of the adjustable roll gap, a material balance across the width is added to the system of equations. Simulations of the model can predict the roll gag, ribbon density, and throughput as a function of feed speed, roll speed, and hydraulic pressure. Throughout the process, it is assumed that the ribbon thickness is equal to the roll gap. The results of the simulation show that ribbon density is affected mainly by changes in pressure, while roll gap is affected by all three input variables. More specifically, simulation results show that an increase of the hydraulic pressure causes an increase of the ribbon density and a decrease of the roll gap width. On the other hand, an increase in the roll speed causes a positive change in the two outputs, but the change in density is not as significant.

The DS of the process is identified using its steady-state performance. The following equations connect the inputs to the outputs of a roller compaction process:

$$\rho_{\text{exit}} = \rho_{\text{in}} \cos\theta_{\text{in}} \left(1 + \frac{h_0}{R} - \cos\theta_{\text{in}}\right)\left(\frac{u_{\text{in}}}{\omega h_0}\right) \qquad (6.8)$$

$$\text{Ph} = \frac{W}{A} \frac{\sigma_{\text{exit}} R}{1+\sin\delta} \int_0^{\alpha} \left[\frac{h_0/R}{\left(1+\left(h_0/R\right)-\cos\theta\right)\cos\theta}\right]^K \cos\theta\, d\theta \qquad (6.9)$$

where ρ_{exit} is the ribbon density; ρ_{in} is the inlet powder density; ω is the angular velocity of the rolls; u_{in} is the velocity of the inlet powder flow; h_0 is the ribbon thickness; $\dot{m}$ is the output throughput; R is the roll radius; W is the roll width; A is the surface; Ph is the applied pressure; $\sigma_{\text{exit}} = C_1 \rho_{\text{exit}}^K$ represents the empirical correlation between the measured stress and the ribbon density with C_1, K the empirical parameters; α is the nip angle; δ is the effective angle of friction; and θ_{in} is the angle of the inlet of the powder. Figure 6.5 shows a schematic representation of the process; however, for further details on the process, the reader should refer to Ref. [54].

The first step of the DS identification is to choose the critical process and material parameters (feed speed, roll speed, pressure, and inlet powder

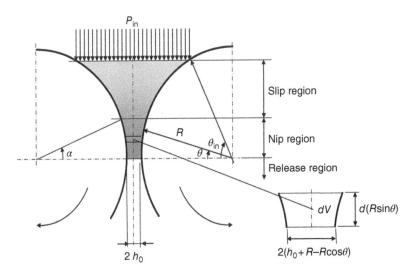

Figure 6.5 Schematic representation of the roller compaction process. *Source:* Hsu [54]. Reproduced with permission of Springer.

density) as inputs of the DS. The second important aspect is the choice of the process model, which in this case is the developed and validated dynamic Johanson's theory model [54]. Finally, the last significant step is to identify what is considered as desired quality, or feasible operation for the specific process. In the following subsections, a deterministic approach for mapping a DS is compared with a stochastic approach, which incorporates uncertainty, and finally, the effects of control are studied for both cases. In all cases, the process model of the underlying process is the same; however, the objective is different, and this fact has a significant effect on the DS.

6.4.2.1 Deterministic Design Space

Following a deterministic DS mapping approach, it is sufficient to choose the upper and lower bounds of the two output variables based on the desired quality specifications of the product. Through mathematical programming, this can be imposed through two additional constraints, which must be strictly satisfied:

$$\rho_{\text{exit}}^{\text{lb}} \leq \rho_{\text{exit}} \leq \rho_{\text{exit}}^{\text{ub}}$$
$$h_0^{\text{lb}} \leq h_0 \leq h_0^{\text{ub}} \tag{6.10}$$

However, by introducing the variable u and solving the minimization problem defined by Equation 6.11 for a range of values within the knowledge space, feasibility can be tested through the sign of variable u. If u is negative or zero, this implies that all constraints are satisfied and thus the process is within the DS. A positive value of u implies that even the minimum possible value of u violates at least one constraint and thus the process is not in the DS:

$$\min_{\rho_{\text{exit}}, h_0, u} u$$

s.t.

$$\rho_{\text{in}} \cos\theta_{\text{in}} \left(1 + \frac{h_0}{R} - \cos\theta_{\text{in}} \right) \left(\frac{u_{\text{in}}}{\omega h_0} \right) = \rho_{\text{exit}}$$

$$\text{Ph} = \frac{W}{A} \frac{\sigma_{\text{exit}} R}{1 + \sin\delta} \int_0^\alpha \left[\frac{h_0 / R}{\left(1 + \left(h_0 / R \right) - \cos\theta \right) \cos\theta} \right]^K \cos\theta \, d\theta \tag{6.11}$$

$$h_0 - h_0^{\text{ub}} \leq u$$

$$h_0^{\text{lb}} - h_0 \leq u$$

$$\rho_{\text{exit}} - \rho_{\text{exit}}^{\text{ub}} \leq u$$

$$\rho_{\text{exit}}^{\text{lb}} - \rho_{\text{exit}} \leq u$$

Consequently, by solving the problem defined by Equation 6.11 for a wider range of operating conditions (knowledge space; Table 6.1), the region within

Table 6.1 Input variables studied ranges (knowledge space) and output variables acceptable ranges.

Variable	Lower bound	Upper bound
Knowledge space		
P_h (MPa)	1	2
u_{in} (cm/s)	3	5
ω (rpm)	5	8
Acceptable ranges		
ρ_{exit} (g/cm^3)	0.9	0.95
h_0 (mm)	2	2.5

which feasible operation is achieved can be identified. It is important to note that the identified DS based on this approach includes regions where both outputs are simultaneously within their acceptable limits. For the roller compactor, the two critical outputs that should be kept within a desired output range are the ribbon density and the ribbon thickness. Thus, if specific upper and lower bounds (Table 6.1) must be strictly satisfied in order to attain a desired output, the acceptable ranges of input variables are shown in Figure 6.6. For the results of this analysis, all parameters are kept at their mean (nominal) conditions; thus their underlying uncertainty is not taken into account.

The DS shown in Figure 6.6 is obtained by solving the problem defined by Equation 6.11 over a grid of points over the three-dimensional space of the three operating conditions and simply identifying whether the feasibility function (u) is positive or negative. The regions bounded by thin gray lines represent the set of conditions that lead to acceptable product in terms of its ribbon density, while the operating conditions within the heavier lines satisfy the thickness quality constraints, and their intersection defines the region where both specifications are met. The results suggest operating the process at low pressures, high inlet flow rates, and low angular velocities if the specific quality specifications must be met. This DS suggests that the process can produce ribbons of desired properties for a range of flow rates and angular velocities; however, as ω increases, u_{in} should also be increased. It should be noted that this DS is strongly dependent on the chosen desired ranges of the output variables, which in this work are chosen mostly for illustration purposes. The simulation of the process is performed in MATLAB, where the fmincon optimization algorithm is used for the solution of problem defined by Equation 6.11.

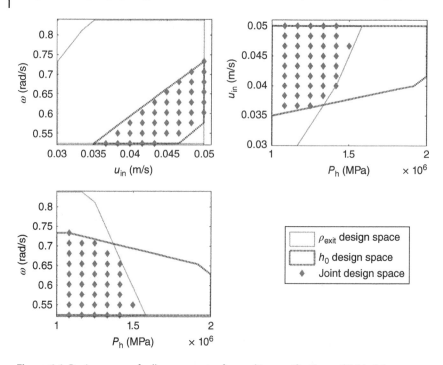

Figure 6.6 Design space of roller compactor for quality specifications of Table 6.1.

6.4.2.2 Stochastic Design Space

Following a stochastic approach—similar to the methodology proposed in [13]—uncertainty of the inlet density is incorporated into the DS. In the present case study, MATLAB is used for simulating the process under different operating conditions in the presence of uncertainty in the material properties. Specifically, it is assumed that the inlet powder density fluctuates as powder enters in the process and follows a normal distribution of mean 0.3 g/cm³ and SD of 0.05. The ability to simulate the process by randomly drawing values from the aforementioned uncertain parameter from its assigned distributions at each simulation allows the evaluation of the level of uncertainty that has propagated to both roll gap and ribbon density. Thus the process is simulated under steady-state conditions, and by studying the predicted density and ribbon throughput, their distribution and the associated parameters can be identified. Based on the obtained steady-state values of ribbon density and throughput for a large amount of random simulations, their distribution type is identified as normal, and thus the mean and SD can be easily calculated. Based on this information, the probability of each output to lie within the predefined acceptable limits can be calculated using cumulative distribution. Finally, the joint probability of both outputs within their acceptable limits is the product of the two calculated probabilities. Consequently, for each point within the knowledge space, the

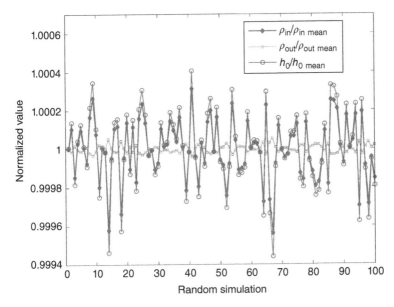

Figure 6.7 Normalized by mean values of inlet density, outlet ribbon density, and ribbon thickness for 100 random simulations.

value of the joint probability can be the figure of merit for defining the reliability or the quantification of the assurance of quality of the DS based on Equation 6.1. An example of the uncertainty profiles of each of the two output variables compared to the inlet variability of the density is shown in Figure 6.7, where it is clear that the ribbon thickness is more sensitive to the inlet density variations compared with the ribbon output density.

In Figure 6.8, the deterministic DS is compared with the stochastic DS, where the effects of the uncertainty of the inlet ribbon density are now taken into account. It can be observed that the first approach results in an overestimated and optimistic DS, which suggests that one can produce ribbons with the desired characteristics operating at regions where the probability of meeting the specifications is only 20–40% due to the uncertainty of the inlet density.

6.4.2.3 Effect of Control Strategies on the Design Space

In order to assess the effect of control in the DS identification procedure, three different control strategies are employed: (i) control of P_h, (ii) control of u_{in}, and (iii) control of ω. In this case, the equalities and inequalities of problem defined by Equation 6.11 must be satisfied, but also one of the operating variables is free to vary based on the values of the remaining input variables in order to satisfy the constraints. In other words, based on the problem posed here, it is investigated whether the control variable can be manipulated in order to drive the process within the DS (negative u). Using the concept of flexibility analysis

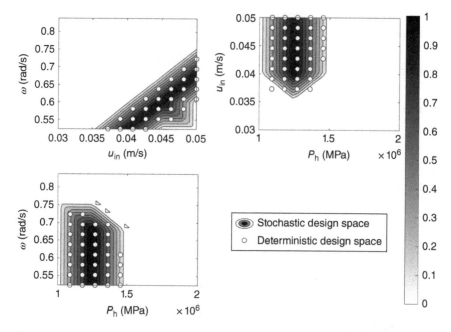

Figure 6.8 Comparison of deterministic and stochastic DS with respect to the operating conditions.

with control variables, the following optimization problem is formed for each case (Equation 6.12):

$$\min_{\rho_{exit},h_0,z,u} u$$

s.t.

$$\rho_{in} \cos\theta_{in} \left(1 + \left(h_0/R\right) - \cos\theta_{in}\right)\left(\frac{u_{in}}{\omega h_0}\right) = \rho_{exit}$$

$$Ph = \frac{W}{A}\frac{\sigma_{exit}R}{1+\sin\delta}\int_0^\alpha \left[\frac{h_0/R}{\left(1+\left(h_0/R\right)-\cos\theta\right)\cos\theta}\right]^K \cos\theta \, d\theta$$

$$h_0 - h_0^{ub} \le u$$

$$h_0^{lb} - h_0 \le u$$

$$\rho_{exit} - \rho_{exit}^{ub} \le u$$

$$\rho_{exit}^{lb} - \rho_{exit} \le u$$

$$z \ge 0$$

(6.12)

where $z = $ Ph for control strategy (i),

$z = u_{in}$ for control strategy (ii),

$z = \omega$ for control strategy (iii)

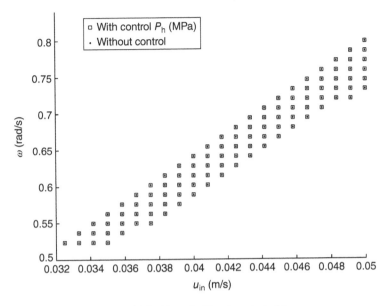

Figure 6.9 Comparison of DS with and without control of P_h.

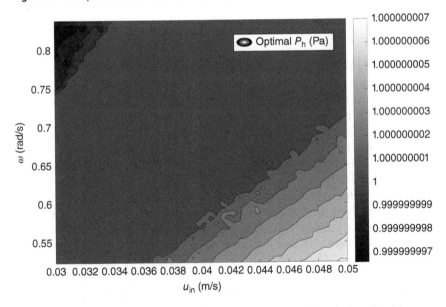

Figure 6.10 Optimal values of P_h as a control variable based on the solution of problem defined by Equation 6.11.

Through the results, it becomes obvious that by manipulating pressure, the DS is not affected at all (Figure 6.9). Figure 6.10 depicts the mapping of the converged optimal values of pressure as a control variable, which fails to enlarge the DS. Similarly, by controlling the inlet powder flow rate (Figure 6.11),

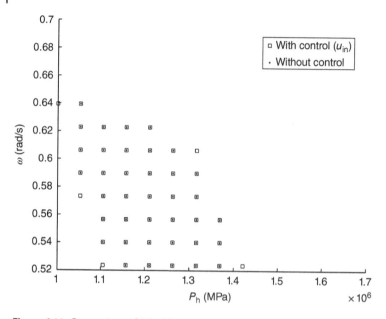

Figure 6.11 Comparison of DS with and without control of flow rate u_{in}.

only a few additional points are added to the DS (Figure 6.12). This is due to the fact that for the current investigated ranges, controlling one of these operating conditions at a time cannot overcome the effects of the remaining two variables, and thus the process fails to produce ribbons of desired properties.

However, if control strategy 3 is employed (Figure 6.13), the DS is enlarged (Figure 6.14). This fact seems reasonable since the angular velocity of the compactor rolls is a variable that has an effect on both output variables [54]. Thus, within this knowledge space, the process gains additional flexibility if the angular velocity is used as a manipulative variable. Following a similar approach, perhaps the effects of additional control strategies can be analyzed (two manipulative variables) in order to verify whether they will lead to the enlargement of the DS with respect to the remaining operating conditions.

Finally, it is of interest to investigate whether the effects of the variability caused by the uncertainty of the raw material properties can be reduced through control. For this analysis, the most effective control strategy is chosen (angular velocity as the manipulative variable). Comparing Figure 6.15 with Figure 6.8, it can be seen that the reliability of the process to meet quality specifications is higher in a larger fraction of the knowledge space when control is employed. Thus, it becomes clear that a successful control strategy can

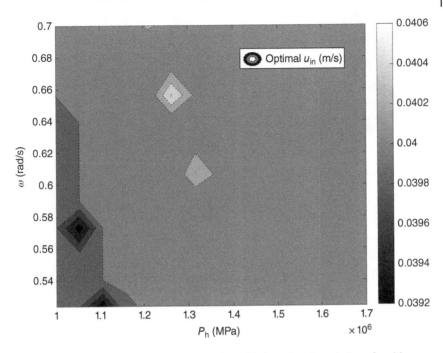

Figure 6.12 Optimal values of u_{in} as a control variable based on the solution of problem defined by Equation 6.11.

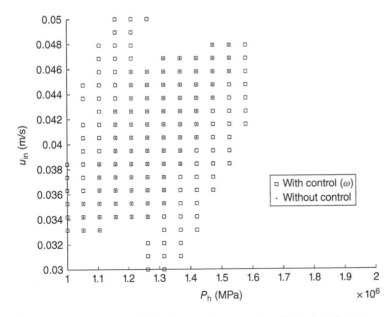

Figure 6.13 Comparison of DS with and without control of angular velocity ω.

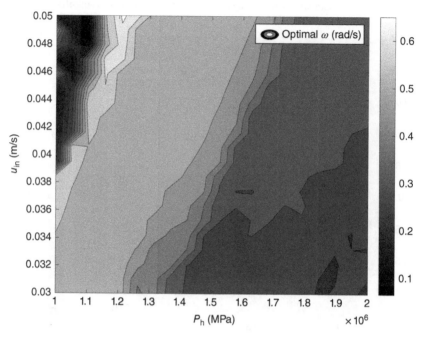

Figure 6.14 Optimal values of ω as a control variable based on the solution of problem defined by Equation 6.11.

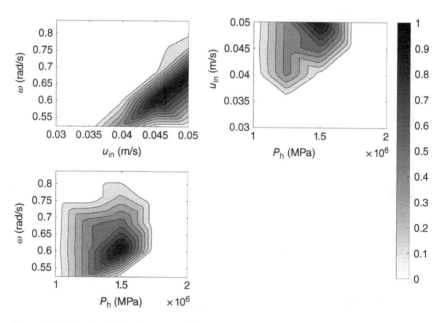

Figure 6.15 Stochastic design space when ω is considered a manipulative variable (control strategy 3).

alleviate the effects of possible uncertainty of the process input variables, allowing the process to have a larger DS.

6.5 Conclusions

We have presented a review of the recent advancements on methods and tools developed for identifying the DS of pharmaceutical processes. The importance and usefulness of the concept of DS in the pharmaceutical industry is emphasized in all of the recent FDA guidance papers, which aim to direct the industry into employing QbD and PAT principles in process and product development. The most important benefit of a detailed and accurate DS is the freedom it offers to operate within an entire region of acceptable conditions by assuring product quality and avoiding time-consuming post-approval procedures.

In this work, the multidisciplinary nature of the DS is presented through a variety of concepts and methods adopted from statistics and process systems engineering and optimization literature. Specifically, process flexibility analysis, which is an established concept in the process systems engineering community, is shown to be an effective tool for the quantification and final presentation of a process DS. The basic steps in identifying a DS include (i) the selection of the critical input variables (operating conditions, design variables, material properties, and control variables), (ii) the formulation of process model, and finally (iii) the identification of desired objective or output quality specifications. The DS can be deterministic or stochastic when the effect of uncertainty is incorporated. The importance of an efficient control strategy is also found to be critical in exploring the flexibility of a process to remain within its DS since oftentimes the process can handle more variation in its input variables when the appropriate variable is manipulated. Consequently, incorporation of control variables in the formulation of the flexibility analysis problem can lead to the identification of the most successful control strategy, namely, the one that leads to the largest DS. In this work it is shown that formulating the correct optimization objective can contribute toward identifying the desired DS of a process including both continuous (operating conditions and material properties) and discrete (design) variables.

In conclusion, even though process DS has been only recently formally introduced in the pharmaceutical industry, it should be realized that it is not a completely unknown concept. Identifying design and operating constraints as well as the necessary control scheme for a process are concepts and ideas that have been extensively studied in the literature, where the developed metrics, theories, and tools can be of great use in pharmaceutical manufacturing. Future work will include identification of the DS of large integrated multicomponent processes and the development of a methodology for dynamic DS identification under uncertainty.

Acknowledgment

The authors would like to acknowledge the support provided by the ERC (NSF-0504497, NSF-ECC 0540855).

References

1 Gernaey KV, Cervera-Padrell AE, and Woodley JM, A perspective on PSE in pharmaceutical process development and innovation. Computers & Chemical Engineering, 2012; 42: p. 15–29.

2 Troup GM and Georgakis C, Process systems engineering tools in the pharmaceutical industry. Computers & Chemical Engineering, 2013; 51: p. 157–171.

3 Boukouvala F, *et al.*, An integrated approach for dynamic flowsheet modeling and sensitivity analysis of a continuous tablet manufacturing process. Computers & Chemical Engineering, 2012; 42: p. 30–47.

4 Oksanen CA and García Muñoz S, Process modeling and control in drug development and manufacturing. Computers & Chemical Engineering, 2010; 34(7): p. 1007–1008.

5 ICH Q8, ICH Harmonized Tripartite Guidelines—*Pharmaceutical Development*, 2006.

6 Garcia T, Cook G, and Nosal R, PQLI key topics—criticality, design space, and control strategy. Journal of Pharmaceutical Innovation, 2008; 3(2): p. 60–68.

7 Lepore J and Spavins J, PQLI design space. Journal of Pharmaceutical Innovation, 2008; 3(2): p. 79–87.

8 Yu L, Pharmaceutical quality by design: Product and process development, understanding, and control. Pharmaceutical Research, 2008; 25(4): p. 781–791.

9 Davis B, Lundsberg L, and Cook G, PQLI control strategy model and concepts. Journal of Pharmaceutical Innovation, 2008; 3(2): p. 95–104.

10 Nosal R and Schultz T, PQLI definition of criticality. Journal of Pharmaceutical Innovation, 2008; 3(2): p. 69–78.

11 MacGregor J and Bruwer M-J, A framework for the development of design and control spaces. Journal of Pharmaceutical Innovation, 2008; 3(1): p. 15–22.

12 García-Muñoz S, Dolph S, and Ward Ii HW, Handling uncertainty in the establishment of a design space for the manufacture of a pharmaceutical product. Computers & Chemical Engineering, 2010; 34(7): p. 1098–1107.

13 Peterson J, A Bayesian approach to the ICH Q8 definition of design space. Journal of Biopharmaceutical Statistics, 2008; 18(5): p. 959–975.

14 Peterson J and Lief K, The ICH Q8 definition of design space: A comparison of the overlapping means and Bayesian predictive approaches. Statistics in Biopharmaceutical Research, 2010; 2: p. 249–259.

15 Cogdill R and Drennen J, Risk-based quality by design (QbD): A Taguchi perspective on the assessment of product quality, and the quantitative linkage of drug product parameters and clinical performance. Journal of Pharmaceutical Innovation, 2008; 3(1): p. 23–29.

16 Lipsanen T, *et al.*, Novel description of a design space for fluidised bed granulation. International Journal of Pharmaceutics, 2007; 345(1–2): p. 101–107.

17 Lebrun P, *et al.*, Development of a new predictive modelling technique to find with confidence equivalence zone and design space of chromatographic analytical methods. Chemometrics and Intelligent Laboratory Systems, 2008; 91(1): p. 4–16.

18 Boukouvala F, Muzzio F, and Ierapetritou M, Design space of pharmaceutical processes using data-driven-based methods. Journal of Pharmaceutical Innovation, 2010; 5(3): p. 119–137.

19 McKenzie P, *et al.*, Can pharmaceutical process development become high tech? AIChE Journal, 2006; 52(12): p. 3990–3994.

20 Peterson J, *et al.*, Statistics in pharmaceutical development and manufacturing. Journal of Quality Technology, 2009; 2: p. 111–147.

21 Peterson J, Miro-Quesada, and del Castillo E, A Bayesian reliability approach to multiple response surface optimization with seemingly unrelated regression models. Quality Technology and Quantitative Management, 2009; 6(4): p. 353–369.

22 Stockdale GW and Cheng A, Finding design space and reliable operating region using a multivariable Bayesian approach with experimental design. Quality Technology and Quantitative Management, 2009; 6(4): p. 391–408.

23 Johnson DB and Bogle IDL, A methodology for the robust evaluation of pharmaceutical processes under uncertainty. Chemical Papers, 2000; 54(6a): p. 398–405.

24 Straub DA and Grossmann IE, Design optimization of stochastic flexibility. Computers & Chemical Engineering, 1993; 17(4): p. 339–354.

25 Sahinidis NV, Optimization under uncertainty: State-of-the-art and opportunities. Computers & Chemical Engineering, 2004; 28(6–7): p. 971–983.

26 Pistikopoulos EN and Mazzuchi TA, A novel flexibility analysis approach for processes with stochastic parameters. Computers & Chemical Engineering, 1990; 14(9): p. 991–1000.

27 Acevedo J and Pistikopoulos EN, Stochastic optimization based algorithms for process synthesis under uncertainty. Computers & Chemical Engineering, 1998; 22(4–5): p. 647–671.

28 Bansal V, Perkins JD, and Pistikopoulos EN, Flexibility analysis and design of dynamic processes with stochastic parameters. Computers & Chemical Engineering, 1998; 22(1): p. S817–S820.

29 Samsatli NJ, Papageorgiou LG, and Shah N, Robustness metrics for dynamic optimization models under parameter uncertainty. AIChE Journal, 1998; 44(9): p. 1993–2006.

30 Grossman IE and Floudas CA, Active constraint strategy for flexibility analysis in chemical processes. Computers & Chemical Engineering, 1987; 11(6): p. 675–693.

31 Halemane KP and Grossmann IE, Optimal process design under uncertainty. AICHE Journal, 1983; 29(3): p. 425.

32 Pistikopoulos EN and Ierapetritou MG, Novel approach for optimal process design under uncertainty. Computers & Chemical Engineering, 1995; 19(10): p. 1089–1110.

33 Swaney RE and Grossmann IE, An index for operational flexibility in chemical process design. Part I: Formulation and theory. AIChE Journal, 1985; 31(4): p. 621–630.

34 Floudas CA and Gumus ZH, Global optimization in design under uncertainty: Feasibility test and flexibility index problems. Industrial & Engineering Chemistry Research, 2001; 40(20): p. 4267–4282.

35 Goyal V and Ierapetritou MG, Determination of operability limits using simplicial approximation. AIChE Journal, 2002; 48(12): p. 2902–2909.

36 Goyal V and Ierapetritou MG, Framework for evaluating the feasibility/operability of nonconvex processes. AIChE Journal, 2003; 49(5): p. 1233–1240.

37 Banerjee I and Ierapetritou MG, Design optimization under parameter uncertainty for general black-box models. Industrial & Engineering Chemistry Research, 2002; 41(26): p. 6687–6697.

38 Banerjee I and Ierapetritou MG, Parametric process synthesis for general nonlinear models. Computers & Chemical Engineering, 2003; 27(10): p. 1499–1512.

39 Banerjee I, Pal S, and Maiti S, Computationally efficient black-box modeling for feasibility analysis. Computers & Chemical Engineering, 2010; 34(9): p. 1515–1521.

40 Boukouvala F and Ierapetritou MG, Feasibility analysis of black-box processes using an adaptive sampling Kriging-based method. Computers & Chemical Engineering, 2012; 36: p. 358–368.

41 Floudas CA, Nonlinear and Mixed-Integer Optimization: Fundamentals and Applications. Oxford: Oxford University Press, 1995.

42 Bernardo FP, *et al.*, Process design under uncertainty: Robustness criteria and value of information. Computer Aided Chemical Engineering, 2003; 16: p. 175–208.

43 Ramachandran R, *et al.*, A quantitative assessment of the influence of primary particle size polydispersity on granule inhomogeneity. Chemical Engineering Science, 2012; 71: p. 104–110.

44 Ramachandran R, *et al.*, Model-based control-loop performance assessment of a continuous direct compaction pharmaceutical process. Journal of Pharmaceutical Innovation, 2011; 6: p. 249–263.

45 Ramachandran R, *et al.*, A mechanistic model for breakage in population balances of granulation: Theoretical kernel development and experimental validation. Chemical Engineering Research and Design, 2009; 87(4): p. 598–614.

46 Sen M and Ramachandran R, A multi-dimensional population balance model approach to continuous powder mixing processes. Advanced Powder Technology, 2013; 24(1): p. 51–59.

47 Sen M, *et al.*, Multi-dimensional population balance modeling and experimental validation of continuous powder mixing processes. Chemical Engineering Science, 2012; 80: p. 349–360.

48 Georgakis C, *et al.*, On the operability of continuous processes. Control Engineering Practice, 2003; 11(8): p. 859–869.

49 Vinson DR and Georgakis C, A new measure of process output controllability. Journal of Process Control, 2000; 10(2–3): p. 185–194.

50 Subramanian S and Georgakis C, Steady-state operability characteristics of reactors. Computers & Chemical Engineering, 2000; 24(2–7): p. 1563–1568.

51 Subramanian S, Uzturk D, and Georgakis C, An optimization-based approach for the operability analysis of continuously stirred tank reactors. Industrial & Engineering Chemistry Research, 2001; 40(20): p. 4238–4252.

52 Subramanian S and Georgakis C, Steady-state operability characteristics of idealized reactors. Chemical Engineering Science, 2001; 56(17): p. 5111–5130.

53 Lima FV, Jia Z, Ierapetritou MG, and Georgakis C, Similarities and differences between the concepts of operability and flexibility: The steady-state case. AIChE Journal, 2010; 56: p. 702–716.

54 Hsu S-H, Reklaitis G, and Venkatasubramanian V, Modeling and control of roller compaction for pharmaceutical manufacturing. Part I: Process dynamics and control framework. Journal of Pharmaceutical Innovation, 2010; 5(1): p. 14–23.

7

Using Quality by Design Principles as a Guide for Designing a Process Control Strategy

Christopher L. Burcham[1], Mark LaPack[2], Joseph R. Martinelli[1], and Neil McCracken[1,3]

[1] Small Molecule Design and Development, Eli Lilly and Company, Indianapolis, IN, USA
[2] Mark LaPack & Associates Consulting, LLC, Lafayette, IN, USA
[3] Bioproduct Research and Development, Eli Lilly and Company, Indianapolis, IN, USA

7.1 Introduction

The main objective in developing pharmaceuticals and new medicines is to help patients live longer, healthier, and symptom-free lives. From a chemistry manufacturing and controls (CMC) perspective, this necessitates delivering high quality active pharmaceutical ingredient (API) that will provide the intended therapeutic benefits. Thus, very tight control of the drug substance quality is paramount and any potential impurities must be kept to extremely low levels. The thresholds for unqualified impurities in the drug substance are outlined in the ICH guidelines [1] and summarized in Table 7.1. Impurities above these levels must be qualified through toxicology tests to demonstrate safety.

It is the responsibility of the CMC development team to outline the relevant potential impurities for a given API. This is not only a federal regulation but also allows the team to plan for implementing appropriate process controls to meet the required impurity limits in the final API. This combination of process understanding and API quality are central to quality by design (QbD) [2]. The discussion herein describes our team's implementation of a QbD approach for the design and development of an impurity control strategy for semagacestat.[1]

Impurities in the final API can result from impurities entering with the raw materials used for the process (i.e., starting material-related impurities) or as reaction by-products formed during the reaction (i.e., process-related impurities). It is important to fully understand the origin and fate of all impurities throughout the synthesis as this allows for appropriate limits and controls to be established.

Comprehensive Quality by Design for Pharmaceutical Product Development and Manufacture,
First Edition. Edited by Gintaras V. Reklaitis, Christine Seymour, and Salvador García-Munoz.
© 2017 American Institute of Chemical Engineers, Inc. Published 2017 by John Wiley & Sons, Inc.

Table 7.1 Thresholds according to ICH guidelines.

Maximum daily dose	Reporting threshold	Identification threshold	Qualification threshold
≤2 g/day	0.05%	0.10% or 1.0 mg/day intake (whichever is lower)	0.15% or 1.0 mg/day intake (whichever is lower)
>2 g/day	0.03%	0.05%	0.05%

An enormous amount of data is required to achieve this level of process understanding. Traditional techniques for offline process monitoring, such as high performance liquid chromatography (HPLC), are the methods by which this type of data is typically acquired. Conversely, many workers have begun to appreciate the power of online analytical tools for process monitoring, such as in situ Fourier transform infrared (FTIR). Process analytical technology (PAT) has been shown to provide valuable tools for understanding chemical processes and for developing predictive process models [3–9]. Utilizing PAT in the laboratory is playing a key role in meeting the QbD principles established by the US Food and Drug Administration (FDA) for pharmaceutical process development [2]. The information obtained from PAT can be used to fully define the process design space that will provide high quality API and enables one to build quality into the process.

The QbD approach presented here comprises several critical steps for the development of a process in which all potential impurities were considered and controlled. A schematic of the flow of this report is depicted in Figure 7.1. The sections in the report follow the linear black arrows beginning at the top left and leading to the definition of the process control strategy at the bottom of the schematic. The actual development path followed was certainly not as linear and involved multiple iterations and considerations, some of which are indicated by the curved gray arrows.

In the sections that follow, the overall synthetic process is first presented followed by a discussion of the impurity genesis and subsequent impurity control limits and specifications. An in-depth description of the reaction step of interest, step 2 (Figure 7.2) of the planned registered sequence (Figure 7.3), is provided, detailing both the deprotection and coupling reactions. Mechanistic models (developed to describe and model the chemistry and physics of the process) are then presented. This is followed by a detailed discussion of how the multiple parameters (rate constants, activation energies, Henry's law constants, and mass transfer coefficients) were experimentally determined. The predictive models were then used to determine the optimal processing conditions necessary to minimize the formation of total and specific impurities. The model was then utilized to probe the design space around the optimal target

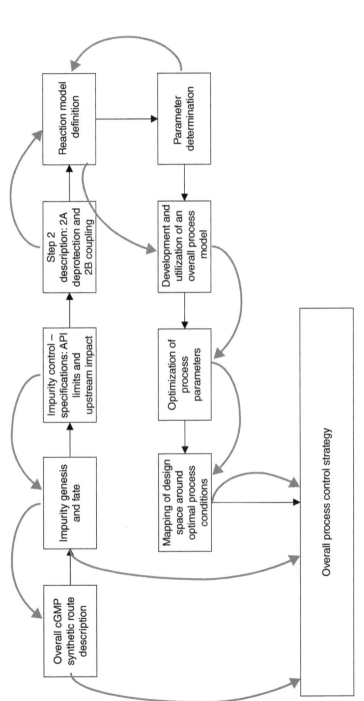

Figure 7.1 Process design and development flow chart.

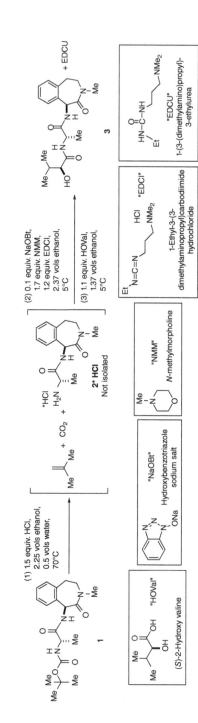

Figure 7.2 Reaction scheme for generating the active pharmaceutical ingredient **3**.

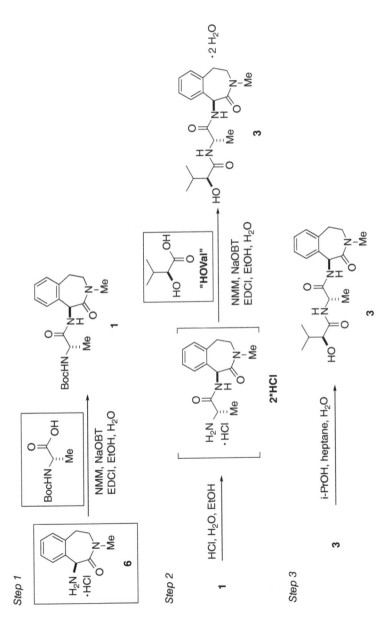

Figure 7.3 Semagacestat GMP sequence.

operational conditions, ultimately culminating in a final discussion about the overall process control strategy. As mentioned, the actual informational and chronological flow was not as linear. The overall cGMP sequence resulted from a consideration of impurity formation and rejection, which was driven by the specifications of impurities in the API and the subsequent upstream impact of the limits in the API to structural analogs or "parents" that form earlier in the intermediate steps, or enter with a given API starting material. Only after the general chemical sequence was established could mechanistic models and elementary reactions steps be defined. Determination of unknown parameters (rate constants, activation energies, Henry's Law constants, and mass transfer coefficients) to utilize the models in a predictive way was then performed; this drove a deeper understanding of the reaction or mass transfer mechanism, resulting in updated mechanistic or physical models. Application of the models and testing with expanded data sets also lead to further model elaboration and development in some instances. The established models allowed for their utilization to probe the optimal processing conditions and the design space around the target conditions. The simulation, in some cases reinforced by designed experiments not presented, were then confirmed through additional experiments at "extremes" or edges of the model predictive space to provide credibility of the predicted results, further establishing the utility and reliability of the overall model. This approach was repeated to some degree for each reaction step. Understanding the process design space around each step and the genesis and fate of all process and starting material-derived impurities leads to an overall control strategy design for the process.

7.2 Chemical Sequence, Impurity Formation, and Control Strategy

7.2.1 Chemical Sequence

As shown in Figure 7.2, the final bond-forming step in the semagacestat process (Figure 7.3) involves two chemical transformations: a protecting group removal followed by a peptide coupling. The N-t-butoxycarbonyl (BOC) amine protecting group is removed under acidic conditions at 70°C (step 2a). The in situ generated HCl salt of **2** is used without further purification in the peptide coupling with (S)-2-hydroxyisovaleric acid (HOVal). The reaction is mediated by 1-ethyl-3-(3-dimethylaminopropyl)carbodiimide (EDCI) HCl at 5°C using the sodium salt of hydroxybenzotriazole (NaOBt) as a catalyst. The product of the reaction sequence is the API, **3**, and 1-(3-(dimethylamino)propyl)-3-ethylurea (EDCU).

The drug substance manufacturing process for semagacestat includes two additional steps, as illustrated in Figure 7.3. The reaction of interest here (step 2) is preceded by another peptide coupling and is followed by a step for crystal form control. In step 1, the aminobenzoazapinone hydrochloride salt (**6**) is reacted with BOC-protected alanine in the presence of *N*-methylmorpholine (NMM) as base and peptide coupling reagents EDCI and NaOBt in ethanol. The product of this reaction, compound **1**, precipitates from the reaction mixture and is isolated by filtration. Step 3 converts the technical grade LY450139 to the desired polymorph (form II) utilizing a seeded isopropanol–heptane crystallization process. LY450139 drug substance is isolated from the reaction mixture by selective crystallization followed by particle size reduction and drying. While these steps are critical to the development of the overall control strategy, the design and development of these steps will not be discussed in detail. Steps 1 and 3 of the GMP sequence will be only included as required for the discussion regarding impurity control and the development of the control strategy.

7.2.2 Impurity Formation

Impurities in the API have their genesis from either impurities entering with reactants or reagents used in the synthesis or as reaction by-products (i.e., process-related impurities) formed during the synthesis. It is important to fully understand the origin and fate of all impurities throughout the synthesis in order to properly design a comprehensive impurity control strategy.

As shown in Figure 7.4, there are several steps that constitute the mechanism for EDCI-mediated peptide coupling reactions. This type of reaction has been studied extensively [10], and it has been shown that the activation of the carboxylic acid by the carbodiimide is the slow or rate-determining step [11]. Despite the obvious differences between this system and the case study presented in the literature, the important features discerned previously are believed to be conserved in the system of interest for this work.

While this coupling reaction seems straightforward, a number of the required manipulations introduce significant opportunity for generating impurities, that is, process-derived impurities. For example, a long hold time at 70°C can result in significant hydrolytic degradation of **2** (Figures 7.5c and 7.6). Additionally, the coupling reagent addition order (and addition rate) can cause the formation of impurities if performed improperly, such as the species shown in Figure 7.5b resulting from the dimerization of HOVal, and subsequent coupling of this dimer. Moreover, things as seemingly innocuous as evolved gases, like CO_2, can have a dramatic effect on the impurity distribution in the product. If entrained/adsorbed CO_2 is not removed prior to the coupling step, a symmetrical urea is formed as a side product (Figure 7.5a). This impurity, an unqualified impurity, has very low solubility. The low solubility results in very little rejection during crystallization of the API. Thus, controlling the formation of this impurity, **4**, became a primary focus of the intense multidisciplinary study presented herein.

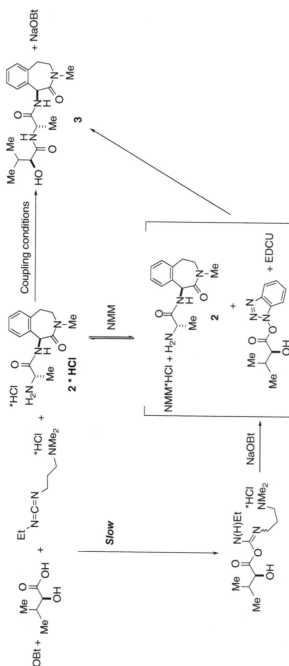

Figure 7.4 Reaction mechanism for the EDCl-mediated peptide coupling.

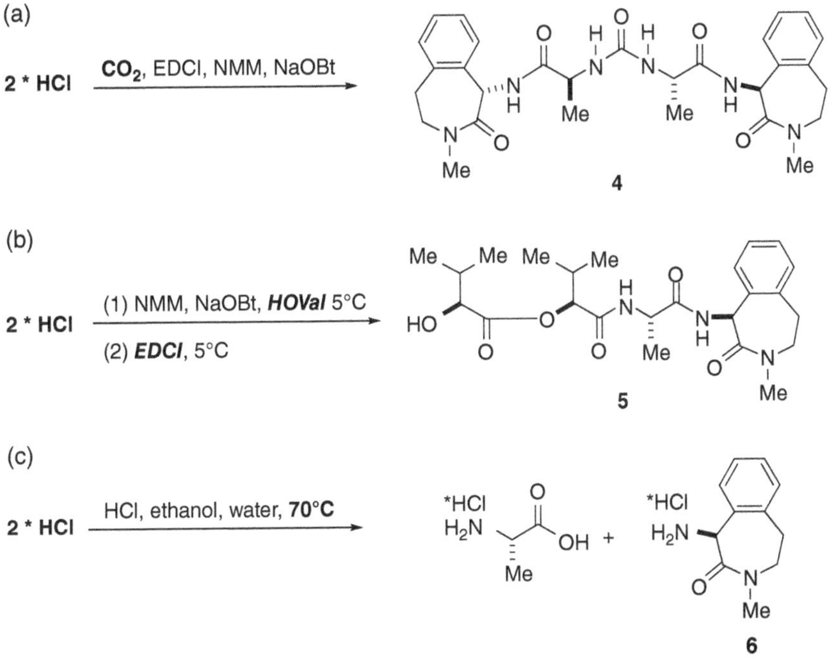

(a)

$2 * HCl$ $\xrightarrow{\text{CO}_2,\ \text{EDCI, NMM, NaOBt}}$

4

(b)

$2 * HCl$ $\xrightarrow[\text{(2) }\textit{EDCI},\ 5°C]{\text{(1) NMM, NaOBt, }\textit{HOVal}\ 5°C}$

5

(c)

$2 * HCl$ $\xrightarrow{\text{HCl, ethanol, water, }70°C}$

6

Figure 7.5 (a–c) Key impurities that can form during step 2.

The symmetrical urea impurity, **4**, discussed earlier was suspected to be derived from dissolved CO_2 liberated during the removal of the BOC protecting group. This hypothesis was tested through a series of control reactions, and it was determined that the impurity, indeed, forms from the reaction of **2** and CO_2 in the presence of EDCI.

Two possible pathways for the formation of **4** are shown in Figure 7.7. It is thought that **2*HCl** will be in equilibrium with the freebase, **2**, in the presence of NMM, which can react with dissolved CO_2 to form a carbamic acid. The carbamic acid can react with EDCI in the same fashion as a carboxylic acid, resulting in an activated o-acylisourea-type intermediate. This species can then undergo reaction with base to form an isocyanate or reaction with NaOBt to form the OBt ester, either of which can react with another equivalent of **2** to form **4**. While the specific pathway has not been discerned, it is clear that depriving the system of CO_2 would prevent the formation of **4**.

BOC-deprotection reactions of this sort are among the most common procedures in pharmaceutical process chemistry [12]. These reactions have been studied widely using PAT [13, 14] and are typically executed without event. This case presents an interesting and unique challenge as the CO_2 produced during the first reaction reacts undesirably in the following reaction to form **4** if not adequately purged prior to the coupling step.

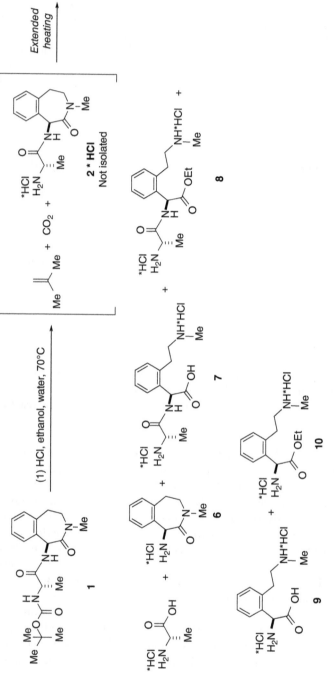

Figure 7.6 Detailed list of impurities resulting from extended heating of **2**.

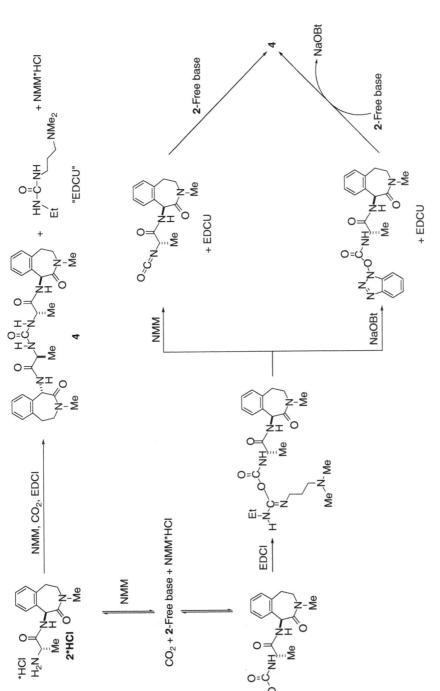

Figure 7.7 Possible pathways for the formation of urea impurity, **4**.

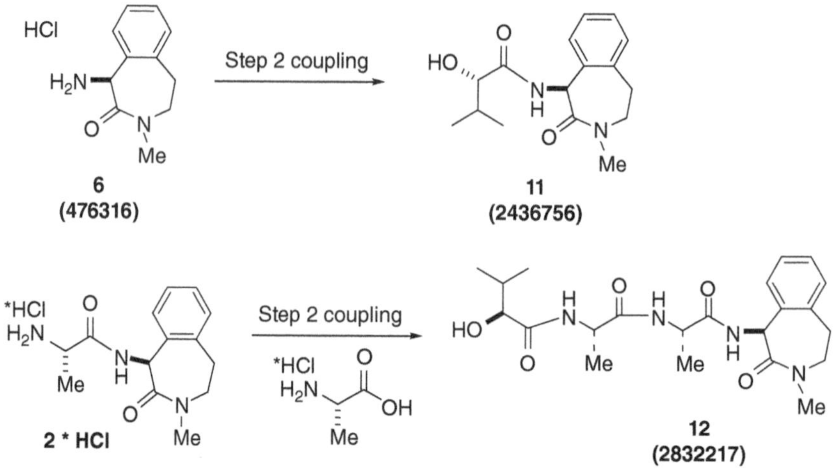

Figure 7.8 Propagation of **6** and alanine through the step 2 to coupling.

In addition to these process-derived impurities, there are several starting material-derived impurities to consider for step 2. These impurities are formed from contaminants in the starting materials used for step 2. Some of the process impurities (species **4–10**) are illustrated in Figures 7.8 and 7.9. It should be noted that residual starting materials also act as impurities further downstream. For example, residual **6** and BOC alanine could enter with **1** or be generated by the degradation pathway described in Figure 7.6. Since all of the degradation impurities shown in Figure 7.6 have amino groups, like **2*HCl**, they can react during the peptide coupling step to generate another set of impurities. Typically, the most troubling of these impurities are those resulting from the coupling of **6** with HOVal and the secondary coupling of residual alanine to result in the impurities shown in Figure 7.8. Similarly, five impurities have been associated with contaminants in hydroxyvaline and could be expected to be present at some level. These species and the impurities resulting from carrying them through the coupling step are illustrated in Figure 7.9.

7.2.3 Control Strategy

Understanding the formation and fate of each impurity in the sequence is important to the controls placed on the process. Understanding the rejection of the impurities through each synthetic step (by way of crystallization and isolation of the desired product) is equally important. Specifications on the API are determined through qualification of known impurities through toxicology studies and maximum daily dosing levels of the drug in the drug product. In the case of semagacestat, two impurities were established in this manner, **5** and **14**, the RSS stereoisomer of the API. All other impurities are limited by

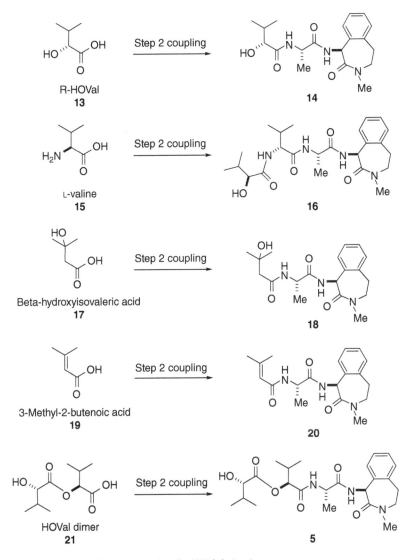

Figure 7.9 Impurities associated with HOVal derivatives.

ICH guidelines [1], which, for a less than 2 g/day dose of the drug substance, do not exceed 0.10% if unidentified or do not exceed 0.15% if identified (or less than 1.0 mg/day intake whichever is lower). Using this as a guide, it is then necessary to determine acceptable limits on impurities entering from previous steps that, when processed under representative conditions, would provide API meeting all quality specifications.

To this end, experiments were conducted to establish limits of impurities in each isolated intermediate in the chemical sequence. The experiments were conducted by adding a known amount of each impurity to the reaction and then determining the final impurity level of the analog formed in the isolated product. It is important to note that these experiments can be done in two ways: either by adding the actual impurity to the reaction mixture just prior to isolation or by adding the impurity precursor at the start of the reaction. Specification levels are set from these "spiking" experiments, either for starting material impurity levels or for in situ levels that then translate to processing constraints. In this fashion spiking experiments were performed for each impurity. Not all impurities were tested at extreme levels to find the edge of failure. In many cases, much lower limits are adequate due to the inherent robustness of the process such that broader ranges are not required. The results of spiking experiments can be visualized as shown in Figure 7.10 for the enantiomer of **6**, **ent-6**.

Figure 7.10 summarizes the spiking studies performed for **ent-6** across the semagacestat chemical sequence. Examining Figure 7.10, one can easily see that both types of spiking experiments were done for step 1. Either adding **ent-6** to the step 1 coupling reaction or adding the diastereomer of **1 (SR-1)** to the crude reaction mixture prior to isolation results in about the same amount of the diastereomer of **1 (SR-1)** in the isolated product from step 1. Since this process was routinely resulting in about 1% of the impurity in **1**, step 2 was spiked at two times that level. Since this level completely purged, we were fairly confident in the processes' ability to handle the routinely observed amount coming out of step 1, which was typically much lower than the 1% seen from the 2% spikes. Finally, the capacity of step 3 was examined for this impurity. It was spiked as the contaminant observed coming out of step 2 and was purged completely from this step as well. The impurity purge plots also offer a way to visualize the level of a given impurity in each intermediate throughout the synthesis as compared with the final specification level.

Working backward from the final specifications for the API with the impurity spiking data as a guide, specifications were set for **6** and **1**. This is best explained by example: consider establishing the specification for an acceptable level of **6** in the step 2 starting material, **1**. If **6** is carried through step 2, it will result in **11** in the final API, which must be limited to no more than 0.10% as both an unspecified and unqualified impurity, conservatively assuming that it is considered an unidentified impurity. With this in mind, we examined how this species propagated through the chemical sequence. As the starting material for step 1, there can be as much as 2% remaining in the step 1 reaction mixture prior to isolation of **1** due to the end of reaction criteria established for step 1. However, after isolation of **1**, this material is routinely well below 0.5%. In fact, an experiment in which the step 1 reaction mixture was laden with 10 area% **6** showed nearly complete rejection; only 0.10% was observed in the final

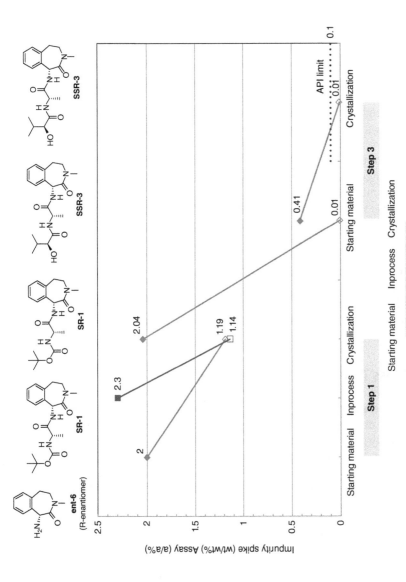

Figure 7.10 Impurity purge plot for **ent-6** across the entire semagacestat GMP sequence.

isolated **1**. Similarly, examining the propagation of this impurity through step 2, a 0.6 wt% (1 mol%) of **6** in the reaction mixture resulted in only 0.05% of the corresponding HOVal-coupled impurity, **11**, in the isolated technical grade **3**. Another point to consider is that **6** can be generated during the BOC removal process. While this pathway of formation is not negligible, it was determined that it could be completely suppressed by control of the reaction. This point will be discussed in greater detail later, but for the sake of setting specifications for **1**, we decided that it would be conservative to build in a small buffer to allow for some in situ generated **6**. Considering these data, along with the goal to have the step 2 product meet the final API specification of 0.1% **11**, we were able to rationalize a specification for **6** in the starting material, setting the specification for **6** at not more than 0.5% by HPLC area. This value was well within the typical limits of step 1, which routinely provides material **6** present at an order of magnitude below this level. Similarly, this amount is within the operating range of step 2 defined by the 0.6 wt% spiking experiment and leaves a small amount of buffer, 0.1%, for any in situ generated **6**.

This process was used to set specifications for each starting material and served as the backbone of our overall control strategy. A critical feature of this approach is to understand both how a specific impurity is generated and how it propagates through the chemical sequence. With the formation mechanisms for the various related impurities understood, and limits established for the various intermediates throughout the chemical sequence, considerations of how to control and minimize their formation can be considered. To this end, mechanistic models were utilized and are described in the remainder of the text.

7.3   Mass Transfer and Reaction Kinetics

To fully understand and predict impurity formation, it was desired to build a mechanistic model of the reaction system. To this end, it was imperative to systematically determine each parameter required for a complete model. The parameters were obtained experimentally and, in certain cases, by more than one experimental method in order to maximize accuracy and minimize error. The methods and techniques employed for obtaining the desired parameters are described in this section.

7.3.1 CO_2 Mass Transfer Model

An understanding of the intrinsic ability of the reaction matrices to absorb/desorb CO_2 was essential to controlling the urea impurity. Gas-to-liquid mass transfer is highly dependent on many process attributes including the reactor agitation rate, reactor size, fill volume, liquid composition, and liquid temperature ([15], pp. 627–628). Due to this high level of system dependence, the mass transfer coefficient is best measured at conditions that closely mimic those of the reaction system. This presented an obvious challenge from the standpoint of generating large amounts of reaction matrices at full production scale (8000 l) before the initial scale-up campaign. Consequently, a method for predicting the mass transfer at production scale was developed:

1) Determine the mass transfer coefficient for CO_2 adsorbing/desorbing in the reaction matrix at laboratory scale.
2) Determine the mass transfer coefficient for CO_2 adsorbing/desorbing in a simpler surrogate matrix (ethanol) at laboratory scale.
3) Determine the mass transfer coefficient for CO_2 adsorbing/desorbing in ethanol at production scale.
4) Develop a relationship for predicting the mass transfer coefficient for CO_2 adsorbing/desorbing in the reaction matrix at production scale utilizing the mass transfer coefficients determined in steps 1–3.

The reactor setup used in this study is described pictorially in Figure 7.11. A number of simplifying assumptions were made in the development of the CO_2 removal model:

1) Dissolved CO_2 does not convert to carbonates or bicarbonates. In the reaction mixture, the presence of HCl supports this assumption [16].
2) The headspace is maintained at constant pressure.
3) The volume of the liquid phase does not change as a result of the evolution of CO_2.
4) The equilibrium condition established between gas and liquid phases is described by Henry's law (a table of symbols is included in the section "Notations"):

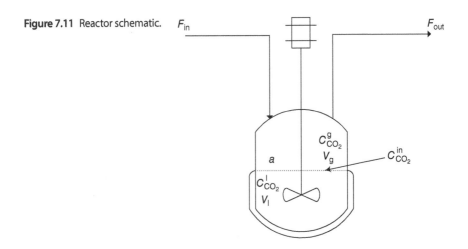

Figure 7.11 Reactor schematic.

$$p_{CO_2} = HC^l_{CO_2} \tag{7.1}$$

5) The composition of the gas phase is uniform (i.e., ideal mixing).
6) The liquid phase is of uniform composition.
7) Mass transfer resistance is limited to the interface and modeled according to

$$j_{CO_2} = -k_l a \left(C^l_{CO_2} - C^{in}_{CO_2} \right) \tag{7.2}$$

Based on these assumptions, the rate of change of CO_2 in the liquid phase is described as follows:

$$\frac{1}{V_l} \frac{dN^l_{CO_2}}{dt} = -k_l a \left(C^l_{CO_2} - C^{in}_{CO_2} \right) + r_{CO_2} \tag{7.3}$$

Both the mass transfer from liquid to gas phases and the generation of CO_2 by the deprotection reaction are accounted for with this expression. For the gas phase, the CO_2 balance includes the mass transfer from the liquid to gas phases and removal through the flow of gas out of the reactor[2]:

$$\frac{dN^g_{CO_2}}{dt} = k_l a V_l \left(C^l_{CO_2} - C^{in}_{CO_2} \right) - C^g_{CO_2} F_{out} + C^i_{CO_2} F_{in} \tag{7.4}$$

An overall mass balance is necessary to determine the total gas flow rate out of the reactor, F_{out}:

$$\frac{dN^g_{tot}}{dt} = \frac{P}{RT} \left(F_{in} - F_{out} \right) - k_l a V_l \left(C^g_{CO_2} - C^{in}_{CO_2} \right) \tag{7.5}$$

Equations 7.3, 7.4 and 7.5 are rewritten using Henry's law and the ideal gas law. Equation 7.5 (with dN^g_{tot}/dt expressed as dP/dt) is rearranged to provide an expression for F_{out}. Upon substitution, the following set of ordinary differential equations (ODEs) describe the concentration of CO_2 in gas and liquid phases:

$$\frac{dN^l_{CO_2}}{dt} = -k_l a \left(N^l_{CO_2} - N^g_{CO_2} \frac{V_l}{V_g} \frac{RT}{H} \right) + V_l r_{CO_2} \tag{7.6}$$

$$\frac{dN^g_{CO_2}}{dt} = k_l a \left(N^l_{CO_2} - N^g_{CO_2} \frac{V_l}{V_g} \frac{RT}{H} \right) + C^i_{CO_2} F_{in}$$
$$- \frac{N^g_{CO_2}}{V_g} \left(F_{in} + k_l a \left(N^l_{CO_2} - N^g_{CO_2} \frac{V_l}{V_g} \frac{RT}{H} \right) \frac{RT}{P} - \frac{V_g}{P} \frac{dP}{dt} \right) \tag{7.7}$$

Many of parameters in the model are operating conditions (T, V_l, V_g, P) and defined. Conversely, the mass transfer coefficient, $k_l a$,[3] and Henry's law coefficient were determined by experiment. This was accomplished by utilizing online measurement techniques to simultaneously measure CO_2 concentrations in gas and liquid phases (void of the reaction). Henry's law constants were obtained under equilibrium conditions with pure CO_2 in the headspace at atmospheric pressure. The uptake of CO_2 in the liquid phase was monitored in real time using attenuated total reflectance Fourier transform infrared (ATR-FTIR) spectroscopy and the change in the headspace by mass spectrometry (MS). The headspace was purged at a known flow rate with varied fractions of CO_2 in argon while monitoring the concentrations in each phase with time. Parameter estimation was performed numerically solving Equations 7.6 and 7.7 while adjusting the mass transfer coefficient until the residuals of the predicted and experimental results were minimized.

7.3.1.1 Determination of Henry's Law Constant

Determining Henry's law constant for carbon dioxide in the reaction matrix was fundamental to the understanding of the mass transfer rate and the reaction rate for the formation of the urea impurity. To ensure accuracy and minimize error in determining these values, two orthogonal methods were utilized at multiple scales.

The first method utilized a Mettler-Toledo MultiMax™ ATR-FTIR automated four-place reactor system with headspace monitoring via an Agilent 5973-MSD™, configured as a 5000A Real-Time Gas Analyzer. Gas streams were then purged through the reactor headspace of different systems and allowed to equilibrate with the liquid contents. The ATR-FTIR and MS data were used to determine Henry's law constants. The general setup is illustrated in Figure 7.12. Henry's law constants for CO_2 in pure ethanol and the reaction matrix determined at various temperatures using this method are presented in Table 7.2.

The second method for determining the solubility was accomplished via a static headspace approach using a pressure-rated HEL AutoMATE® reactor. After determining the vessel volume, the vessel was charged and the liquid degassed by pulling vacuum on the system. The headspace was then pressurized with carbon dioxide (agitation off) and sealed. At this point, the agitation was started and the pressure decay in the headspace monitored until equilibrium was established. Using the ideal gas law, the mass of carbon dioxide dissolved in the matrix was determined and Henry's law constant calculated. This approach was used to determine Henry's law constant for several variations of the solvent and reaction matrices:

1) Deprotection solvent matrix (ethanol and water) without compound **2** and HCl (Table 7.3)
2) Deprotection solvent matrix (ethanol and water) with compound **2** and 0.5 equivalents of HCl (the same matrix as measured by IR and MS)
3) Deprotection matrix including HCl

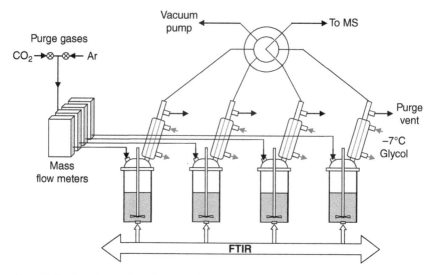

Figure 7.12 Experimental configuration for multidimensional analysis of parallel processes at small scale.

Table 7.2 Henry's law coefficient as determined using MS and FTIR (bold) and by pressure decay experiments.

	H (kPa l/mol)			
T (°C)	Ethanol (system 1)	Ethanol and water (system 2)	Deprotection reaction matrix (system 3)	Deprotection reaction matrix with excess HCl (system 4)
5	**543**	1200	**1105**, 1400	1300
20	**802**		**1593**	
40	**1538**	2300	**2360**, 2600	2600
75	**2062**		**8621**	

4) Coupling matrix without compound **2** or EDCI (discussed in Section 7.3.2.3)
5) Coupling matrix without EDCI (discussed in Section 7.3.2.3)

Good agreement was observed for Henry's law constants determined using both methods. Henry's law constants determined in the deprotection reaction matrix were used in the subsequent models since the values covered the broadest temperature range.

Table 7.3 Matrix compositions.

Material	Composition (wt%)			
	Ethanol (system 1)	Ethanol and water (system 2)	Deprotection reaction matrix[a](system 3)	Deprotection reaction matrix with excess HCl (system 4)
Compound 2	0	0	24	23.7
3A ethanol	100	68	52	51.2
Water[b]	0	32	24	23.6
HCl	0	0	0	1.5
NMM	0	0	0	0
NaOBt	0	0	0	0

[a] The matrix used for the MS/FTIR experiment differed slightly: 21.9% compound 2, 52.6% 3A ethanol, 21.3% water, and 4.2% HCl.
[b] Including water from HCl.

7.3.1.2 Determination of the Mass Transfer Coefficient

Various methods have been reviewed in the literature for measuring the mass transfer coefficients [17–19]. Ethanol was initially used to develop the methodology as well as to calibrate the ATR-FTIR. The adsorption, equilibration, and desorption cycle utilized the following experimental sequence:

1) Establish a baseline: A known volume of solution was charged to the reactor and the headspace purged with argon (Ar) until baseline values were obtained with the ATR-FTIR and the MS in liquid and gas, respectively.
2) Adsorption: The gas flow was switched from Ar to a known flow rate of CO_2 until the argon was fully purged.
3) Desorption: The flow was switched back to Ar until CO_2 was fully purged and the system returned to baseline in both gas and liquid.

Initially, this methodology was applied at 50 ml scale. The adsorption/desorption curve for ethanol and for the reaction matrix each at 5°C is shown in Figure 7.13. In the case of ethanol, the concentration of CO_2 in the liquid phase tracks that in the gas phase quite well indicating little resistance to mass transfer. However, in the reaction matrix, mass transfer is more limited as evidenced by the lag in the changes to the liquid-phase CO_2 concentration. The mass transfer coefficients are determined by solving and regressing Equations 7.6 and 7.7 with the appropriate initial conditions (i.e., 100% CO_2 in the inlet and no CO_2 in the system, or no CO_2 in the inlet, and equilibrium values in the system) until residuals of the predicted and experimental results are minimized. The values determined in ethanol and the reaction matrix are provided in Table 7.4.

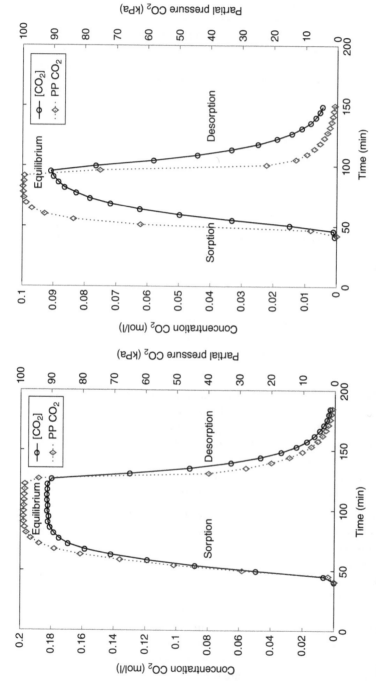

Figure 7.13 Adsorption/desorption curve for CO_2 in ethanol (left) and the reaction matrix (right) at 5°C.

Table 7.4 Mass transfer coefficients for CO_2 in ethanol and the deprotection reaction matrix.

	k_la (min^{-1})	
T (°C)	Ethanol	Reaction matrix
5	0.154	0.114
20	0.264	0.131
40	0.380	0.253
75	3.183	3.074

The natural log of the mass transfer coefficient for the two systems is plotted versus 1/T (Figure 7.14). The fit allows for the mass transfer coefficients to be estimated as a function of temperature for the 50 ml scale system.[4] Values measured at 75°C were not included since the rate of mass transfer was too high to be measured with any certainty. The high mass transfer measured at 75°C is likely due to the system being near reflux.

Mass transfer is inherently scale and equipment dependent. Even with geometrically similar reactor systems, mass transfer coefficients are difficult to predict [20, 21]. A survey of available literature was conducted to find correlations for gas–liquid mass transfer constants at various scales. Unfortunately, the majority of the data available ([15], pp. 627–634, [22–24]) is for gas sparging systems. To ensure successful transfer of the process to a larger scale, such as that used for commercialization, a scale-up methodology was developed as it was not possible to measure the mass transfer coefficient in the reaction matrix at scale prior to the start of production.

Given the lack of precedence in the literature as to how to estimate the mass transfer of the reaction matrix at a larger scale, the following ratio, obtained from the results in Figure 7.14, was assumed to hold:

$$\frac{\left(k_la\left(T,\text{scale1}\right)\right)_{\text{mixture}}}{\left(k_la\left(T,\text{scale1}\right)\right)_{\text{ethanol}}} = \frac{\left(k_la\left(T,\text{scale2}\right)\right)_{\text{mixture}}}{\left(k_la\left(T,\text{scale2}\right)\right)_{\text{ethanol}}} \tag{7.8}$$

Using this rationale, the mass transfer coefficient for CO_2 in the reaction matrix was estimated at a different scale by measuring the mass transfer coefficient for CO_2 in ethanol at the desired scale with the same mixing condition for both matrices.[5] Prior to scale up to 8000 l, this approach was tested at an intermediate scale (8 l) and found to hold.

The equipment configuration used to measure the mass transfer coefficient for CO_2 in ethanol at 8000 l scale is illustrated in Figure 7.15. Accounting for

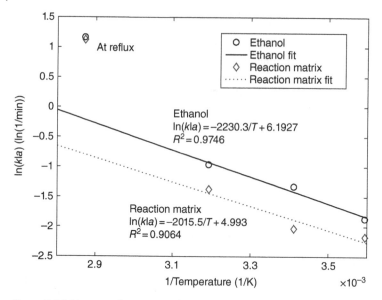

Figure 7.14 Mass transfer rates in ethanol and the reaction mixture.

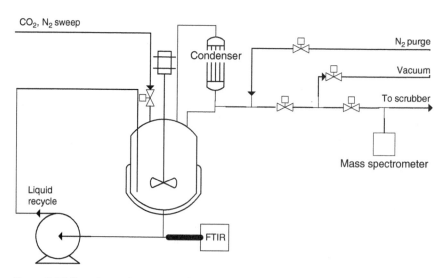

Figure 7.15 Experimental setup used for determining the k_la at 8000 l scale.

the nominal volume and the overhead pipe lengths and diameters, the total volume was estimated to be 8550 l. As with the smaller-scale experiments, a mass spectrometer was interfaced to the vent of the reactor. The ATR-FTIR probe was installed in the recycle loop. Ethanol (9521) was charged to the reactor and heated to 40°C. The reactor was fully purged with CO_2 under agitation

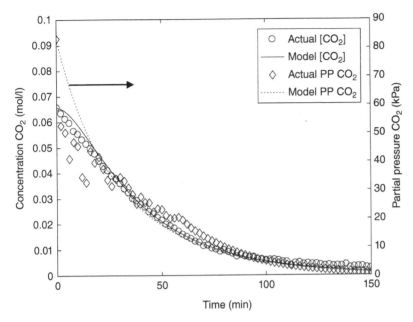

Figure 7.16 Desorption of CO_2 from ethanol at 8550 l scale, 40°C, liquid volume = 950 l, gas volume = 7600 l, purge gas flow = 260 l/min. Black arrow indicates secondary axis for partial pressure.

while acquiring FTIR and MS data. Once equilibrium was achieved, the reactor was purged with nitrogen at a rate of 260 l/min until CO_2 was no longer detected in either phase.

The desorption data and model fit are shown in Figure 7.16. The mass transfer coefficient for CO_2 from ethanol is determined to be $0.0617 \, min^{-1}$. From this and Equation 7.8, the mass transfer coefficient in the reaction mixture at 40°C is estimated to be $0.0369 \, min^{-1}$.

7.3.2 Reaction Kinetics

Understanding the mass transfer of CO_2 as well as the generation rate of the impurities, intermediate, and the desired product was essential to the control strategy for the process. From the measured Henry's law constants, it is evident that the lowest solubility of CO_2 in the reaction matrix is at the highest temperature, as expected. Therefore, the optimal temperature to remove CO_2 from the system would be at the deprotection reaction temperature. However, the desired intermediate, compound 2, can react under the deprotection conditions, resulting in undesired impurities (Figure 7.17). To fully minimize impurity formation, the reaction kinetics of the deprotection reaction and associated impurities had to be determined to ascertain the optimal CO_2 desorption temperature. The applicable reaction scheme is shown in Figure 7.17.

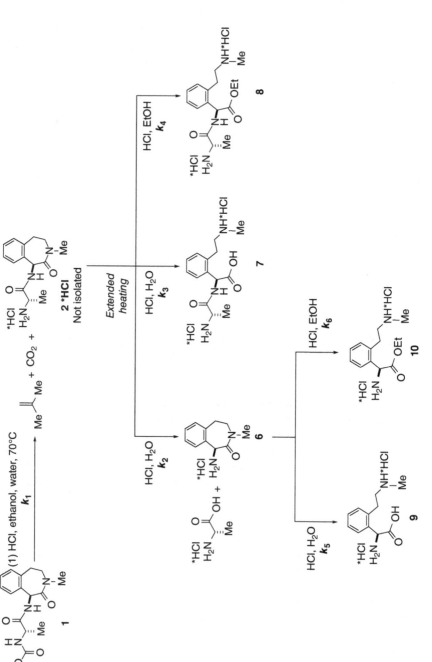

Figure 7.17 Elementary chemical expression for intermediate **2** formation and associated degradation impurities.

7.3.2.1 Deprotection Reaction Kinetics

The deprotection reaction model was developed by first proposing the pathway and describing the pathway via systems of Arrhenius and mass transfer expressions. The deprotection and degradation pathways are shown in Figure 7.17. The corresponding rate expressions used in the model are given by Equations 7.9–7.18. In Equation 7.9, the surface area of the solids is changing as the solids are dissolving. This is accounted for in Section 7.3.2.2:

$$\frac{dm_{\text{solid}}}{dt} = -k_{\text{sL}}a_{\text{solid}}\left(C_1^{\text{interface}} - C_1^{\text{bulk}}\right)\frac{M_{\text{soln}}}{\rho_{\text{soln}}}$$
$$= -k_{\text{sL}}a_{\text{solid}}\left(M_{\text{soln}}\frac{C_1^{\text{sat}}(T)}{\rho_{\text{soln}}} - MW_1 N_1\right) \tag{7.9}$$

$$\frac{dN_1}{dt} = -k_1^{\circ}e^{\frac{-Ea_1}{RT}}N_1\frac{N_{\text{HCl}}}{M_{\text{soln}}} + k_{\text{sL}}a_{\text{solid}}\left(\frac{M_{\text{soln}}C_1^{\text{sat}}(T)}{MW_1 \rho_s} - N_1\right) \tag{7.10}$$

$$\frac{dN_2}{dt} = -k_2^{\circ}e^{\frac{-Ea_2}{RT}}N_2\frac{N_{\text{HCl}}}{M_{\text{soln}}} - k_3^{\circ}e^{\frac{-Ea_3}{RT}}N_2\frac{N_{\text{HCl}}}{M_{\text{soln}}} - k_4^{\circ}e^{\frac{-Ea_4}{RT}}N_2\frac{N_{\text{HCl}}}{M_{\text{soln}}}$$
$$+ k_1^{\circ}e^{\frac{-Ea_1}{RT}}N_1\frac{N_{\text{HCl}}}{M_{\text{soln}}} \tag{7.11}$$

$$\frac{dN_6}{dt} = -k_5^{\circ}e^{\frac{-Ea_5}{RT}}N_6\frac{N_{\text{HCl}}}{M_{\text{soln}}} - k_6^{\circ}e^{\frac{-Ea_6}{RT}}N_6\frac{N_{\text{HCl}}}{M_{\text{soln}}} + k_2^{\circ}e^{\frac{-Ea_2}{RT}}N_2\frac{N_{\text{HCl}}}{M_{\text{soln}}} \tag{7.12}$$

$$\frac{dN_7}{dt} = k_3^{\circ}e^{\frac{-Ea_3}{RT}}N_2\frac{N_{\text{HCl}}}{M_{\text{soln}}} \tag{7.13}$$

$$\frac{dN_8}{dt} = k_4^{\circ}e^{\frac{-Ea_4}{RT}}N_2\frac{N_{\text{HCl}}}{M_{\text{soln}}} \tag{7.14}$$

$$\frac{dN_9}{dt} = k_5^{\circ}e^{\frac{-Ea_5}{RT}}N_6\frac{N_{\text{HCl}}}{M_{\text{soln}}} \tag{7.15}$$

$$\frac{dN_{10}}{dt} = k_6^{\circ}e^{\frac{-Ea_6}{RT}}N_6\frac{N_{\text{HCl}}}{M_{\text{soln}}} \tag{7.16}$$

$$\frac{dN_{\text{HCl}}}{dt} = -k_1^{\circ}e^{\frac{-Ea_1}{RT}}N_1\frac{N_{\text{HCl}}}{M_{\text{soln}}} - k_2^{\circ}e^{\frac{-Ea_2}{RT}}N_2\frac{N_{\text{HCl}}}{M_{\text{soln}}} - k_3^{\circ}e^{\frac{-Ea_3}{RT}}N_2\frac{N_{\text{HCl}}}{M_{\text{soln}}}$$
$$- k_4^{\circ}e^{\frac{-Ea_4}{RT}}N_2\frac{N_{\text{HCl}}}{M_{\text{soln}}} - k_5^{\circ}e^{\frac{-Ea_5}{RT}}N_6\frac{N_{\text{HCl}}}{M_{\text{soln}}} - k_6^{\circ}e^{\frac{-Ea_6}{RT}}N_6\frac{N_{\text{HCl}}}{M_{\text{soln}}} \tag{7.17}$$

$$\frac{dM_{soln}}{dt} = -\left(k_1^\circ e^{\frac{-Ea_1}{RT}} N_2 \frac{N_{HCl}}{M_{soln}} \right) \left(MW_{CO_2} + MW_{isobutylene} \right) \qquad (7.18)$$

To obtain the rate constants for each of the six reactions, four experiments were conducted: deprotection of **1** at two temperatures and the degradation of **6** at two temperatures[6] analyzing the reaction matrix over time.[7] The rate constants were determined by minimizing the overall sum of square error between the actual and predicted values for the various components. The series of ODEs (Equations 7.9–7.18) were solved in MATLAB® (version 7.5.0 R2007B and version 7.11.0 R2010b). The two constants at the two temperatures were used to calculate the pre-exponential factor and activation energy (Arrhenius constants).

The rate of decomposition of **6** (rates constants k_5 and k_6) was first determined. These rate constants were used in the determination of the various rate constants associated with the deprotection of **1** and subsequent degradation of **2*HCl**. The model prediction and experimental results for the decomposition of **6** are shown in Figures 7.18 and 7.19 with the numerical results tabulated in Table 7.5.[8]

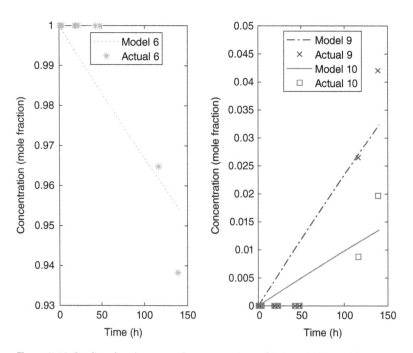

Figure 7.18 Predicted and measured concentrations of **6**, **9**, and **10** at 56°C.

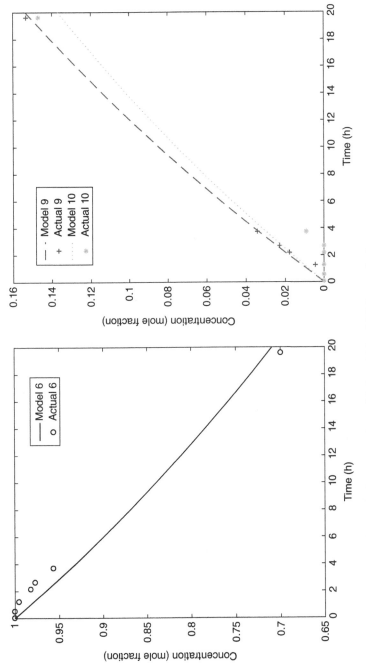

Figure 7.19 Predicted and measured concentrations of **6**, **9**, and **10** at 80°C (near reflux).

Table 7.5 Summary of **6** degradation kinetic parameters.

Temperature (°C)	80	56
k_5 (g solution/mol/h)	4.69	0.333
k_6 (g solution/mol/h)	4.18	0.139
Sum of squared error **6** prediction	1.97E–03	8.90E–04
Sum of squared error **9** prediction	8.83E–05	4.06E–04
Sum of squared error **10** prediction	1.51E–03	9.93E–05

Next, the kinetics of formation of **2*HCl** and the subsequent degradation reactions were determined by applying the derived kinetics for the degradation of **6** in the parameter estimation. Since gas is generated in the formation of **2*HCl**, the rate equation was modified to account for the mass of the entire system. For simplicity, it was assumed (and confirmed) that essentially all of the carbon dioxide and isobutylene are liberated from the reaction as they are generated.

Initially, the experiments were run dilute so that all the starting material, **1,** was in solution avoiding mass transfer in the parameter estimation. However, when the rate constants were then used to model standard reaction conditions, the model predicted differing respective amounts of the impurities than experimentally observed. Specifically, the levels of the three main impurities in the dilute (8 vol 3A ethanol with respect to **1**) experiment were **8** > **7** > **6**. In the case of a more concentrated (2.26 vol 3A ethanol) experiment, the amounts of **7** and **8** reversed places in relative ending concentration, indicating that a reaction order shift occurs with respect to **8** when the reaction is run too dilute. In the standard experiment, water acts as the primary nucleophile, but when the amount of ethanol is increased from 2.25 volumes to 8 volumes, the sheer abundance of ethanol overwhelms its relatively poor nucleophilicity, resulting in **8** becoming a significant contributor to the impurities.

Reactions were then run at standard concentrations and these experiments were used to conduct parameter estimation. In the case of these estimations, the solid dissolution mass transfer equations had to also be included (see Section 7.3.2.2), and consequently the dissolution mass transfer coefficients were estimated. The rate constants along with their resulting predictions compared against actual data are shown in Figures 7.20 and 7.21, with the rate constants tabulated in Table 7.6. The mass transfer coefficients for these two experiments were also calculated. In both cases, the constant was approximately 2.0×10^{-3} ($g_{sfl}/cm^2 \times h$).

The Arrhenius constants (activation energy, Ea_i, and the pre-exponential, k_i^o) were calculated using the rate constants determined at different temperatures (Table 7.7).

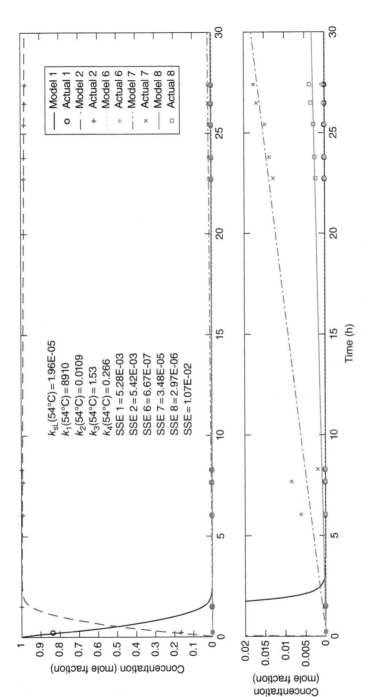

Figure 7.20 Predicted and measured concentrations of **1** at 54°C. Compounds **6** and **7** are near 0.

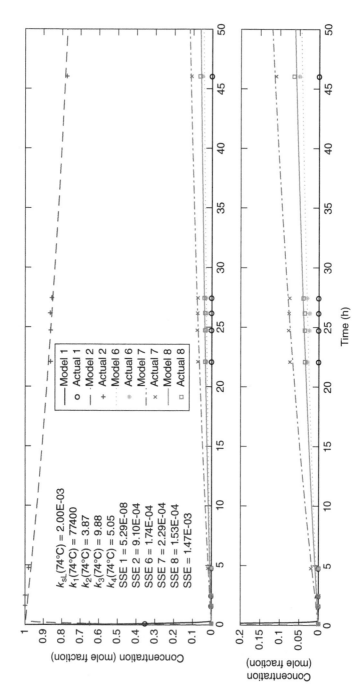

Figure 7.21 Predicted and measured concentrations of **1** at 74°C. Compound **1** is near 0.

Table 7.6 Summary of **1** deprotection kinetic parameters.

Temperature (°C)	54	74
k_{sL} (g sfl/cm^2 h)	2.83E−03	2.00E−03
k_1 (g solution/mol/h)	7830	77 400
k_2 (g solution/mol/h)	0.0806	3.87
k_3 (g solution/mol/h)	1.53	9.88
k_4 (g solution/mol/h)	0.263	5.05
Sum of squared error **1** prediction	6.66E−03	5.29E−08
Sum of squared error **2*HCl** prediction	6.80E−03	9.10E−04
Sum of squared error **6** prediction	7.17E−07	1.74E−04
Sum of squared error **7** prediction	3.47E−05	2.29E−04
Sum of squared error **8** prediction	3.03E−05	1.53E−04

Table 7.7 Calculated deprotection reaction Arrhenius constants.

	Ea_i (J/mol)	k_i^0 (g solution/mol/h)
k_1	1.08E+05	1.43E+21
k_2	1.83E+05	1.20E+28
k_3	8.80E+04	1.74E+14
k_4	1.39E+05	4.85E+21
k_5	1.06E+05	2.62E+16
k_6	1.37E+05	7.66E+20

7.3.2.2 Calculation of Dissolution Constants

For the reaction of **1** to form **2*HCl** to occur, **1** must first be solubilized. In the case where the reaction rate is considerably higher than the rate of **1** dissolution, the rate of mass transfer has the ability to control the overall rate. In the event that the dissolution rate is considerably faster than the reaction rate, the overall reaction is controlled by the reaction rate and the reaction takes place more in the bulk liquid. Based on prior work, mass transfer was known to be rate limiting. As a result it was necessary to account for the solid dissolution mass transfer rate in the kinetic model.

The dissolution process is modeled as a system of solid particles surrounded by a thin film of stagnant liquid ([15], p. 566). Compound **1** must diffuse across this film before it comes into contact with the bulk liquid and react. The film thickness dictates the rate of mass transfer and is effected by a number of system parameters, with one being agitation conditions, including the vessel geometry and agitation rate, both very scale dependent. The rate of dissolution is modeled according to (i.e., Equation 7.9)[9]

$$
\begin{aligned}
\frac{dm_{\text{solid}}}{dt} &= -k_{\text{sL}} a_{\text{solid}} \left(C_1^{\text{interface}} - C_1^{\text{bulk}} \right) \frac{M_{\text{soln}}}{\rho_{\text{soln}}} \\
&= -k_{\text{sL}} a_{\text{solid}} \left(M_{\text{soln}} \frac{C_1^{\text{sat}}(T)}{\rho_{\text{soln}}} - \text{MW}_1 N_1 \right)
\end{aligned}
\tag{7.19}
$$

The concentration of solute at the particle surface is assumed to be equal to the saturation concentration at the temperature of the reaction.

The surface area of the solids changes as the crystals dissolve. Assuming that the crystal dissolves equally along both major and minor axes (Figure 7.22), the ratio of $x{:}y$ is maintained. As a result, defining an aspect ratio, $\phi = y/x$, and assuming that the particles are monodisperse, the surface area is

$$
a_{\text{solid}} = n_{\text{p}} \left(2x^2 + 4x^2 \phi \right)
\tag{7.20}
$$

The mass of solids is determined from the total volumes of solids according to

$$
m_{\text{solid}} = n_{\text{p}} \phi x^3 \rho_{\text{s}}
\tag{7.21}
$$

The number of particles is determined from the initial mass of solids, $m_{\text{solids}}^{\text{o}}$, of initial particle size, x_{o}^3, charged according to

$$
n_{\text{p}} = \frac{m_{\text{solid}}^{\text{o}}}{\phi x_{\text{o}}^3 \rho_{\text{s}}}
\tag{7.22}
$$

Equations 7.19, 7.20, and 7.21 are combined to provide a differential equation for the mass of solids with time:

$$
\frac{dm_{\text{solid}}}{dt} = -2 \frac{(1 + 2\phi)\left(m_{\text{solids}}^{\text{o}} \right)^{1/3}}{x_{\text{o}} \phi \rho_{\text{s}}} \left(m_{\text{solids}} \right)^{2/3} k_{\text{sL}} \left(M_{\text{soln}} \frac{C_1^{\text{sat}}}{\rho_{\text{soln}}} - \text{MW}_1 N_1 \right)
\tag{7.23}
$$

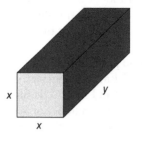

Figure 7.22 Crystal dimensions.

The particle aspect ratio is determined from specific surface area and particle size analysis, and the initial particle size by particle size analysis, and it is found to be 6.7. The equilibrium solubility was measured for the solvent composition as a function of temperature. The solubility was fit using a Van't Hoff correlation according to

$$C_1^*(T) = 3.25 \times 10^9 \exp\left(\frac{-8405.3}{T}\right) \text{mol} / \text{mol soln} \qquad (7.24)$$

The solubility of compound **1** in the reaction matrix was used with Equation 7.23 to determine the dissolution rate.

7.3.2.3 Coupling Reaction Kinetics

For the urea formation, **4**, the determination of the reaction rate constant was not as direct as for the deprotection step. Typically, reaction rates, mechanisms or reaction pathways, rate equations, and associated rate constants can be derived using experimental data at a variety of reaction concentrations and temperatures. However, since the reaction chemistry for the urea formation in the coupling reaction involves carbon dioxide that easily transfers from the liquid to the gas phase, the reactions had to be conducted in a sealed vessel. The following describes the experiments conducted to measure the rate of formation of the urea:

1) Determination of the Henry's law coefficient and the mass transfer coefficient for CO_2 into the coupling reaction matrix (different from the deprotection reaction matrix as described in the previous section)
2) Determination of the desired reaction rate constant
3) Determination of the rate of urea, compound **4**, formation under controlled reactor headspace conditions

Once the kinetic parameters were determined, the subsequent reaction models were used to predict the desired product quality and associated limits on allowable levels of in situ carbon dioxide prior to the coupling reaction.

The gas–liquid mass transfer coefficients for the coupling system were determined using a method modified from one that has been described previously [25]. However, the mass transfer coefficient could not be measured in the actual coupling system since the reaction would progress. As a result the mass transfer coefficient was measured without EDCI so that the reaction would not occur (Table 7.8). The mass transfer and Henry's law coefficients determined for the coupling matrix without EDCI were used in the subsequent process simulations. The measured mass transfer coefficients for both matrices are shown in Table 7.9. The Henry's law coefficient was determined as described in Section 7.3.1.1 using the pressure decay method. The measured values are shown in Table 7.10.

The coupling reaction kinetics were determined by isolating the desired reaction from the undesired reaction. The pathways of interest are shown in Figure 7.4 for the desired product, **3**, and the urea formation, **4**, in Figure 7.7.

Table 7.8 Matrix compositions used to determine the mass transfer and Henry's law coefficients in the coupling reaction.

	Composition (wt%)	
Material	Coupling matrix without compound 2 or EDCI	Coupling matrix without EDCI
Compound 2	0	13
3A ethanol	67	58
Water[a]	23	20
NMM	9	8
NaOBt	1	1

[a] Including water from HCl.

Table 7.9 Step 2 matrix calculated k_la values.

	k_la (min^{-1})	
T (°C)	Coupling matrix without compound 2 or EDCI	Coupling matrix without EDCI
5	6.2	8.5
40	20	No data

Table 7.10 Henry's law coefficient as determined by pressure decay experiments.

	H (kPa l/mol)	
T (°C)	Coupling matrix without compound 2 or EDCI	Coupling matrix without EDCI
5	490	410
40	1600	No data

From the literature, the rate-limiting step is assumed to be the reaction of the carboxylic acid, in this case HOVal, with EDCI [11]. The reaction of the *o*-acylisourea with NaOBt and the subsequent reaction of this ester with compound **2** are much faster than the reaction of EDCI with HOVal. To describe the overall reaction kinetics, it is assumed that the formation of the desired product, **3**, can be described as the reaction of compound **2** with

HOVal described as shown in Equation 7.26. The balance for the deprotected intermediate, compound **2**, is given by Equation 7.25. Equations 7.28 and 7.29 were used to close the mass balance. The formation of the urea impurity, **4**, is described by Equation 7.27. Equations 7.30 and 7.31 describe the balance of CO_2 in the reaction matrix and the reactor headspace, respectively, and were derived in Section 7.3.1. These expressions have been simplified for the conditions of the reaction system. It should be noted that elementary steps for each of these reactions have not been rigorously modeled. Instead, a simplified model has been assumed that preserves the important, that is, rate-limiting, components of the system and is limited to the stoichiometry described herein:

$$\frac{dN_2}{dt} = -k_7 N_2 \frac{N_{HOVal}}{M_{soln}} - 2k_8 N^l_{CO_2} \frac{N_2^2}{M_{soln}} \tag{7.25}$$

$$\frac{dN_3}{dt} = k_7 N_2 \frac{N_{HOVal}}{M_{soln}} \tag{7.26}$$

$$\frac{dN_4}{dt} = k_8 N^l_{CO_2} \frac{N_2^2}{M_{soln}^2} \tag{7.27}$$

$$\frac{dN_{HOVal}}{dt} = F_{HOVal} - k_7 N_2 \frac{N_{HOVal}}{M_{soln}} \tag{7.28}$$

$$\frac{dM_{soln}}{dt} = \frac{F_{HOVal}}{x_{HOVal}} MW_{HOVal} - k_1 a MW_{CO_2} \left(N^l_{CO_2} - N^g_{CO_2} \frac{M_{soln}}{\rho_{soln} V_g} \frac{RT}{H} \right) \tag{7.29}$$

$$\frac{dN^l_{CO_2}}{dt} = -k_1 a \left(N^l_{CO_2} - N^g_{CO_2} \frac{M_{soln}}{\rho_{soln} V_g} \frac{RT}{H} \right) - k_8 N^l_{CO_2} \frac{N_2^2}{M_{soln}^2} \tag{7.30}$$

$$\frac{dN^g_{CO_2}}{dt} = k_1 a \left(N^l_{CO_2} - N^g_{CO_2} \frac{V_l}{V_g} \frac{RT}{H} \right)$$
$$- \frac{N^g_{CO_2}}{V_g} \left(k_1 a \left(N^l_{CO_2} - N^g_{CO_2} \frac{V_l}{V_g} \frac{RT}{H} \right) \frac{RT}{P} - \frac{V_g}{P} \frac{dP}{dt} \right) \tag{7.31}$$

The rate constant, k_7, for the API formation, **3**, was determined by conducting the coupling reaction at 5°C after first fully purging the system of CO_2. The reaction was initiated by adding all the reactants simultaneously and then monitoring the concentration of the reactant, **2**, and the product, **3**, by HPLC with time. The system of ODEs was solved in MATLAB.[10] The rate constant k_7 was determined by minimizing the residuals between the modeled concentration and experimental values. The system of ODEs was solved subject to the initial conditions and constraints as given in Equation 7.32:

$$N_2(t=0) = N_2^0, \quad N_{HOVal}(t=0) = N_{HOVal}^0,$$
$$N_3(t=0) = 0, \quad M_{soln}(t=0) = M_{soln}^0$$
$$N_{CO_2}^l(t=0) = 0, \quad N_{CO_2}^g(t=0) = 0, \tag{7.32}$$
$$N_4(t=0) = 0, \quad P(t=0) = P^o, \quad \dot{F}_{HOVal} = 0$$

The rate constant that provided the best fit to this experimental data was 1.0×10^4 g solution/(mol × h). The measured product concentration (crosses) is compared with the predicted concentration data over time in Figure 7.23.

The rate constant for the undesired reaction forming the urea impurity, **4**, was determined using two separate methods due to the complexity of reaction monitoring. In one method the uptake of CO_2 from the headspace was monitored, and in the other, compound **4** formation was monitored with known spikes of CO_2. The rate constant was regressed in a similar manner as done for the desired reaction with the initial conditions:

$$N_2(t=0) = N_2^0, \quad N_{HOVal}(t=0) = N_{HOVal}^0,$$
$$N_3(t=0) = 0, \quad M_{soln}(t=0) = M_{soln}^0,$$
$$N_{CO_2}^l(t=0) = C_{CO_2}^{sat} V_l, \quad N_{CO_2}^g(t=0) = N_{CO_2}^{g,0}, \tag{7.33}$$
$$N_4(t=0) = 0, \quad P(t=0) = P^o, \quad \dot{F}_{HOVal} = 0$$

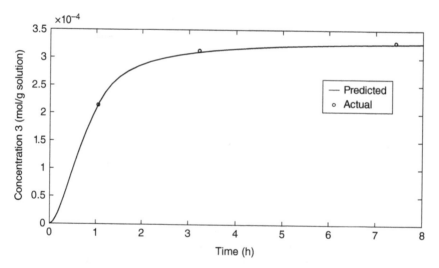

Figure 7.23 Coupling reaction kinetic data for the desired product, compound **3**, formation, experimental (open circle) versus predictive (line).

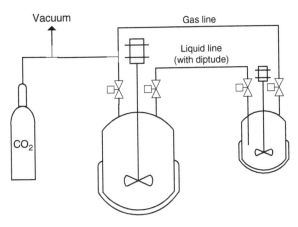

Figure 7.24 Compound **4** reaction kinetics equipment configuration.

The CO_2 uptake studies carried out from the headspace were performed using HEL pressure-rated reactors. The setup for the urea reaction studies involved the use of two reactors (300 and 100 ml volumes) as shown in Figure 7.24. The experiments were conducted according to the following:

1) Charge the coupling reaction matrix to the 300 ml reactor.
2) Charge the NMM/EDCI/NaOBt mixture to the 100 ml reactor. Position the dip tube in the 100 ml reactor so that the desired amount will transfer to the 300 ml reactor when the 100 ml reactor is pressured.
3) Cool both solutions to 5°C and purge the headspace of the reactors using carbon dioxide.
4) Stop agitation on both vessels.
5) Pressurize the 100 ml reactor to the desired experiment pressure and allow the liquid in the 100 ml reactor (EDCI/NMM solution) to transfer to the 300 ml reactor as fast as possible.
6) Turn on the agitation in the 300 ml reactor to the desired set point.

The rate constant for the formation of compound **4** that provided the best fit to this experimental data was 2×10^6 (g solution/mol)^2/h. The measured headspace pressure (crosses) is compared with the predicted pressure over time in Figure 7.25. This method, however, resulted in the formation of a thick white slurry of pure urea during the course of the study. The increase in the slurry viscosity over the course of the reaction results in a decrease of the mass transfer coefficient over the extent of the reaction. This effect was not accounted for in the model.

Because the slurry viscosity and consequently the mass transfer coefficient changed during the progression of the reaction, it was suspected that there

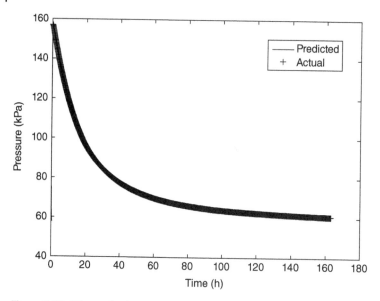

Figure 7.25 CO_2 uptake during the urea formation reaction (+ = actual, line = predicted).

was potential error in the obtained value of the rate constant k_8 using the pressure decay method. Consequently, an alternate method was used to determine the rate constant for the urea reaction. Known amounts of CO_2 were added to the headspace of a coupling experiment in a sealed system (a 2.3 l, 1 l nominal Mettler-Toledo RC1-E reactor) as follows:

1) Add coupling reaction matrix in the reactor and cool to 5°C.
2) Charge NaOBt to a pressure-equalizing dropping funnel; charge EDCI and NMM solution to a pressure-equalizing dropping funnel; charge HOVal solution to a syringe pump for a timed addition or an addition funnel for bolus addition; keep all solutions under nitrogen and sealed from the main reactor headspace.
3) Allow the reaction solution to equilibrate with a gas headspace of known carbon dioxide concentration for several hours; monitor by in situ ATR-FTIR; then seal the reactor.
4) Within the sealed system, add the NaOBt solution to the reactor from the addition funnels.
5) Add the EDCI and NMM solution to the reactor from the addition funnel.
6) Add the HOVal solution to the coupling solution over a specified time period using a syringe pump (or dropping funnel for bolus addition).
7) Once the reaction has completed, warm the mixture to 40°C and analyze the product.

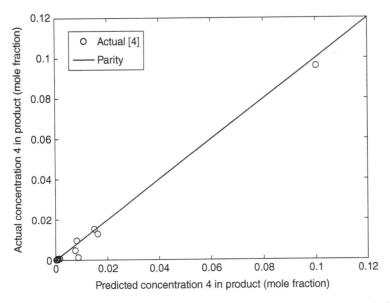

Figure 7.26 Comparison of actual and predicted urea concentration at the end of the reaction when spiked with various amounts of CO_2 in the headspace.

Three CO_2 spiking experiments were used to generate the rate constant, and eight were used to experimentally confirm the method. The comparison of the actual and predicted data is shown in Figure 7.26. The rate constant was determined by measuring the ratio of urea to desired product formed. Knowing the rate constant of the desired reaction, the unknown rate constant was easily determined. The rate constant obtained using this method was 5.5×10^6 g solution^2/(mol^{2}h). The rate constant for the urea formation obtained by the two different methods were in good agreement considering the limitations of each method, 5.5×10^6 versus 2×10^6, further reinforcing the accuracy of the model.

7.4 Optimal Processing Conditions

With all kinetic and mass transfer coefficients determined, it was possible to fully simulate the process. The overall step 2 process model consists of the deprotection reaction model; step 2a, the CO_2 mass transfer model; and the coupling reaction model, step 2b. The step 2 model could then be used to minimize the overall impurity profile and to optimize the process efficiency. In doing so, the conditions used to remove CO_2 prior to initiating the coupling reaction are needed to be designed with two primary factors in mind. First, based on the thermodynamics of the system being studied, the solubility of carbon dioxide is lower at higher temperatures. At the same time, the degradation of the post-deprotection intermediate, **2*HCl**, is an issue at higher temperatures. **2*HCl** has been shown to degrade through a cascade of degradation reactions (see Figure 7.17). The degassing temperature was probed by simulating these two counteracting mechanisms using a combination of the mathematical models describing the deprotection reaction: degassing and coupling reaction. The modeled system was simulated at a series of degassing temperatures, and the output (product impurity load) was used as a constraint to estimate the optimal degassing temperature and degassing regiment.

7.4.1 Use of Combined Models

The three models (deprotection and coupling kinetic models and the mass transfer model) were aggregated to form a "macro-expression" of the step chemistry. The macro-expression consisted of simultaneously solving Equations 7.6, 7.7 (CO_2 mass transfer), 7.9–7.18 (cascade of deprotection reactions), and 7.23 (compound **1** dissolution expression) subject to the appropriate initial conditions (concentration of compound **1** in suspension) and processing conditions (temperature and pressure with time). The output (calculated concentrations of compound **2** and CO_2) was used as the initial conditions simultaneously solving Equations 7.25–7.31, again subject to the appropriate processing conditions (temperature and pressure with time). Two process variables within the degassing step, temperature and nitrogen flow rate through the reactor headspace, were varied simultaneously. Specifically, the temperature of the degassing step was varied from 5° to 60°C, and the nitrogen sweep rate during the degassing steps was varied from 0 to 50000 l/h for an 8550 l reaction vessel. The program used the MATLAB ode23tb solver to solve the series of ODEs for the predicted impurity profile. The predicted total impurity profiles are shown in Figure 7.27.

Increasing the nitrogen flow rate through the headspace decreases the total impurity level at low flow rates but plateaus at higher flow rates as expected. Once all CO_2 is purged from the system, the urea impurity is no longer formed, and impurities formed are only those from continued degradation of the desired intermediate, compound **2**. The effect of temperature exhibits a saddle point at the low nitrogen sweep rates. At higher rates, impurity formation is minimized at temperatures below 40°C. From this analysis, the optimal degassing conditions were determined to be 20–40°C and a nitrogen sweep rate of 2000 l/h or more.

Since the levels of individual impurities are just as important as the total impurities of total related substances (TRS), the individual impurities were also analyzed. Each of these impurities must meet an acceptable in situ limit.

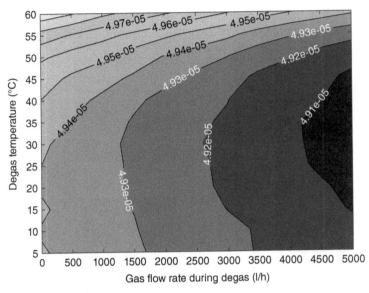

Figure 7.27 Predicted impurity levels as a function of degassing temperature and headspace degassing flow rate.

The limits are based on the potential for product to not meet specifications based upon the previous experiments. It was determined that the predicted levels were within acceptable ranges in the proposed design space. The contour plots predicting each of the related impurities are shown in Figure 7.28.

A degassing temperature of 40°C was chosen as the condition to degas the reaction matrix prior to starting the coupling reaction at 5°C. The rationale for CO_2 removal at the higher end of the optimal range was to perform this operation under conditions where CO_2 would have the lowest solubility in the liquid phase. With the optimal temperature defined, various purge options were considered.

7.4.2 Carbon Dioxide Removal Process Options

The aforementioned simulation considered a constant sweep of nitrogen through the headspace of the reactor. However the model assumes ideal mixing in the reactor headspace—likely a poor assumption in the headspace of a commercial-scale (8550l) reactor (without agitation in the headspace). It was desired to sparge the solution of CO_2 by injecting nitrogen subsurface allowing for intimate contact of gas and liquid phases at the point of CO_2 generation. However, due to the agitator design and vessel configuration, it was not possible to install a sparge ring under the agitator. Alternatives considered were a constant

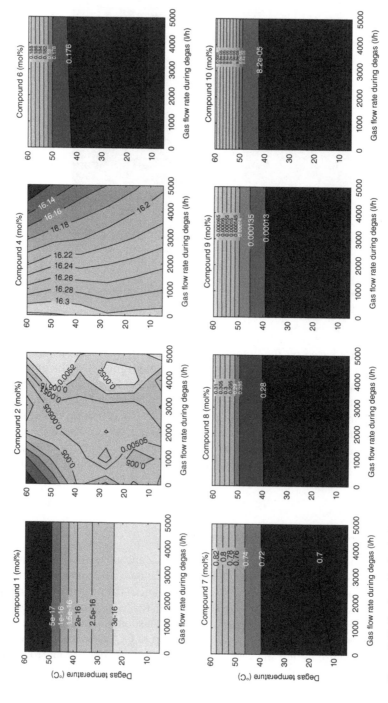

Figure 7.28 Predicted individual impurity levels as a function of degassing temperature and headspace degassing flow rate.

nitrogen sweep through the reactor headspace, pressurization/depressurization cycles, vacuum/pressurization cycles, or combinations of these.

At commercial scale (8550 l), approximately 680 mol of CO_2 is generated by the deprotection exceeding the gaseous molar CO_2 capacity (280 mol) of the reactor and reactor overhead at the reaction conditions (101 kPa, 70°C). Constraints were placed on the process from the manufacturing facility. It was desired to perform not only the deprotection reaction and the coupling reaction in the same vessel but also the crystallization of the product. The crystallization process is an antisolvent crystallization in which the final volume at the end of crystallization is almost three times that of the volume at the start of crystallization. As a result, the volume at each phase of the process is highly constrained: 13% fill volume at the end of the deprotection reaction, 32% fill volume at the end of the coupling reaction, and 90% fill volume at the end of crystallization. Therefore, the headspace-to-liquid volume ratio at the end of the deprotection reaction is undesirably high.

To proceed forward, the various removal operations were simulated with the combined deprotection model and mass transfer model. The deprotection reaction was simulated with a 1 h reaction held at 70°C and then cooled to 40°C over 60 min.

The first option considered was the utilization of three vacuum cycles, pulling vacuum to 33 kPa followed by re-pressurization to 101 kPa at 40°C. The vacuum pressure is limited to 33 kPa to avoid bringing the solvent matrix to reflux. The results of the simulation are shown in Figure 7.29. The top graph plots the concentration of CO_2 in the liquid phase and the solubility of CO_2 in the liquid phase if in equilibrium with the headspace CO_2 concentration over the course of the reaction. As the reaction begins, the liquid-phase concentration of CO_2 quickly rises above the solubility limit. The model does not account for nucleation of gas bubbles in the liquid phase. The supersaturation is only relieved by mass transfer of CO_2 between phases, which is why the depicted liquid-phase concentration exceeds the solubility limit. At the end of the reaction, the headspace is almost entirely filled with evolved CO_2 (see the second panel). The third panel shows the mole fraction of CO_2 in each phase (normalized by the total moles of evolved CO_2), and the last the temperature and pressure profiles over the course of the reaction and subsequent pressure manipulations. The phases are in equilibrium at the end of each vacuum cycle as a result of the high mass transfer rates. The vacuum cycle effectively dilutes the headspace concentration by 66% with each cycle.

Additional CO_2 removal conditions considered included the following: option 2, three pressure and three vacuum cycles (Figure 7.30); option 3, a constant headspace purge (70 l/min) with three vacuum cycles (Figure 7.31); option 4, a constant headspace purge with three pressure cycles and three vacuum cycles (Figure 7.32); and option 5, a constant headspace purge with

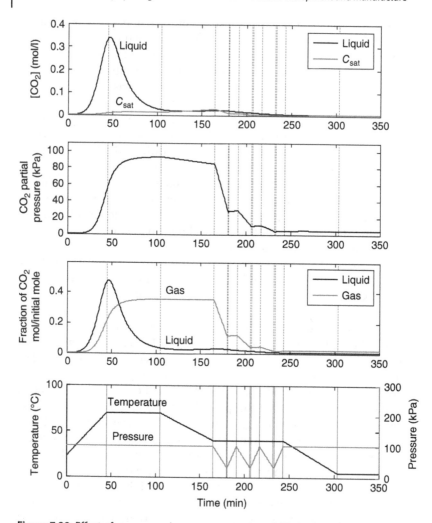

Figure 7.29 Effect of vacuum cycles on concentration of CO_2 in the reactor prior to the coupling reaction (option 1).

four vacuum cycles (Figure 7.33). CO_2 concentrations after cooldown to 5°C are summarized in Table 7.11.

The upper limit for the total CO_2 in the gas phase (at the same ratio of liquid to gas volumes) was determined through experiments to be 1.0% (mol/mol compound **1**) and the coupling reaction kinetic model (discussed in Section 7.5). Under these conditions, levels of compound **4** below the acceptable limit of

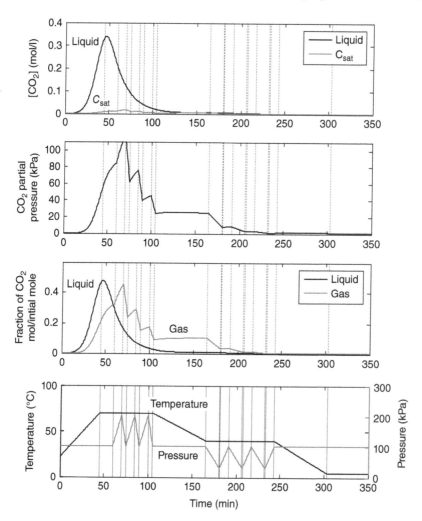

Figure 7.30 Effect of pressure and vacuum cycles on concentration of CO_2 in the reactor prior to the coupling reaction (option 2).

0.10% (wt/wt) were observed in the isolated product. Of the five options considered (Table 7.11), options 2 and 4 result in levels of CO_2 below the upper limit with option 3 just above the limit. Both options 2 and 4 utilize a pressure purge cycles during the reaction, followed by vacuum purge cycles at 40°C— option 4 incorporates a headspace sweep, while option 2 does not. The utilization of a reasonable headspace sweep, although flow rate dependent, offers

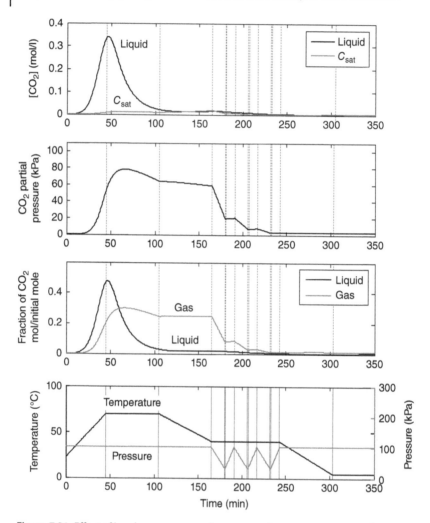

Figure 7.31 Effect of headspace purge and vacuum cycles on concentration of CO_2 in the reactor prior to the coupling reaction (option 3).

only a modest decrease in the CO_2 level at the end of the removal process. The utilization of additional pressure purge cycles (option 5) without vacuum cycles did not reduce the CO_2 levels below the target. The reason for the result is due to the requirement that the reaction time must not exceed 60 min. This required the pressure cycles to begin earlier in the reaction when large amounts of CO_2 are already off-gassing since the reaction is run under

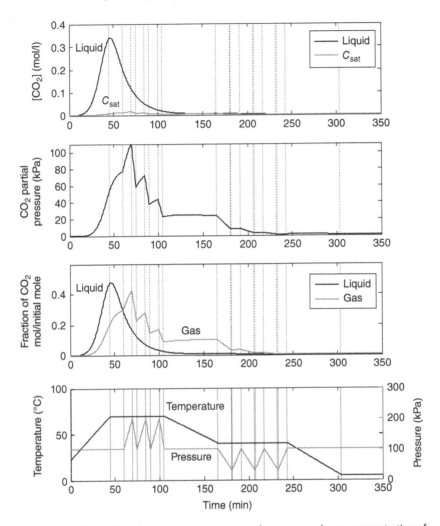

Figure 7.32 Effect of headspace purge, vacuum, and pressure cycles on concentration of CO_2 in the reactor prior to the coupling reaction (option 4).

constant pressure. In the end, option 2 was selected as the appropriate CO_2 removal process. It is important to note that if the ratio of headspace volume to liquid volume were to change, the results would also change. If the ratio were to decrease, that is, a smaller headspace and larger liquid volume, the total amount of CO_2 in the reactor system would also decrease. However, the mass transfer rate would be expected to decrease and would need to be considered.

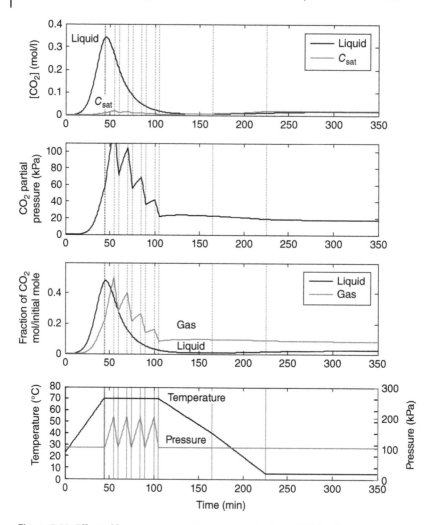

Figure 7.33 Effect of four pressure cycles on concentration of CO_2 in the reactor prior to the coupling reaction (option 5).

7.5 Predicted Product Quality under Varied Processing Conditions

Given the deprotection model, the residual levels of CO_2 post-degassing, and the formation of **4** relative to **3**, the design space as it applies to all process parameters could be simulated to determine the impact on the product quality. The impact of hydroxyvaline addition time and the level of CO_2 prior to

Table 7.11 Summary of total mole fraction of CO_2 in the reaction system (as a fraction of the total amount of compound 1 at the start of the reaction) for the various CO_2 removal options considered.

		Total mole fraction CO_2 of initial 1				
Conditions	Option	1	2	3	4	5
	Headspace sweep	No	No	Yes	Yes	No
	Pressure purge (cycles)	No	Yes (3)	No	Yes (3)	Yes (4)
	Vacuum purge (cycles)	Yes (3)	Yes (3)	Yes (3)	Yes (3)	No
Process point	End of reaction	39%	62%	28%	59%	71%
	End of pressure cycle 1	—	36%	—	35%	56%
	End of pressure cycle 2	—	21%	—	20%	33%
	End of pressure cycle 3	—	11.8%	—	11.4%	19%
	End of pressure cycle 4	—	—	—	—	10.9%
	Cool to 40°C	39%	11.4%	27%	11.0%	10.5%
	End of vacuum cycle 1	15%	4.4%	10.2%	4.2%	—
	End of vacuum cycle 2	6.0%	1.8%	4.2%	1.8%	—
	End of vacuum cycle 3	2.6%	0.81%	1.9%	0.78%	—
	Cool to 5°C	2.6%	0.81%	1.9%	0.78%	10.7%
	Hold at 5°C—total	**2.6%**	**0.81%**	**1.9%**	**0.78%**	**10.8%**
	Hold at 5°C—gas phase	*2.1%*	*0.63%*	*1.5%*	*0.61%*	*8.4%*
	Hold at 5°C–liquid phase	*0.57%*	*0.17%*	*0.40%*	*0.17%*	*2.5%*

the start of the coupling reaction were modeled using the previously derived reaction and mass transfer models. The urea formation is shown in Figure 7.34. The results of the simulation allowed for an estimation of the maximum allowable levels of urea formation at the end of the degassing operation.

Utilization of the combined model finds that shorter addition times of hydroxyvaline, result in lower levels of compound **4** in the product. However too short of an addition time leads to the formation of another impurity (an analog of a hydroxyvaline dimer, compound **5**, shown in Figure 7.5). A 1 h addition of hydroxyvaline coupled with an initial CO_2 level of less than 15 ppm in solution at the start of the coupling reaction results in product meeting the compound **4** limit of less than 0.10% (wt/wt). These conditions also provide a product void of the hydroxyvaline dimer analog, **5**.

7.5.1 Virtual Execution of PAR and Design Space Experiments

One approach that has been used to prepare a new process submission is the use of laboratory models to generate process information. Specifically, univariate

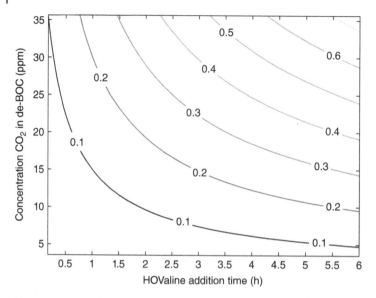

Figure 7.34 Mass fraction of urea, compound 4, in the product as a function of CO_2 concentration in solution at the start of the coupling reaction and as a function of the hydroxyvaline addition rate.

experiments have been used to determine proven acceptable ranges (PAR) and multivariate experiments to approximate the design space of a limited number of process variables. The execution of this work has typically begun with the estimation of PAR by running single experiments (via the laboratory model) at ± six standard deviations (or sigma) [26–29] of the set point and determining whether or not the product meets intermediate or API quality attributes. Experiments that pass are considered to have low or medium risk where medium is assigned from scientific judgment based on whether or not the variable could potentially interact with another process variable to produce a failing product. Experiments that fail to meet the quality attribute over the range of ± six-sigma in the PAR studies are assigned as high risk and are also designated as potential critical process parameters since they may impact a critical quality attribute outside of an acceptable range. High- and medium-risk variables are then studied in multivariate experiments to fully explore the design space.

Another method of determining the PAR and design space is one that takes advantage of the flexibility of experimentally confirmed mathematical models. Once the model described in Section 7.3.2.1 was created, it was able to be used in this approach. Specifically, this method involved ascertaining the PAR and design space by running the model under the various process conditions and not through more experiments in the laboratory. This approach offers significant time savings, the ability to explore multiple interactions, and a mathematical representation of the design space.

Table 7.12 Deprotection chemistry process parameters.

Parameter	Sigma[a]	– 6*Sigma	Center	+6*Sigma
Total 3A alcohol (vol)	0.0225	2.12	2.25	2.39
Total purified water (vol)	0.00564	0.530	0.564	0.598
Compound 1 (eq)	0.010	0.94	1.00	1.06
Concentrated HCl (eq)	0.015	1.41	1.50	1.59
Stir time prior to heating (min)	10	0	0	60
DeBOC heating hold time (min)	10	0	60	120
Heat-up ramp rate (°C/min)	0.0037	1.034	1.056	1.078
Deprotection heating temperature (°C)	2	58	70	82
Cooldown ramp rate (°C/min)	0.0045	0.473	0.5	0.527
Cooldown hold time (min)	10	0	0	60
Cooldown temperature (°C)	2	28	40	52

[a] Note that σ for volume or weight = 1% of the charge, temperature = 2°C, manual time = 10 min, and rates are based on the calculated propagation of error for the two terms (i.e., temperature and time).

7.5.1.1 Process Parameters

The first step in simulating the perturbation of process variables was to determine the potential ranges for each process variable in the deprotection chemistry. Limits were calculated using a ± six-sigma estimation from the process center point. The one-sigma values were determined using accumulated data on the capability of processing equipment. The calculated values are shown in Table 7.12.

7.5.2 Acceptable In Situ Values

The model created for the deprotection process predicts the in situ impurity profile prior to execution of the coupling chemistry (Section 7.3.2.3). In order to correlate the in situ impurity profile to the final isolated product, deprotection reaction solution that was stressed at 75–76°C for 3.5 h was forward processed through the coupling process and the dry product was isolated. This product was determined to be acceptable, and consequently, the following impurities at their associated in situ levels were determined to be acceptable for the step 2a reaction in situ data point (Table 7.13). It must be noted that these levels of impurities are considered to be the *limit of knowledge* and not necessarily the *limit of failure* for the process.

Table 7.13 Deprotection process acceptable in situ impurity levels.[a]

Compound	Area % by HPLC
2	96.3
6	1.2
7	2.2
8	0.3

[a] Compounds 1, 9, and 10 were not observed in the impurity validation experiments, and consequently, the criteria for each of these compounds were conservatively set at a limit of 0.1%.

7.5.3   PAR Simulation

A total of 22 PAR experiments were conducted virtually using the deprotection model. The process conditions considered are described in Table 7.14. The results from these experiments are shown in Table 7.15. The results highlighted in gray indicate those experiments that had **1** levels or **8** levels that had not been demonstrated to be acceptable and consequently are conditions that need further study to determine if they are acceptable. The termination of the project did not allow for the experiments to be completed. However, if the project would have continued, these conditions would indeed have been run experimentally. Further experimental confirmation of the model should also be performed by selecting sets of process conditions at extremes in predicted impurities and conducting experiments under those conditions to further verify the model predictability.

7.5.4   Design Space Simulation: Interactions

Based on the results described in Section 7.3.2.1, the two process parameters that did not meet the acceptable impurity levels were the deprotection reaction hold time and the deprotection reaction temperature. These parameters were consequently used to conduct virtual multivariate experiments in order to ascertain the design space for the high-risk process variables. The other parameters in this process step were run at their center point values. The results for total impurity level (TRS, sum of **1**, **6**, **7**, **8**, **9**, and **10**) versus time and temperature are shown in Figure 7.35.

These plots show that the high TRS is associated with two sources of impurities: (i) unreached **1** in the low hold time/low temperature regions and (ii) impurities resulting in degradation of **2** due to high temperature and high reaction hold time.

Examining TRS alone does not show which regions in the design space would be expected to pass based on the demonstrated impurity profile (Table 7.13).

Table 7.14 PAR conditions.

Experiment	3A alcohol (vol)	Purified water (vol)	Cpd 2 (eq)	HCl (eq)	Mass solution (g)	HCl initial (mol/g)	Initial temperature (°C)	Stir time prior to heat (h)	Heat-up ramp rate (°C/min)	De-BOC heating hold time (h)	De-BOC heating temperature (°C)	Cool-down ramp rate (°C/min)	Cooldown hold time (h)	Cooldown temperature (°C)
1	2.115	0.5641	1	1.5	3.64	0.00114	22.5	0	1.056	1	70	0.5	0	40
2	2.385	0.5641	1	1.5	3.86	0.00108	22.5	0	1.056	1	70	0.5	0	40
3	2.25	0.5303	1	1.5	3.72	0.00112	22.5	0	1.056	1	70	0.5	0	40
4	2.25	0.5979	1	1.5	3.78	0.00110	22.5	0	1.056	1	70	0.5	0	40
5	2.25	0.5641	0.94	1.5	3.67	0.00106	22.5	0	1.056	1	70	0.5	0	40
6	2.25	0.5641	1.06	1.5	3.84	0.00115	22.5	0	1.056	1	70	0.5	0	40
7	2.25	0.5641	1	1.410	3.73	0.00105	22.5	0	1.056	1	70	0.5	0	40
8	2.25	0.5641	1	1.590	3.78	0.00117	22.5	0	1.056	1	70	0.5	0	40
9	2.25	0.5641	1	1.5	3.75	0.00111	34.5	0	1.056	1	70	0.5	0	40
10	2.25	0.5641	1	1.5	3.75	0.00111	10.5	0	1.056	1	70	0.5	0	40
11	2.25	0.5641	1	1.5	3.75	0.00111	22.5	1	1.056	1	70	0.5	0	40
12	2.25	0.5641	1	1.5	3.75	0.00111	22.5	0	1.078	1	70	0.5	0	40
13	2.25	0.5641	1	1.5	3.75	0.00111	22.5	0	1.034	1	70	0.5	0	40
14	2.25	0.5641	1	1.5	3.75	0.00111	22.5	0	1.056	2	70	0.5	0	40
15	2.25	0.5641	1	1.5	3.75	0.00111	22.5	0	1.056	0	70	0.5	0	40
16	2.25	0.5641	1	1.5	3.75	0.00111	22.5	0	1.056	1	58	0.5	0	40
17	2.25	0.5641	1	1.5	3.75	0.00111	22.5	0	1.056	1	82	0.5	0	40

(Continued)

Table 7.14 (Continued)

Experiment	3A alcohol (vol)	Purified water (vol)	Cpd 2 (eq)	HCl (eq)	Mass solution (g)	HCl initial (mol/g)	Initial temperature (°C)	Stir time prior to heat (h)	Heat-up ramp rate (°C/min)	De-BOC heating hold time (h)	De-BOC heating temperature (°C)	Cool-down ramp rate (°C/min)	Cooldown hold time (h)	Cooldown temperature (°C)
18	2.25	0.5641	1	1.5	3.75	0.00111	22.5	0	1.056	1	70	0.527	0	40
19	2.25	0.5641	1	1.5	3.75	0.00111	22.5	0	1.056	1	70	0.473	0	40
20	2.25	0.5641	1	1.5	3.75	0.00111	22.5	0	1.056	1	70	0.5	1	40
21	2.25	0.5641	1	1.5	3.75	0.00111	22.5	0	1.056	1	70	0.5	0	28
22	2.25	0.5641	1	1.5	3.75	0.00111	22.5	0	1.056	1	70	0.5	0	52

Table 7.15 PAR simulation results (area %).

Experiment	Cpd 1	Cpd 2 (limit = 96.3)	Cpd 6 (limit = 1.2)	Cpd 7 (limit = 2.2)	Cpd 8 (limit = 0.3)
1	0.00	99.51	0.09	0.29	0.11
2	0.00	99.54	0.09	0.27	0.10
3	0.00	99.52	0.09	0.28	0.11
4	0.00	99.53	0.09	0.28	0.11
5	0.00	99.56	0.08	0.26	0.10
6	0.00	99.52	0.09	0.29	0.11
7	0.00	99.60	0.08	0.24	0.09
8	0.00	99.45	0.10	0.32	0.12
9	0.00	99.54	0.09	0.28	0.10
10	0.00	99.53	0.09	0.28	0.11
11	0.00	99.53	0.09	0.28	0.11
12	0.00	99.54	0.09	0.27	0.10
13	0.00	99.53	0.09	0.28	0.11
14	0.00	99.18	0.16	0.47	0.19
15	0.68	99.19	0.02	0.09	0.03
16	1.32	98.57	0.01	0.08	0.02
17	0.00	97.90	0.75	0.80	0.55
18	0.00	99.53	0.09	0.28	0.11
19	0.00	99.52	0.09	0.28	0.11
20	0.00	99.51	0.09	0.29	0.11
21	0.00	99.52	0.09	0.28	0.11
22	0.00	99.53	0.09	0.27	0.11

In order to understand the portion of the design space that meets the in situ requirements described in Table 7.13, each impurity must be individually considered (Figure 7.36).

The acceptable space where all impurity levels (**2, 6, 7,** and **8**) meet the in situ specification is indicated by the filled region of Figure 7.37. In the area outside the filled space, at least one impurity is above an acceptable level based on the known edge of acceptability. Experiments should be conducted in the area outside the known acceptable limits to determine if the higher levels of impurities would match the model predictions and also meet

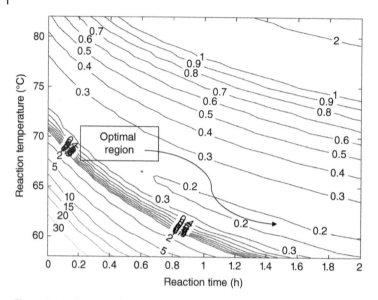

Figure 7.35 Contour plot of total impurity level (TRS) as a function of the reaction time and temperature.

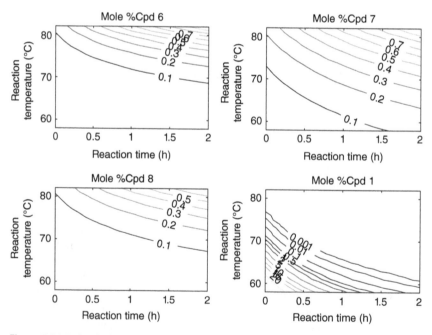

Figure 7.36 Individual impurities levels at the end of the step 2a reaction as a function of reaction temperature and reaction time.

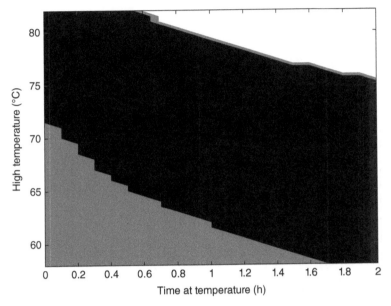

Figure 7.37 Acceptable operational region for the deprotection reaction as a function of reaction time and reaction temperature. Areas outside of the filled region do not indicate an edge of failure, but instead a limit of knowledge. Unreacted starting material is represented by the region in gray, Compound 8 in excess by the region in white at the top right.

product specifications if carried forward in the synthesis and isolation. If the product generated under these conditions meets the quality specifications of the API, then the range of knowledge would be expanded to account for higher in situ impurity levels.

7.5.5 Design Space Simulation: Screening Design Experiment and Multifactor Experiment Simulation and Data Analysis

An alternate method for determining design space is one that involves conducting experiments based on a statistical design (design of experiments (DOE)). The experimental design can involve full factorial or screening experiments (that estimate where main effects or minimal levels of interactions could be present) followed by full factorial or design space experiments based on the results from the screening analysis.[11]

A custom screening design was established in JMP® using all 12 of the process variables at their six-sigma ranges.[12] This design was simulated using the model, and the simulation results (total impurities calculated) were analyzed within JMP using a regression fit model. The variables in Table 7.16 were determined to have a statistically significant impact on impurities' response

Table 7.16 Screening design terms and relative statistical significance.

Term	Estimate	Standard error	t ratio		Probability > |t|
DeBOC heating hold time (h) × DeBOC heating temperature (°C)	11.477052	0.775163	14.81	++++++++++++	<0.0001
DeBOC heating hold time (h)(0, 2)	−8.933728	0.743917	−12.01	− − − − − − − − − − − −	<0.0001
DeBOC heating temperature (°C)(58, 82)	−8.539475	0.751393	−11.36	− − − − − − − − − − −	<0.0001
DeBOC heating hold time (h) × DeBOC heating hold time (h)	9.7841156	2.650773	3.69	++	0.0007
DeBOC heating temperature (°C) × cooldown hold time (h)	2.4384966	0.772911	3.15	+++	0.0031
DeBOC heating hold time (h) × cooldown hold time (h)	2.256227	0.793233	2.84	++	0.0071
Cooldown hold time (h) (0, 1)	−2.092625	0.742193	−2.82	− −	0.0076
DeBOC heating temperature (°C) × DeBOC heating temperature (°C)	7.1835479	2.5614	2.8	++	0.0079
Compound **1** (eq) (0.94, 1.06)	1.8432129	0.750999	2.45	++	0.0188

(*p*-value < 0.05). The fit of the regression model is shown in Figure 7.38. Based on this, the interactions italicized in Table 7.16 were shown to impact the level of simulated impurity. These three interactions were then modeled in three multivariate simulations (Figure 7.39).

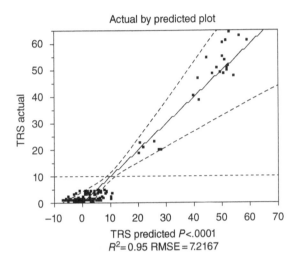

Figure 7.38 Fit model summary for screening design.

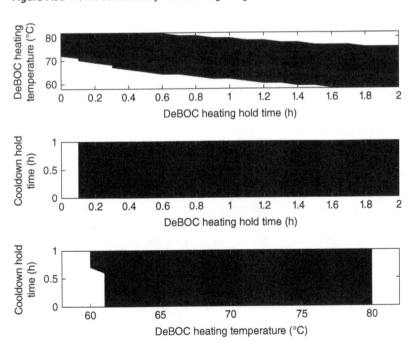

Figure 7.39 Screening design multivariate experiment contour plots.

Figure 7.39 shows the combinations of reaction process parameters (in black) where the process is expected to produce acceptable product (based on the limit of knowledge). These plots show that the design space around the reaction temperature and the reaction duration is the most limited among acceptable processing parameters. Hold time at the cooldown temperature interacting with the reaction temperature is the next most limited. The reaction hold time and cooldown hold time design space has a small region in the ± six-sigma process range that could produce undesirable product.

7.5.6 Confirmation of the Design Space with Experiment

Because any model is limited in the phenomena that it can describe, it is important to note that it is prudent to confirm corner regions of the modeled design space with additional experiments at laboratory scale. Termination of further development of this project prevented these activities from occurring. The reaction models presented and the associated kinetic parameters determined are scale independent. However, the mass transfer parameters associated with the dissolution of compound 1 and the absorption/desorption of CO_2 across the gas–liquid interface are scale dependent. Confirmation of these parameters at process scale or with changes to equipment sets should also be considered.

7.6 Conclusions

The development of the deprotection and peptide coupling reactions described here presents a complete picture of the elements of QbD outlined by the FDA. The project team was able to leverage the multiphase online monitoring power of PAT in combination with traditional offline techniques to construct mechanistic process models capable of fully simulating the chemical and physical processes encompassed by the final chemical bond-forming step of the synthetic sequence. The experimental approach and interpretation can be complex, both in experimental setup and in data analysis, but the results can be exceptionally useful and versatile. Predictive models developed in this fashion allow for accelerated process development by enabling rapid in silico experimentation. This leads to less trial and error in the lab and allows for greatly reduced uncertainty with scale-up. Not only can this speed up the rate of process development but also lessens the material demand for experimentation. The QbD paradigm established by the FDA calls for processes with built-in quality. The processes resulting from development efforts employing predictive modeling necessarily involve an in-depth understanding of the complete process. Thus, the design process builds in quality by bringing the science of process development to bear directly on the control strategy. The establishment of impurity specifications and operational conditions that ensure product meeting these specifications is key element in defining the control strategy. This approach ensures quality product when operating in the defined design space.

Acknowledgments

A number of individuals have been associated with the development of the semagacestat process and it is impossible to list them all. The early synthetic work on the project was performed by Drs. David Mitchell and Radhe Vaid with engineering support from Jeff Vicenzi, Amanda McDaniel, and Robert Towsley. More recently David Varie provided chemistry oversight to the project. The guidance and support of these individuals are appreciated. Analytical support was provided by Dr. Michael Watson, Curtis Miller, Carolyn Stobba-Wiley, and Mary Kay McCauley. Consultations with Dr. Bernard McGarvey in the development and solutions of the kinetic models and with Chad Wolfe for statistical analysis are acknowledged. The laboratory assistance of Amanda Parrish is greatly recognized. Finally the authors would like to thank Drs. Kevin Seibert and Shanthi Sethuraman for their insightful and thought-provoking comments while reviewing the manuscript.

Notation

Acronyms

API	active pharmaceutical ingredient
FDA	Food and Drug Administration
FTIR	Fourier transform infrared
HPLC	high pressure liquid chromatography
MS	mass spectrometry
PAR	proven acceptable ranges
PAT	process analytical technology
QbD	quality by design
TRS	total related substances

Symbols

ϕ	particle aspect ratio, m/m
ρ_s	crystal density of compound 1, g/cm^3

ρ_{soln}	density of the reaction matrix, g/cm^3
a	area of mass transfer at the gas–liquid interface, m^2
a_{solid}	total surface area of solid compound **1**, m^2
$C_{CO_2}^l$	concentration of CO_2 in the liquid phase in the reactor, mol/l
C_1^{bulk}	concentration of compound **1** in the bulk, mol/l
$C_{CO_2}^g$	concentration of CO_2 in the gas phase in the reactor, mol/l
$C_{CO_2}^i$	concentration of CO_2 at the inlet gas stream to the reactor, mol/l
$C_{CO_2}^{in}$	concentration of CO_2 at the gas–liquid interface, mol/l
$C_1^{interface}$	concentration of CO_2 at the solid–liquid interface, g/l
C_1^{sat}	equilibrium solubility of compound **1**, g/l
Ea$_i$	activation energy of reaction i, J/mol K
F^{in}	total volumetric gas flow rate entering the reactor, l/min
F_{HOVal}	mass flow rate of hydroxyvaline into the reactor, g/min
F^{out}	total volumetric gas flow rate exiting the reactor, l/min
H	Henry's law constant, l kPa/mol
j_{CO_2}	mass transfer rate of CO_2 from the liquid to the gas phase, mol/min
k_l	mass transfer coefficient at the gas–liquid interface, min^{-1} m^{-2}
k_i	reaction rate constant of reaction i, g/mol min
k_i^o	Arrhenius reaction rate constant pre-factor of reaction i, g/mol min
k_{sL}	mass transfer coefficient at the solid–liquid interface, min^{-1} m^{-2}
m_{solid}	mass of compound **1** as a solid, g
m_{solid}^o	initial mass of compound **1** as a solid, g
M_{soln}	total mass of liquid, g
MW$_i$	molecular weight of compound i, g/mol
n_p	number of particles of solid compound i, unitless
$N_{CO_2}^l$	moles of CO_2 in the liquid phase in the reactor, mol
$N_{CO_2}^g$	moles of CO_2 in the headspace gas in the reactor, mol
N_{total}^g	total moles of gas in the reactor, mol
N_i	moles of compound i in the reactor, mol
N_i^o	initial moles of compound i in the reactor, mol
p_{CO_2}	partial pressure of CO_2, kPa
P	pressure in the reactor, kPa
R	ideal gas constant, 8.314 l kPa/mol K
r_{CO_2}	rate of CO_2 generation, mol/min
t	time, min
T	temperature, K
V_l	volume of the liquid phase in the reactor, l
V_g	volume of the gas phase (headspace) in the reactor, l
x	particle size, m
x_o	initial particle size, m
x_{HOVal}	mass fraction of HOVal in feed solution to the reactor, g/g

Notes

1 Semagacestat was developed by Eli Lilly and Company for the treatment of Alzheimer's disease but has since been halted in its investigation due to undesirable interim clinical results.

2 A generalized form of the CO_2 balance is derived that includes CO_2 entering the reactor from an external source. This form of the expression is used in the determination of the mass transfer coefficients from experimental results.

3 The area term was lumped into the mass transfer coefficient as there was a sufficient amount of swirling in the reactor and as such using the cross-sectional area of the reactor was not appropriate.

4 The same mixing conditions (defined as the same reactor and agitator geometry, rotation rate and fill volume) used to determine the mass transfer coefficient were used in all subsequent experiments.

5 Mixing conditions are defined as the same reactor and agitator geometry, rotation rate, and fill volume.

6 The temperature dependence of the reactions is assumed to follow an Arrhenius model. Confirmation experiments were conducted to test this assumption and found it to be appropriate over the range of temperatures examined.

7 These reactions were run by first heating up the reaction solution (prior to HCl addition) to their respective reaction temperatures. HCl was then charged in a bolus addition.

8 The SSE was calculated by first subtracting each actual data point from its predicted levels by the model. Each difference was then squared. Finally, all of the squared values were summed to create one value.

9 The units for area [15] are in terms of area per volume of solid-free liquid. The concentration terms are also in terms of volume of liquid.

10 Equations 7.25, 7.26, 7.28, and 7.29 were used in this determination.

11 It is also possible to consider the simulation a full factorial DOE using all 12 process variables. This, however, leads to 3^{12} (531 441) experiments. This approach was not pursued since JMP® version 7.0.2 can only process up to 10 000 experiments.

12 A full factorial design was not used in JMP® due to the limitations of the JMP® software to create a design of this size.

References

1 ICH Harmonized Tripartite Guideline: Impurities in New Drug Substances, Q3a(R2), Current Step 4 Version. Brussels: International Conference on Harmonization of Technical Requirements for Registration of Pharmaceuticals for Human Use, October 25, 2006.

2 Pharmaceutical Quality for the 21st Century: A Risk-Based Approach. Rockville, MD: U.S. Food and Drug Administration, May 2007.

3 am Ende DJ, Preigh MJ. Process optimization with in situ technologies. Current Opinion in Drug Discovery & Development. 2000; 3: 699–706.

4 am Ende DJ, Clifford PJ, DeAntonis DM, SantaMaria C, Brenek SJ. Preparation of Grignard reagents: FTIR and calorimetric investigation for safe scale-up. Organic Process Research & Development. 1999; 3: 319–329.

5 Argentine MD, Braden TM, Czarnik J, Conder EW, Dunlap SE, Fennell JW, LaPack MA, Rothhaar RR, Scherer RB, Schmid CR, Vicenzi JT, Wei JG, Werner JA. The role of new technologies in defining a manufacturing process for PPAR agonist LY518674. Organic Process Research & Development. 2009; 13: 131–143.

6 Littler B. PAT tools in early chemical development. Chemistry Today. 2008; 26: 20–22.

7 McConnell JR, Barton KP, LaPack MA, DesJardin MA. Streamlining process R&D using multidimensional analytical technology. Organic Process Research & Development. 2002; 6: 700–705.

8 Sistare F, St. Pierre Berry L, Mojica C. Process analytical technology: An investment in process knowledge. Organic Process Research & Development. 2005; 9: 332–336.

9 Yu H, Richey RN, Stout JR, LaPack MA, Gu R, Khau VV, Frank SA, Ott JP, Miller RD, Carr MA, Zhang TY. Development of a practical synthesis of DPP IV inhibitor LY2497282. Organic Process Research & Development. 2008; 12: 218–225.

10 Han S-Y, Kim YA. Recent development of peptide coupling reagents in organic synthesis. Tetrahedron. 2004; 60: 2447–2467.

11 Chan LC, Cox BG. Kinetics of amide formation through carbodiimide/ N-hydroxybenzotriazole (HOBt) couplings. Journal of Organic Chemistry. 2007; 72: 8863–8869.

12 Dugger RW, Ragan JA, Brown Ripin DH. Survey of GMP bulk reactions run in a research facility between 1985 and 2002. Organic Process Research & Development. 2005; 9: 253–258.

13 Dias EL, Hettenbach KW, am Ende DJ. Minimizing isobutylene emissions from large scale tert-butoxycarbonyl deprotections. Organic Process Research & Development. 2005; 9: 39–44.

14 Doherty S, LaPack M, Garrett A. Mass spectrometry: Another tool from the PAT toolbox. European Pharmaceutical Review. 2005; 4: 63–71.

15 Paul EL, Atiemo-Obeng VA, Kresta SM. Handbook of Industrial Mixing—Science and Practice. Hoboken, NJ: John Wiley & Sons, Inc., 2004.

16 Burt EE, Rau AH. The determination of the level of bicarbonate, or carbon dioxide in aqueous solutions. Drug Development and Industrial Pharmacy. 1994; 20: 2955–2964.

17 Poughon L, Duchez D, Cornett JF, Dussap CG. $k_L a$ determination: Comparative study for a gas mass balance method. Bioprocess and Biosystems Engineering. 2003; 25: 341–348.

18 Linek V, Sinkule J, Benes P. Critical assessment of gassing-in methods for measuring $k_L a$ in fermentors. Biotechnology and Bioengineering. 1991; 38: 323–330.

19 Van't Reit K. Review of measuring methods and results in non-viscous gas-liquid mass transfer in stirred vessels. Industrial & Engineering Chemistry Process Design and Development. 1979; 18: 357–364.

20 Chandrasekharan K, Calderbank PH. Further observations on the scale-up of aerated mixing vessels. Chemical Engineering Science. 1981; 36: 819–823.

21 Martin M, Montes FJ, Galan MA. Physical explanation of the empirical coefficients of gas-liquid mass transfer equations. Chemical Engineering Science. 2009; 64: 410–425.

22 Fujasová M, Linek V, Moucha T. Mass transfer correlations for multiple-impeller gas–liquid contactors. Analysis of the effect of axial dispersion in gas and liquid phases on "local" $k_L a$ values measured by the dynamic pressure method in individual stages of the vessel. Chemical Engineering Science. 2007; 62: 1650–1669.

23 Kapic A, Heindel TJ. Correlating gas-liquid mass transfer in a stirred-tank reactor. Chemical Engineering Research and Design. 2006; 84: 239–245.

24 Garcia-Ochoa FF, Gomez E. Theoretical prediction of gas–liquid mass transfer coefficient, specific area and hold-up in sparged stirred tanks. Chemical Engineering Science. 2004; 59: 2489–2501

25 Machado RM. Fundamentals of mass transfer and kinetics for the hydrogenation of nitrobenzene to aniline. Mettler Toledo. 2007; 01-2007.

26 Mitchell JD, Abhinava K, Griffiths KL, McGarvey B, Seibert KD, Sethuraman S. Unit operations characterization using historical manufacturing performance. Industrial & Engineering Chemistry Research. 2008; 47: 6612–6621.

27 Seibert KD, Sethuraman S, Wolfe C. The Use of Routine Process Capability in the Determination of Proven Acceptable Ranges (Pars) Critical Process Parameters (Cpps) and Mapping of the Design Space for an API Process. Salt Lake City, UT: AIChE Annual Meeting, November 5–9, 2007.

28 Mitchell JD, McGarvey B, Abhinava K, Seibert K, Sethuraman S, Griffiths K. Synchronized Approach to Developing Optimal Control and Design Spaces. Salt Lake City, UT: AICHE Annual Meeting, November 5–9, 2007.

29 Seibert KD, Sethuraman S, Mitchell JD, Griffiths KL, McGarvey B. The use of routine process capability for the determination of process parameter criticality in small-molecule API synthesis. Journal of Pharmaceutical Innovation. 2008; 7: 105–112.

8

A Strategy for Tablet Active Film Coating Formulation Development Using a Content Uniformity Model and Quality by Design Principles

Wei Chen[1], Jennifer Wang[1], Divyakant Desai[1], Shih-Ying Chang[1], San Kiang[1], and Olav Lyngberg[2]

[1] Drug Product Science and Technology, Bristol-Myers Squibb Company, New Brunswick, NJ, USA
[2] Chemical Development, Research and Development, Bristol-Myers Squibb Company, New Brunswick, NJ, USA

8.1 Introduction

Since its development in the 1960s, nonfunctional or cosmetic tablet aqueous film coating is routinely used to provide a unique product image, to differentiate tablet strengths, to mask the taste of an active molecule, or to protect it against photolytic degradation. For many oncology drug products, it is also used to protect caregivers against potential toxicity of the drug. Normally, cosmetic coating itself is not an indispensable aspect of a tablet formulation. Nonfunctional or cosmetic coating is applied to core tablets. It neither contains any active pharmaceutical ingredients (APIs) nor controls the release of the drug. In a tablet active film coating process, the API is included in the coating layer instead of in the core tablet or in addition to the core tablet. An active film coating technology has become one of the formulation approaches to address certain dosage form requirements. These requirements include (i) stabilization of a drug molecule and (ii) development of a fixed-dose combination (FDC) product with the ability to control the drug release.

For many drug products, sugar beads are coated with active coating layer and filled into capsules. Depending on the desired combination and required drug release profiles, the coating can also include release-controlling polymers. The beads are assayed for the drug content and their fill weight is adjusted to obtain capsules with desired strength. The approach involving the active coating of sugar beads is well known in the pharmaceutical industry. Many commercial products have been developed using this approach. On the other hand, active coating of tablets is rare and more challenging. Generally, it is easier to develop

Comprehensive Quality by Design for Pharmaceutical Product Development and Manufacture,
First Edition. Edited by Gintaras V. Reklaitis, Christine Seymour, and Salvador García-Munoz.
© 2017 American Institute of Chemical Engineers, Inc. Published 2017 by John Wiley & Sons, Inc.

an active coating process for water-soluble API compared with water-insoluble one. For water-soluble compound, it can be dissolved in the aqueous coating solution or suspension, followed by spraying on the core tablets. In the case of the water-insoluble API, the particle size needs to be fine enough so that the spray guns are not clogged. Moreover, API particles should stay suspended during the coating process to keep the suspension homogeneous to get satisfactory content uniformity. Thus, the challenges involving active coating of tablets are many as described in the following text.

Two examples are provided to show how a tablet active coating approach can be used to stabilize a drug molecule.

Peliglitazar, a PPAR α/γ agonist, was used in the treatment of dyslipidemia and diabetes. The compound was very potent requiring a daily dose as low as 0.5 mg. The molecule can undergo an acid- as well as a base-catalyzed degradation. It was reported that tablets manufactured using an active film coating approach where the API was incorporated into the coating material and spray coated on inert tablet cores showed superior stability compared with tablets manufactured using other approaches such as dry or wet granulation formulations [1]. The stability enhancement observed in the active coating approach is attributed to the higher drug to excipient ratio in the film coated material compared with that in the dry or wet granulated formulations [1].

A tablet formulation was also successfully developed using a three-layer film coating process for another compound. The compound that is sensitive to common pharmaceutical unit operations cannot be used to prepare the tablet. In the active film coating approach, as shown in Figure 8.1, the API was sandwiched between two inert coating layers. With all three layers spray coated onto inert core tablets, the innermost layer kept API apart from the excipients of the inert core tablet. The outer layer protected the API from the environment. This approach helped in stabilizing the API molecule by avoiding common pharmaceutical unit operations, such as granulation, milling, and tablet compression, in which mechanical stress is much involved [2]. The final drug product achieved a 36-month shelf-life.

Active coating technology is often used to prepare a FDC tablet formulations. For example, in Claritin-D™, loratadine and pseudoephedrine sulfate are

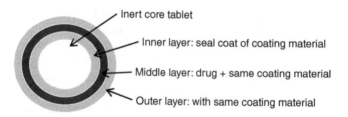

Figure 8.1 Schematic of an active film coated tablet.

coated on the extended-release pseudoephedrine core tablets [3]. Upon oral administration, the coating layer dissolves immediately to release loratadine and pseudoephedrine, providing an initial dose followed by the extended release of pseudoephedrine from the tablet core matrix [3]. Similar approach was also adopted for Advicor™ where lovastatin was coated on extended-release niacin core tablets [4]. One important advantage of the active coating technology is that the different release profiles of two drugs can be maintained. There are FDC tablet formulations where the main objective is to keep two actives apart to minimize their chemical interactions. In such cases, either one active could be included in the core tablets followed by the active coating of the other active. Alternatively, both actives can be coated on inert cores by applying two separate coating layers separated by the coating layer containing no active, if needed.

For traditional tablet formulations developed using direct compression, wet granulation, or dry granulation technology, API is weighed and mixed with other excipients as a part of the manufacturing process. In the active coating approach, API is sprayed on the core tablet. There are two main technical challenges in this approach. First, it is how to determine the coating end point so that the tablets will reach the target potency when the coating process is stopped. Second, it is how to make sure that all the coated tablets will have satisfactory tablet-to-tablet API content uniformity given the inherent variability in the coating operation. The content uniformity issue is more pronounced for the active coated tablets compared with active coated beads because the former has relatively smaller surface area.

The coating end point is determined either based on the amount of suspension sprayed or the weight gain of the core tablets for traditional cosmetic film coating. This approach has been further modified for the active coating. During the active film coating, tablet samples are taken periodically and analyzed not only for weight gain but also for the amount of API deposited on the core tablets by performing an in-process assay. As shown in Figure 8.2, when the coating conditions, especially the spray rate, are kept constant during the entire coating process, a linear relationship was observed between the actual and theoretical API amount deposited on the core tablets [2]. Normally, when 70–90% of the coating liquid is sprayed, the amount of API deposition is analytically determined. Based on that information, additional amount of suspension needed to get the target potency is determined.

Achieving tablet content uniformity is another challenge for an active film coating process. For the traditional tablet formulations, the amount of API needed for the batch is weighed and mixed with the formulation excipients followed by dry or wet granulation. The granulation step locks in the API with the rest of the excipients. Since the API is intimately mixed with the excipients, the proper control on the weight variation of the tablet ensures satisfactory API content uniformity between tablets. In the active coating, the API is

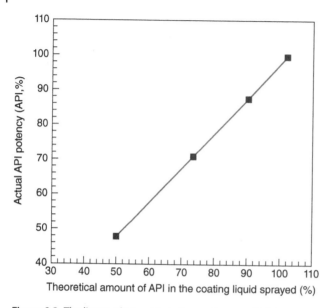

Figure 8.2 The linear relationship between the actual API potency of coated tablets and the theoretical amount of API in the coating liquid sprayed.

sprayed along with the coating materials on core tablets. The factors governing the content uniformity of active coated tablets are very much different from those for the traditional tablets. Control on the spraying operation is vital for the content uniformity. Essentially, whenever tablets come into the spray zone, the API and coating materials get sprayed and deposited on them. Nevertheless, the underlying mechanism of the coating process is complicated, and the predictability of the coating performance is limited due to the large number of process variables in the operation. In active film coating, the API content uniformity of the tablets becomes very important in order to meet the required regulatory standards. It is therefore necessary to be able to understand the factors and coating mechanisms that impact content uniformity.

In a pan coating process as shown in Figure 8.3, the tablets are put inside a perforated pan coater. When the pan rotates, the tablets cascade through the spray zone and are coated by an atomized solution or suspension from one or multiple spray nozzles at the center of the coater. The tablets then leave the spray zone and go back to the drying zone where the solvent is evaporated by the inlet hot air flowing through the bed. As the pan continues to rotate, the tablets travel between the spray zone and drying zone, and this process repeats itself until the end point is reached. The tablets are analyzed for the API potency to determine if the potency meets the required specification. The content uniformity of the tablets is affected by factors including the movement of

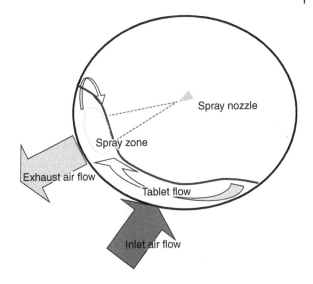

Figure 8.3 Schematic of a perforated pan coating process.

tablets, the projected surface area of the tablets in the spray zone, and the spray quality. The modeling efforts presented here will address how they are related.

8.2 Content Uniformity Model Development

The coating modeling started with Sherony [5]. He first proposed the two-zone model concept for the particle spray coating process in a fluidized bed by visualizing the equipment into the spray and drying zones. The particles are assumed to move between these two zones. It was found that the content uniformity is inversely proportional to the square root of the coating time. Later Cheng and Turton [6, 7] found similar results using the surface renewal theory that the content uniformity is inversely proportional to the coating time and the main contribution to the overall content uniformity was from the mass coated-per-pass distribution (76–86%). Joglekar *et al.* [8] further developed a random theory-based model to predict coat weight variability in the pan coating. They found that the model was explicit in some process parameters and an average number of coating events per tablet can be used as a scale-up factor. Wassgren *et al.* [9–11] recently systematically studied the tablet mixing behavior and content uniformity in the rotating pan from the random theory and DEM model. They defined model parameters, such as segregation value, fraction residence time, and appearance frequency, to characterize the tablet mixing and content uniformity. Pandey *et al.* [12] used a Monte Carlo approach to simulate the weight variability in a pan coating process. Fichana *et al.* [13] used DEM modeling and experimentation to investigate tablet mixing in a pan coater. They studied the tablet residence time distribution (RTD) in the spray zone to estimate the spray deposition and weight variability.

The previously described studies presented useful information for understanding and improving tablet coating operation design, however, models that can be used to quantitatively predict tablet content uniformity are lacking. Denis *et al.* [14] presented the surface renewal model, but they depended on empirical data to fit parameters used in their model. Yamane *et al.* [15] used DEM computer simulations to investigate surface residence time, which was limited to only 600 tablets and very small pan radius to tablet diameter ratios. Kalbag *et al.* [9, 10] and Freireich *et al.* [11] used DEM model to evaluate the tablet mixing and then predict the content uniformity; their model heavily relied on the segregation value (one of model parameters) that was difficult to obtain. It also required intensive computations for a commercial-scale pan coater.

In order to overcome the difficulties in obtaining model parameters and high computational capacity requirements, a mathematical model [16] based on a two-zone concept using RTD theory for calculating the relative standard deviation (%RSD) of coated tablets has been developed. This model will be called the RSD model in the chapter. Three independently measureable parameters related to process parameters—tablet velocity, tablet number density, and spray zone width—were integrated into the model. This model, verified by experimental data from a wide range of coater and tablet sizes, was used as a process design tool to expedite the selection of operating conditions to achieve the desired %RSD from laboratory- to commercial-scale pan coaters. It can also be an important quality by design (QbD) tool in establishing the design space of an active film coating process.

8.2.1 Principles of the Model

The model was developed based on the following assumptions [16]:

1) The spray is uniform and the spray rate is constant throughout the spray zone.
2) Tablets are free flowing and well mixed in the coating pan.
3) The spray zone is much smaller than the drying zone.
4) Tablets only acquire coating material from the spray zone (the coated material is not transferred between tablets).

The model calculates the tablet content uniformity using the RTD of tablets in the spray zone. Variations in residence time of tablets within the spray zone are dependent on the mixing of tablets in the drying zone. If the spray rate is constant, the %RSD can be calculated based on the RTD in the spray zone. The %RSD is defined by the following equation:

$$%RSD = \frac{\sqrt{\sum(m_i - \bar{m})^2 / N}}{\bar{m}} = \frac{\sigma_m}{\bar{m}} \tag{8.1}$$

Equation 8.1 shows the equality between mass-based and time-based RSD calculation, that is, $\%\text{RSD} = \sigma_m/\bar{m} = \sigma_t/\bar{t}$, if the spray rate r is fixed. With this prerequisite condition, we can calculate the standard deviation and mean residence time from the RTD to represent %RSD. Thus, the predicted %RSD for the tablets is obtained via calculation of the RTD of the tablets in the spray zone. Assume that there are n tablets passing through the spray zone and the spray rate is constant. After tablets pass through the spray zone, the tablet weight gain from the spray will be the fractional residence time multiplied by the spray rate $t_i \times r$ where r is the spray rate, t_i is the fractional residence time in the spray zone for each individual tablet. The mean tablet weight gain can be calculated by Equation 8.2:

$$\bar{m} = \frac{1}{n}\sum_{i=1}^{n} m_i = \frac{1}{n}\sum_{i=1}^{n} t_i r = \frac{r}{n}\sum_{i=1}^{n} t_i = r\bar{t} \tag{8.2}$$

The variance of tablet weight gain is

$$\sigma_m = \sqrt{\frac{\sum_{i=1}^{n}\left(m_i - \bar{m}\right)^2}{n}} = \sqrt{\frac{\sum_{i=1}^{n}\left(t_i - \bar{t}\right)^2}{n} \times r^2} = r\sigma_t \tag{8.3}$$

Based on the definition of %RSD, we can see

$$\%\text{RSD} = \frac{\sqrt{\sum\left(m_i - \bar{m}\right)^2/n}}{\bar{m}} = \frac{\sigma_m}{\bar{m}} = \frac{\sigma_t}{\bar{t}} \tag{8.4}$$

We also can see the equality between mass-based and time-based RSD calculation from Equation 8.3. In Equation 8.3, $1/n$ is the probability distribution function of the tablet residence time. In this case, $1/n$ means uniform distribution function for a discrete non-flow system. For a continuous flow system, the residence time of fluid elements or tablets in the spray and drying zone is considered as a stochastic process and can be described by a probability density function or the RTD density function ($E(t)$ function) from Danckwerts [17]. We can use this function to calculate the mean and variance of the tablet coated mass and %RSD. The goal of this modeling approach is to obtain a probability density function $E(t)$ or a RTD density function by using ideal flow patterns.

8.2.2 Total Residence Time and Fractional Residence Time

A schematic representation of the tablet movement in a perforated pan coating process is shown in Figure 8.3. It shows that the pan can be divided into two regions: a "spray zone" and a "drying zone." In the spray zone, the tablets move quickly over the surface of the bulk bed of tablets and are impacted by spray droplets applied in a narrow band near the top of the tablet bed. In the drying zone, the residual moisture on the tablets is removed and the motion of the pan and baffles mixes the bulk of the tablet bed. We define the spray zone plus the drying zone as the total zone. The fractional residence time is the tablet residence time in the spray zone, while the total residence time is the tablet residence time in the total zone. In order to obtain %RSD, we need to know the RTD density function in the spray zone or in the drying zone or in the total zone and the relationship between the mean and variance of the residence time and the RTD density function between the two zones. Here we use the fundamental stochastic process theory to derive these relations.

The total number of tablets N_t divided by the total residence time t in the total zone is equal to the number of tablets in spray zone n divided by the fractional residence time that each tablet spends in the spray zone, $\Delta\tau$. This is a "conservation of tablet flow rate" concept that must always hold true:

$$\frac{n}{\Delta\tau} = \frac{N_t}{t} \tag{8.5}$$

For a certain coating process, n and N_t are fixed and we consider t and $\Delta\tau$ as stochastically random variables. According to the fundamental theorem [18] of stochastic processes, the density function of fractional RTD $\Delta\tau$ is given by the following equation:

$$E(\Delta\tau) = \frac{1}{|d\Delta\tau/dt|} E(t) = \frac{N_t}{n} E\left(\frac{N_t}{n}\Delta\tau\right) \tag{8.6}$$

If we know the density function of total residence time distribution t, the density function of the fractional residence time distribution $\Delta\tau$ can be obtained using Equation 8.6. By using this equation, we can find the mean fractional residence time and variance as follows:

$$\Delta\bar{\tau} = \int_{-\infty}^{\infty} \Delta\tau E(\Delta\tau)d\Delta\tau = \int_{-\infty}^{\infty} \Delta\tau \frac{N_t}{n} E\left(\frac{N_t}{n}\Delta\tau\right)d\Delta\tau = \frac{n}{N_t}\bar{t} \tag{8.7}$$

$$\sigma_{\Delta\tau}^2 = \int_{-\infty}^{\infty} (\Delta\tau - \Delta\bar{\tau})^2 E(\Delta\tau)d\Delta\tau = \left(\frac{n}{N_t}\right)^2 \int_{-\infty}^{\infty} (t-\bar{t})^2 E(t)dt = \left(\frac{n}{N_t}\right)^2 \sigma_t^2 \tag{8.8}$$

By dividing square root of Equation 8.8 by Equation 8.7, we can obtain

$$\frac{\sigma_{\Delta\tau}}{\Delta\bar{\tau}} = \frac{\sigma_t}{\bar{t}} = \%RSD \tag{8.9}$$

From Equation 8.9, we can see that $\sigma_{\Delta\tau}/\Delta\bar{\tau}$ in the spray zone is equal to the $\sigma_t/\bar{t}$ in the total zone. As long as we know the $\sigma_t/\bar{t}$ in the total zone, we can calculate $\%RSD = \sigma_{\Delta\tau}/\Delta\bar{\tau}$ in the spray zone no matter how different the size of the two zones. Now we can convert the calculation of %RSD from $\sigma_{\Delta\tau}/\overline{\Delta\tau}$ in the spray zone into a calculation of %RSD from $\sigma_t/\overline{t}$ in the total zone.

8.2.3 The RSD Model Derivation

The model is conceptually illustrated by Figures 8.4 and 8.5. Well-known modeling techniques from chemical reaction engineering [19] classifies reactors into modules based on idealized reactor models such as the plug flow reactor (PFR) or the continuous stirred-tank reactor (CSTR). Since the number of tablets in spray zone is small compared with the number of tablets in the drying zone, the variance of the RTD of the tablets in the spray zone can be neglected compared with that in the drying zone. The PFR is therefore selected as the idealized reactor most closely resembling the spray zone and the CSTR as the idealized reactor most closely resembling the drying zone.

The RSD model equation was derived by considering a unit step (Heaviside function) increase in the concentration of a tracer "A" into a long series of PFRs, each followed by a CSTR as shown in Figure 8.5. At $t = 0$, the inlet concentration of A undergoes a step increase from 0 to C_{A0}. Each PFR represents one cycle of tablets flowing through the spray zone. Each CSTR represents one cycle of tablets through the drying zone. We obtain the mass balance equation for the first CSTR shown in Equation 8.10 ($u(t - t_p)$ is a unit step function):

$$t_m \frac{dC_{A1}}{dt} + C_{A1} = u(t - t_p) \tag{8.10}$$

Figure 8.4 Analogy between a perforated pan coater and ideal chemical reactors.

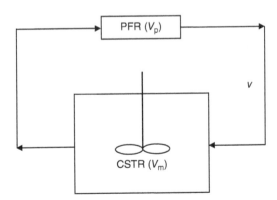

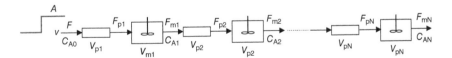

Figure 8.5 Schematic of the RSD model development.

This equation is subject to the initial condition $C_{A1}\big|_{t=t_p} = 0$ and can be solved by using Laplace transforms to give

$$C_{A1} = F_{m1}(t) = 1 - \exp-\frac{(t-t_p)}{t_m} \tag{8.11}$$

where $F_{m1}(t)$ is a cumulative distribution density function, t_p is the mean residence time in PFR, and t_m is the mean residence time in CSTR. In a similar manner, we also can obtain $F(t)$ for the second CSTR and third CSTR:

$$C_{A2} = F_{m2}(t) = 1 - \left(1 + \frac{(t-2t_p)}{t_m}\right)\exp-\frac{(t-2t_p)}{t_m} \tag{8.12}$$

$$C_{A3} = F_{m3}(t) = 1 - \left(1 + \frac{(t-3t_p)}{t_m} + \frac{(t-3t_p)^2}{2t_m^2}\right)\exp-\frac{(t-3t_p)}{t_m} \tag{8.13}$$

and for the Nth CSTR

$$C_{AN} = F(t) = F_{mN}(t) = 1 - \left(\sum_{k=1}^{N} \frac{\left((t-Nt_p)/t_m\right)^{k-1}}{(k-1)!}\right)\exp-\frac{(t-Nt_p)}{t_m} \tag{8.14}$$

$$E(t) = \frac{dF(t)}{dt} = \frac{(t-Nt_p)^{(N-1)}\exp-(t-Nt_p)/t_m}{\Gamma(N)t_m^N} \tag{8.15}$$

where $E(t)$ is the RTD density function that describes the fluid motion characteristics in a mixing tank. We define τ and θ as

$$\tau = Nt_m, \quad \theta = \frac{(t-Nt_p)}{Nt_m} \tag{8.16}$$

Rewriting Equation 8.15 with τ and θ,

$$E(t) = \frac{N^N \theta^{(N-1)}\exp-N\theta}{\Gamma(N)\tau} \tag{8.17}$$

$$E(\theta) = \tau E(t) = \frac{N^N \theta^{(N-1)}\exp-N\theta}{\Gamma(N)} \tag{8.18}$$

$$\sigma^2 = \int_0^\infty \theta^2 E(\theta)d\theta - 1 = \frac{\Gamma(2+N)}{N^2(N-1)!} - 1 \tag{8.19}$$

$$\Gamma(N) = \int_0^\infty t^{N-1}e^{-t}dt \tag{8.20}$$

σ^2 is the variance of dimensionless RTD $E(\theta)$:

$$\sigma^2 = \frac{\sigma_t^2}{\tau^2} \tag{8.21}$$

$$\%\text{RSD} = \frac{\sigma_{\Delta\tau}}{\Delta\bar{\tau}} = \frac{\sigma_t}{\tau} = \sigma = \sqrt{\frac{\Gamma(2+N)}{N^2(N-1)!} - 1} \tag{8.22}$$

N is the total number of tablet cycles between the two zones and is a function of the coating time t, and t_m the mean residence time in the drying zone as shown in Equation 8.22.

Equation 8.23 shows the relation between N and the coating time:

$$N = \frac{t}{t_m} \tag{8.23}$$

The relationship between the mean residence time in the drying zone and the number of tablets and the tablet flow rate is given by Equation 8.24:

$$t_m = \frac{V_m}{f} \tag{8.24}$$

where V_m is the total number of tablets in the drying zone. Since the number of the tablets in the spray zone is relatively small compared with the number of tablets in the drying zone, the total number of tablets in the pan N_t is close to the total number of tablets in the drying zone. f (#/s) is the tablet flow rate that, in turn, is the product of the tablet velocity v (cm/s), tablet number density ρ (#/cm^2), and the spray zone width w (cm), all of which are process conditions that can be measured using camera systems with image analysis. Equation 8.25 shows the relation of the tablet flow rate to the measureable model parameters:

$$f = v \times \rho \times w \tag{8.25}$$

Normally the total number of tablets in the pan is fixed, and a target %RSD can be achieved by adjusting the tablet flow rate f(#/s) and the coating time t (min).

When N is an integral, Equation 8.22 can be simplified by Equation 8.26: %RSD for the Nth CSTR in a simplified format:

$$\%\text{RSD} = \sqrt{\frac{1}{N}} = \sqrt{\frac{1}{t/t_m}} = \sqrt{\frac{V_m}{t \times f}} = \sqrt{\frac{V_m}{t \times v \times \rho \times w}} \tag{8.26}$$

According to the definition of $E(\Delta\tau)$, we can obtain the distribution of the assay of the coated material (API) from Equation 8.27:

$$C(\Delta\tau) = M \times E(\Delta\tau) = \frac{M \times N_t}{n} E(t) \tag{8.27}$$

where M is the total coated mass on the tablet at the coating time t.

8.2.4 Model Parameters and Their Measurements

The %RSD of the API content uniformity of the coated tablets, as indicated in Equation 8.22 (or the simplified version, Equation 8.26), is only a function of the number of tablet cycles, N, and inversely proportional to the square root of the total number of cycles that the tablets going between the spray zone and the drying zone. The larger the number of tablet cycles, the better the content uniformity of the coated tablet. The number of cycles is a function of the total number of tablets in the drying zone, spray zone width, tablet velocity, tablet number density, and total coating time. This can be used to understand the links between process parameters and measurable conditions of the system. These links are shown in Figure 8.6. The tablet flow rate in the first column is characterized by the product of the spray zone width, tablet velocity, and tablet number density as shown in the second column. The tablet flow rate is an indication of the coating performance: the higher the tablet flow rate, the larger the number of tablet cycles for a certain amount of tablets and coating time, and the better the content uniformity (i.e., lower %RSD). These model parameters can then be linked to the actual controllable process parameters as shown in the very right column. For example, the tablet velocity can be linked to pan speed, pan load level, tablet size, density, and surface roughness. In this way, the changes of the process parameters can be consolidated to the change of one model parameter. As long as the three model parameters (spray zone width, tablet velocity, and tablet number density) can be measured, the content uniformity profile then can be determined with the model. Alternatively, the minimum coating time can be calculated for a desired target content uniformity given the number of tablets in the pan the total coating time.

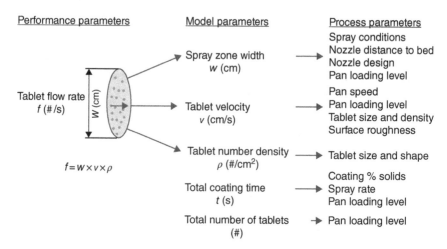

Figure 8.6 Links between model parameters and measurable conditions of the system.

8.2.4.1 Tablet Velocity

The tablet velocity as one of the model input parameters can be measured by the in-line Pro-Watch® imaging system (OSEIR Inc., Tampere, Finland). The system has an eight-bit grayscale CCD camera with a stroboscopic LED ring light source. The camera and light source are integrated into a single stainless steel enclosure to replace a spray nozzle during the measurement. The system continuously acquires dual image pairs in the adjustable time interval of 400–800 µs.

The measured images are processed in real time using a specially tuned spatial cross-correlation-based algorithm to extract the displacement of each tablet in view. The velocity of each tablet is then derived from the displacement data and time interval of two consecutive images. The ability to use a submillisecond pulse interval ensures that the matching tablets in the images are identified correctly regardless of their orientation and rotation properties. System outputs include the mean and standard deviation of radial and axial velocity of tablets and the number of tablets per unit viewing area.

The typical viewing area of the system is 10 cm × 8 cm and the focal distance can be adjusted to fit different sizes of coating pans. Before each experiment, the camera is first calibrated to focus on the upper portion of tablet bed cascading surface of the bed, which is the region that contains the area where the tablets are sprayed during a typical coating operation. Figure 8.7 illustrates an installation of the imaging system in a BFC-400 commercial coater. The velocity of tablets of various sizes and shapes at 50 and 400 kg scales were measured. Figures 8.8 and 8.9 show the tablet image and velocity vector on the surface of the pan coater, respectively. A representative relationship of the tablet velocity versus pan speed is shown in Figure 8.10. Generally, the tablet velocity increases with an increase of the pan speed.

Alexander *et al.* [20] and Levin [21] analyzed the relationships between particle velocity, vessel radius, and rotation speed using Rayleigh's method. They provided an empirical correlation (Equation 8.28) for the rotating cylinders of 6.3, 9.5, 14.5, and 24.5 cm diameter filled to 50% of capacity:

$$V = kR\omega^{2/3} \left(\frac{g}{d} \right)^{1/6} \tag{8.28}$$

where V is the particle velocity, k is a constant, R is the pan radius, ω is the pan angular speed, g is the gravity, and d is the particle diameter. The tablet velocity data in 5, 10, and 40 kg batch were from Muller and Kleinebudde [22]; the data cover a range of pan radii and pan speeds. The tablet velocity at 50 and 400 kg batch was measured by an in-line imaging device (Pro-Watch®, OSEIR, Finland) for A-type tablets. Combining these data and Muller and Kleinebudde's data, we can obtain an empirical correlation given in Equation 8.29. Results in Figure 8.11 show good correlation between the experimental data and fitted

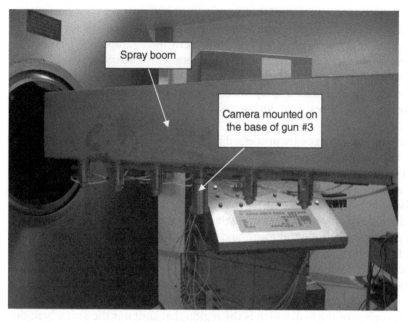

Figure 8.7 Installation of a Pro-Watch® camera in a commercial pan coater.

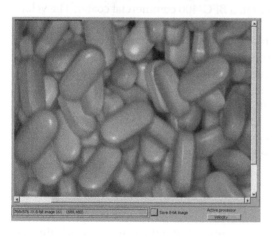

Figure 8.8 Tablet imagine (number density) on the surface of the pan coater.

Equation 8.29. The equation was used in 200 kg batch to extrapolate the tablet velocities with 18.9% (v/v) fill level:

$$V = 2.6075 V_T^{2/3} R^{1/3} d_p^{1/6} \tag{8.29}$$

where V_T is the peripheral velocity of pan coater and d_p is the equivalent diameter of the tablets. The good correlation between the tablet velocity and

Figure 8.9 Tablet velocity vector on the surface of the pan coater.

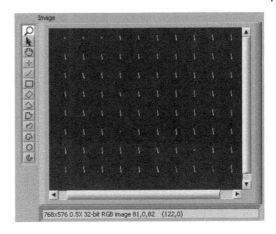

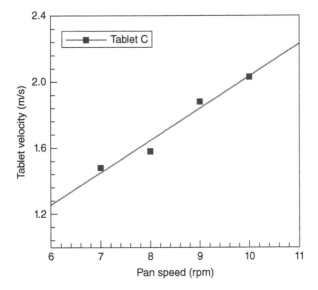

Figure 8.10 Representative tablet velocity versus pan speed in a commercial pan coater.

coating process variables suggested that Equation 8.29 can be used to calculate the tablet velocity in a perforated pan coating operation.

8.2.4.2 Tablet Number Density

In an early study of the pan coating process, Shukla *et al.* [23] proposed a scale-up approach for the pan coating process by using the number of passes of the tablets through the spray zone. A tablet number density was used to calculate

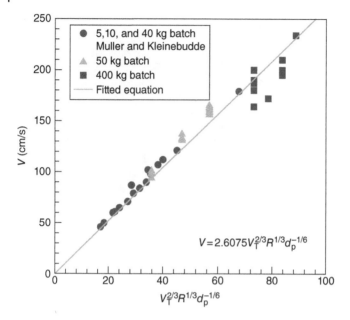

Figure 8.11 Empirical correlation of the tablet velocity, peripheral velocity (pan speed), and pan radius and particle size.

the number of passes of the tablets in this approach. Muller and Kleinebudde [24] assumed that the projection area was the half of the tablet surface and calculated the tablet number density by dividing this area by the projection area of one tablet. Pandey *et al.* [25] described a similar concept that was a project surface area in tablet motion measurements. In this area, the tablet can "see" the spray droplet. The idea of both approaches is to calculate how fast the coating will occur in the spray zone. In our study, the tablet number density is measured by the SprayWatch® camera system and used to calculate the tablet flow rate. The tablet number density can be extracted from the image as shown in Figure 8.8. The measured tablet number density values for the various weight, size, and shape tablets ranged in 0.2–3.0 tablet/cm^2.

8.2.4.3 Spray Zone Width

The spray droplet size and droplet size distribution are key qualities during the coating process, while the spray zone width is associated with the spatial droplet distribution. The SprayWatch® imaging system (OSEIR, Tampere, Finland) was used to measure the droplet volume flux profile to independently determine the spray zone width [26, 27]. The imaging system had been customized to measure the spray characteristics of typical suspensions used in the pharmaceutical industry. The detailed configuration of this system and the measurement principles are described in the following paragraphs.

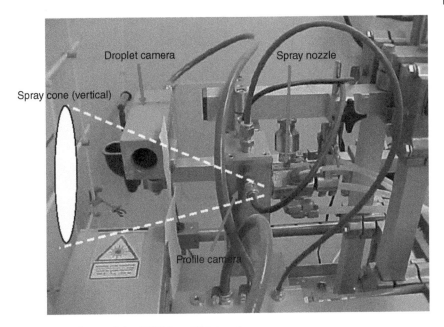

Figure 8.12 Set up of the OSIER SprayWatch system.

The spray nozzle studied was a model 930/7-1S35 from Düsen-Schlick GmBh (Untersiemau, Germany). Bore sizes of 1.2 and 0.8 mm were tested. The aqueous suspension used for spray coating contained Opadry® II white from Colorcon (West Point, PA). The solid content in the liquid suspension ranged from 4.2 to 13.2 wt%.

The measuring system consists of two cameras. A CCD camera with microscope lens and a diode laser light source, shown in Figure 8.12, were used to measure the droplet size and velocity distribution. The measurement area was 6.40 × 4.82 mm^2 around the central line of the spray zone. This camera system was mounted on a metal frame with a stepper motor in order to move it along the spray zone with a measured increment. This arrangement allowed measurement of the spatial distributions of droplet size and velocity within the spray zone. The test nozzle was mounted so that its tip was 17 cm away from the measuring camera. This distance was selected to immolate a typical nozzle-to-tablet bed distance of 15–20 cm. A typical image captured with this setup is shown in Figure 8.13.

Droplet velocities were derived from measurements using the PTV algorithm. The laser was triggered to generate three pulses at a fixed interval to capture triplet images corresponding to a single particle. A real-time image processing tool (LabVIEW, National Instruments, TX, USA) was applied to enhance droplet images and then to identify those droplet tracks formed by

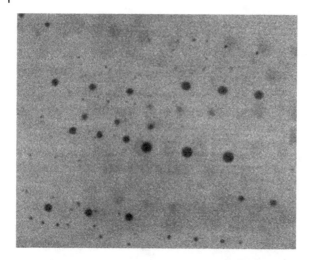

Figure 8.13 A typical image of three-pulse backlight droplets for velocity and size analysis.

the triplet images. The corresponding velocity of each track was calculated from measured displacement and the known pulse interval.

The droplet size was derived from the circumference of the enhanced droplet images. There were usually more than a thousand droplets recorded in each image; distribution of velocities and sizes can therefore be derived. Based on the measured droplet size and velocity distribution, the droplet characteristics including volume flux (unit volume per unit area and time, $nl/s.mm^2$), number density (number/mm^2), volume average size (D90, D50, and D10), and standard deviation were available.

The system was configured to measure a droplet size range of 8–120 µm and velocities up to 60 m/s.

The spray profiles were recorded by a second camera mounted at the proximity of the nozzle tip. This camera was illuminated by a matrix of LED at NIR wavelength and focused through a wide-angle lens (Figure 8.14). The cone angle was derived from the measured angle between the cone borderlines. The borderlines were detected through software by differentiating the intensities between spray and dry air. Figure 8.14 shows a typical image from the profile camera.

Typically, the spray zone width ranges from 8 to 20 cm per gun at a gun-to-bed distance from 15 to 25 cm shown in Figure 8.15. When a two-fluid nozzle is used in the spray system, the ratios of the atomization air flow rate (AA) to the spray rate (SR) and atomization air flow rate to pattern air flow rate (PA) are key parameters that affect spray characteristics. With the proper selection of the AA/SR and AA/PA mass flow ratios, it is possible to generate sprays with consistent spatial distribution of volume flux with minimal variation in the mean droplet size. Then, a certain spray zone width with a uniform volume flux can

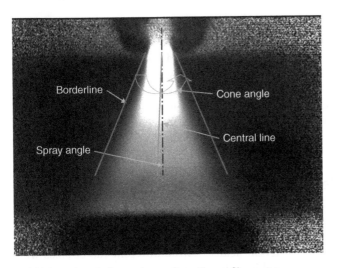

Figure 8.14 A typical spray image from the profile camera.

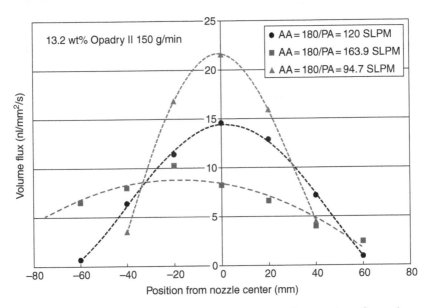

Figure 8.15 Effect of AA/PA ratio on the spatial distribution of spray volume flux and spray zone width (SLPM, standard liter per minute).

be controlled by AA/SR and AA/PA mass flow ratios. Typically after the spray is fully atomized, the droplet size and its distribution may have little effect on the content uniformity, but they can affect the efficiency of the pan coater. Fine droplets may decrease the coating efficiency due to the spray drying with a certain gun-to-bed distance, while coarser droplets may increase the coating efficiency. The selection of the AA/SR and AA/PA mass flow ratios should take into account both the spray zone width-related content uniformity and the efficiency of the pan coater. The coating efficiency is the ratio of amount sprayed accounted on tablet (potency or weight gain) to the actual amount sprayed. The efficiency is considered 100% when the entire amount sprayed gets deposited on the tablets. With proper control of all the parameters mentioned here, it is not difficult to get the coating efficiency of at least 93–95%.

8.3 RSD Model Validation and Sensitivity Analysis for Model Parameters

The RSD model was validated using experimental coating runs for different size tablets and for different scales. Three different tablet sizes were used in these coating runs. Tablet A is a standard concave tablet 5/16" with 200 mg weight. Tablet B is a biconvex-shaped tablet with dimensions 0.748" × 0.364" with 1024 mg weight. Tablet C is also a biconvex-shaped tablet with dimensions 0.913" × 0.429" with 1450 mg weight. When not mentioned, the coating runs were carried out using Tablet A. Three different size perforated pan coaters (BFC-50, BFC-200, and BFC-400, Bohle LLC, Ennigerloh, Germany) were used in the study. The key operating parameters are listed in Table 8.1.

The dimensions of the BFC series coaters can be found in the paper by Chen *et al.* [26] Schlick (Düsen-Schlick GmbH, Germany) two-fluid spray guns with 012 nozzles (inside diameter 1.2 mm) were used for all experiments. In all runs, the coating aqueous suspension contained 3–15% API and coating material Opadry®, which was sourced from Colorcon Inc. Tablet samples were collected throughout the active coating process at different time points corresponding to 20, 40, 60, 80, and 100% mg of target active weight gain. The 30 tablets were assayed for API content to determine the content uniformity or %RSD at each collection time point in each run. The content uniformity data from 10 experimental runs in a BFC-50 pan coater with a 50 kg batch size, 6 experimental

Table 8.1 Key operational parameters for various size pan coaters.

Pan coater	Pan load (kg)	Pan speed (rpm)	Spray rate (ml/min/gun)	Inlet air flow rate (CFM)
BFC-50	40–65	9–18	65–110	500–800
BFC-200	200	10–15	70–120	1200–1500
BFC-400	300–500	8–10	60–100	2000–2900

runs in a BFC-200 with a 200 kg batch size, and 15 experimental runs in a BFC-400 with a 400 kg batch size were used to verify the model prediction and check the sensitivity of the model parameters.

8.3.1 Model Validation

The model predictions were compared with experimental data from the process by plotting the model %RSD and experimental %RSD values on the same graph as a function of the number of tablet cycles through the spray zone as shown in Figure 8.16. Figure 8.16 shows the %RSD as a function of the number of tablet cycles as predicted by the model and measured experimentally, demonstrating that there is a good agreement between the model predicted results and experimental results from about 30 batches including various core tablet weights (200–1450 mg), sizes and shapes (concave, or biconvex-shaped), strengths in the active coat (2.5–12.5 mg), and pan load (50–400 kg) with different scales, but the same type, of perforated pan coaters. One of the key insights gained in the course of this work was these variations can all be captured by one variable—the number of tablet cycles—as shown in Equation 8.26.

In addition to the %RSD, the potency distribution of the 30 assayed tablets can also be calculated by Equation 8.27. A comparison between the model predictions and experimental results for the potency distribution in a 200 kg batch

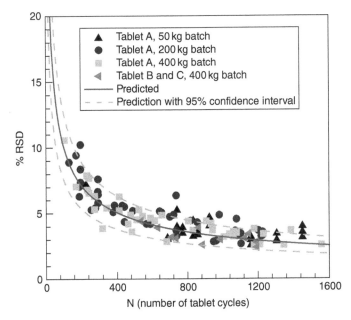

Figure 8.16 Comparison between model predicted and experimental content uniformity.

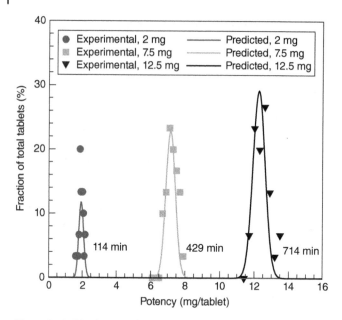

Figure 8.17 Distribution of potency predicted by the model and compared with the measured values in the 200 kg batch.

is shown in Figure 8.17. The model predictions and experimental results for the potency distribution from three selected sample points at beginning, middle, and end of a coating run agree well with each other. Differences between the model prediction and experimental data do exist, especially at the early stages of the coating run. These differences may be due to the experimental sampling at this time as large variations in content uniformity exist at the beginning of the coating run. These differences could also be explained by the precision of the experimental measurements or the analysis of the data as it was made into a histogram, potentially without sufficient bins for such a small mass.

The agreement between the experimental and predicted results shown in Figures 8.16 and 8.17 confirms the observation that there is a common mechanism for the coating process such that one model is able to describe this coating operation at all scales and for any pan coater operated under similar mechanisms. Overall, the percent relative error between the experimental and predicted %RSDs from 165 data points (including results during coating and at the end coating) is 13%. These comparisons suggest that the model can be used as a predictive tool to describe the operating parameters of an active film coating process in a pan coater.

Measuring content uniformity in %RSD, the model mechanistically links the tablet active film coating variables and the final product critical quality attributes together. The model can be used to evaluate how changes in the coating process and conditions will affect the content uniformity (%RSD). By changing the input parameters including the spray zone width, tablet velocity, and the tablet number density, we can calculate the %RSD under different process scenarios. While the effects of the process variables on the content uniformity of the film coated tablets will be discussed in more detail in the next section, a general discussion of the sensitivity of the model parameters is summarized as follows.

8.3.2 Effect of Spray Zone Width on Content Uniformity

From Equation 8.26, it can be seen that by increasing the spray zone width while keeping the other model parameters fixed, the %RSD can be reduced. This simulation result was confirmed by experimental data as shown in Figure 8.18. Figure 8.18 shows the effect of the spray zone width on the %RSD: as the spray zone width increased from 12 to 18 cm/gun by increasing the pattern air flow rate, the %RSD decreased from 4.3 to 3.2% at a coating time of 800 min in a 400 kg batch. Keeping the spray zone width at a maximum is the best strategy in pan coater operations to improve the tablet content uniformity. It was also noticed that if the coating time extends into the plateau zone in the %RSD versus N curve, this spray zone width effect may be less significant. For example, when the spray zone width changed from 12 to 18 cm, the %RSD

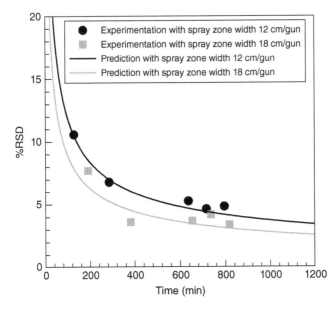

Figure 8.18 Effects of spray zone width on tablet content uniformity.

values changed from 4.3 to 3.2% and 2.7 to 2.1% for the coating time of 800 and 2000 minutes, respectively. In fact, when the coating time increases, the spray rate has to be reduced and it will become increasingly difficult to maintain the maximum spray zone width. Other factors such as the gun-to-bed distance and pan load may also affect the spray zone width.

8.3.3 Effect of Tablet Velocity on Content Uniformity

The effect of the tablet velocity on the content uniformity is shown in Figure 8.19. A tablet velocity change from −10 to 10% at the center velocity of 119 cm/s would cause the %RSD to change from 4.1 to 3.7% at the same coating time of 300 min in a 50 kg batch. This is an approximately 10% change of %RSD caused by a 20% change of tablet velocity. The experimental data (Figure 8.19) from a 50 kg batch operating at the pan speeds of 15 and 18 rpm confirmed this simulation result. Although the pan speed increased from 15 to 18 rpm, the %RSD was only reduced from 4.1 to 3.8%. Similar results are obtained for the larger pan coaters. It should be emphasized that it is important to keep the tablets in the well-mixed region for pan coater operations. Within the well-mixed region [28], tablet velocity has less influence on the %RSD. This is especially true in lager pan coaters. At lower pan speeds, the tablet-to-tablet contacting time may increase, and if the spray rate is relatively high, this may lead to a "wet" condition in the coater. This condition is likely to cause API to transfer

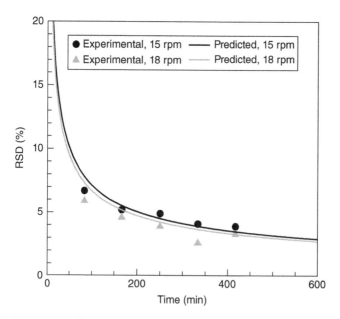

Figure 8.19 Effects of tablet velocity on tablet content uniformity.

between the tablets, leading to poor content uniformity. At higher pan speeds, the tablet mechanical strength becomes more important due to the possibility of broken tablets and excessive tablet attrition.

8.3.4 Effect of Tablet Size on Content Uniformity

Tablet number density is affected by the tablet weight, size, and shape. Smaller tablets in more symmetrical shape tend to result in a higher number density. For a 200 mg round tablet, a number density change of 20%, from −10 to 10% of its center point, will cause a 10% of %RSD change from 3.4 to 3.1% at a coating time of 800 min. It is difficult to design experiments in which the %RSD sensitivity to the tablet size can be studied independently. Tablet size change leads to a different total number of tablets in the coater bed. Normally, in order to keep the tablets in the well-mixed regime [28], the tablet fill volume fraction should be kept fairly constant. For example, the tablet fill volume fraction is usually kept at 18.9% [24] in BFC series coaters. The ratio of the total number of tablets to tablet number density can be further defined as the equivalent tablet surface area (ETSA). It is the total top surface area occupied by the tablets of the entire pan load when they are randomly spread into one layer on a flat surface. The ETSA of smaller tablets is higher than larger tablets. A reduction of ETSA is beneficial for a lower %RSD given the same coating time. For example, when the tablet weight changed from 200 to 1450 mg, the total number of tablets changed from 2 000 000 to 241 379 with the same fill volume fraction in the coater. With the tablet number density changes from 1.9 to 0.57 tablet/cm^2, or the ETSA change from 1 052 631 to 424 963 cm^2, the RSD value will change from 3.2 to 2.4% at a coating time of 800 min along with other fixed process parameters. The experimental data at a pan speed of 10 rpm shown in Figure 8.20 confirmed these simulation results.

8.3.5 Effect of Pan Load on Content Uniformity

Similar to tablet size, pan load can affect %RSD in three different ways through changes in three model parameters: the total number of tablets, tablet velocity, and the spray zone width. The tablet velocity variation due to changes in the pan load is limited. The total number of tablets is easy to calculate based on the weight of the tablet. The spray zone width variation can be calculated using Figure 8.21, assuming that the spray zone width is a baseline under the spray gun. For example, when the pan load increases from 400 to 500 kg, the gun-to-bed distance is reduced by about 6 cm. If the gun is in a fixed location and the spray cone has the same shape, based on the similarity of the triangle, the spray zone width will be reduced by 3.5 cm. Correspondingly, the %RSD changes from 4.9 to 6.1% at a coating time of 700 min. The experimental data also verified the model prediction as shown in Figure 8.22.

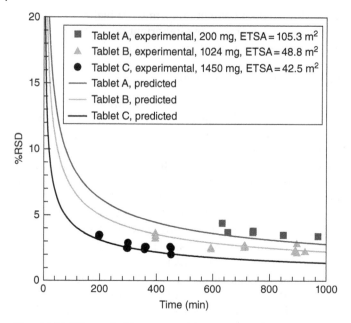

Figure 8.20 Effects of tablet size on tablet content uniformity.

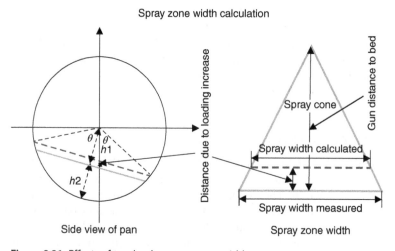

Figure 8.21 Effects of pan load on spray zone width.

8.3.6 Effect of Coating Time on Content Uniformity

As shown in Equation 8.26, the %RSD is inversely proportional to the square root of the coating time. Therefore, lower %RSD can be achieved by increasing the coating time. The longer coating time provides ample opportunity to the

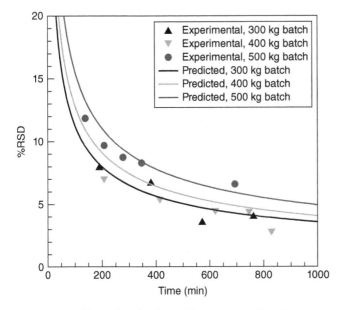

Figure 8.22 Effects of pan load on tablet content uniformity.

tablets to come in front of a spray zone. The coating time is also related to the spray rate, which is determined by the amount of API to be coated on the tablet, and the solid concentration of the API suspension. The coating time can be adjusted by changing the spray rate and solid concentration. One key consideration is the need to maintain the maximum spray zone width with a uniform spray. Longer coating times are always the first option for achieving lower %RSD using a smaller spray rate. However, it should be emphasized that in order to maintain the maximum spray zone width with a smaller spray rate, the nozzle setup has to be adjusted. Usually finer droplets are generated by a smaller spray rate and may have negative impact on efficiency in the pan coater.

8.4 Model-Based Design Space Establishment for Tablet Active Film Coating

A tablet active coating process development, scale-up, process design space exploration, and technology transfer (from one type of perforated coater to another) can all be guided by utilizing the RSD model predictions. The number of experimental runs may be greatly reduced, leading to a shorter and more economical way of developing, optimizing, and troubleshooting an active film coating process using a perforated pan coater.

The current definition of design space is "The multidimensional combination and interaction of input variables and process parameters that have been demonstrated to provide assurance of quality" [29–32]. It has been widely recognized [33–37] that QbD is a systematic approach that is carried out during pharmaceutical development to establish the design space, ensuring the quality of the drug product. QbD begins with the identification of the desired dosage form and its target product profile. From this target product profile, a list of product critical quality attributes can be developed. From process development point of view, QbD requires an understanding of how the process variables, such as input material attributes and process parameters, influence the quality of the final drug product. When product critical quality attributes are defined to meet target product profile, it is important to design a manufacturing process to produce a final product that has these physical, chemical, biological, or microbiological quality attributes. For a tablet that requires an active film coating process to incorporate the API into the dosage form, the tablet-to-tablet content uniformity is usually one of the product's critical quality attributes that needs to be addressed during process development.

8.4.1 Establish a Model-Based Process Design Space at a Defined Scale

The process development and scale-up of active coating process often starts with the knowledge base created by routinely used cosmetic film coating process, which has been studied in the pharmaceutical industry extensively [13, 38, 39]. It is known that the characteristics of a film coated tablet are linearly correlated with the efficiency with which water is removed from the coating process [40–42]. An environmental equivalency factor (EEF) [40] approach is particularly useful for film coating process development. EEF is derived from a first-principles model built upon the coupling of heat and mass transfer in evaporative mass transfer. EEF is an indicator of the relative rate of water evaporation from the tablet bed surface. It is a single quantity that is indicative of the drying rate of a film coating process by determining the thermodynamic and mass transfer relationships. EEF is defined as the ratio of the surface areas over which heat is convectively transferred from the drying air to the tablet bed surface to how moisture is transferred from the liquid film at the tablet surface into the pass airstream. In a perforated pan coater, the thermodynamic conditions of an aqueous film coating process may be characterized by an EEF that incorporates many process variables such as the inlet air temperature, air humidity and air flow rate, the solid concentration of the coating liquid, the coating liquid spray rate, the atomization air flow rate, and the temperature of the exhaust air and its humidity. According to the Thermodynamic Analysis of Aqueous Film Coating (TAAC) program [43], the range of EEF values fall between 1.0 and 5.2, with 3.3 being a typical production value. An EEF value of 1.0 indicates that the tablet bed is completely saturated with the coating liquid,

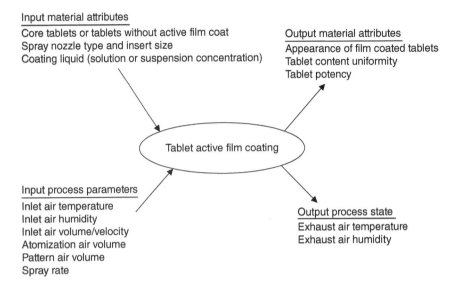

Input material attributes
Core tablets or tablets without active film coat
Spray nozzle type and insert size
Coating liquid (solution or suspension concentration)

Output material attributes
Appearance of film coated tablets
Tablet content uniformity
Tablet potency

Tablet active film coating

Input process parameters
Inlet air temperature
Inlet air humidity
Inlet air volume/velocity
Atomization air volume
Pattern air volume
Spray rate

Output process state
Exhaust air temperature
Exhaust air humidity

Figure 8.23 Process map for tablet active film coating.

while higher values indicate drier tablet bed. The EEF is very useful in developing and optimizing an aqueous film coating process. It is an effective scaling factor when scaling up a coating operation from lab to production scale or transferring the operation between coating pans of different sizes or from different suppliers.

A process map illustrating the relationship between the input attributes, process parameters, output attributes, and the quality of the final drug product in a tablet active film coating process is presented in Figure 8.23. Once a general tablet film coating process is set up with one type of film coating equipment, such as a perforated pan coater, the next step is to establish a design space for an active film coating process at a desired scale. The establishment of the design space is a way to prove that an understanding of the process has been gained. A design of experiment (DOE) approach is usually the most effective starting point for developing a design space for a unit operation or process to adequately address the relationship between the input variables and the final product quality attributes; now with the help of RSD model and understating the mechanism of tablet film coating, the process variables controlling the content uniformity can be narrowed down to two main model: parameters of the spray zone width and tablet velocity as shown in Figure 8.24. We can see that the first three process variables in Figure 8.24 can be lumped into the one model parameter, which is the spray zone width. This measure of the quality of the spray can be independently characterized by a camera system. The next four process variables in Figure 8.24

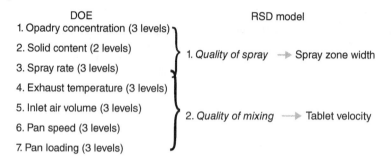

Figure 8.24 Correlation between DOE factors and RSD model parameters.

are related to the tablet movement or tablet velocity. These items can be lumped into one model parameter, the tablet velocity, which is a measure of the quality of the tablet mixing. This tablet velocity also can also be independently measured by a camera system. In other words, we can determine the quality of spray and quality of mixing before conducting any pan coating experiments. As long as the model parameters are linked with the process conditions determined by the camera systems, the content uniformity can be predicted. The model facilitates specification of an operating range with greater accuracy. The following is an example to demonstrate a design space by RSD model (Figure 8.25).

As illustrated by the RSD model in Figure 8.25, when other parameters are held constant, the total length of coating time becomes critical in determining the content uniformity of the tablet. Lower %RSD can be achieved by increasing the length of coating time, which is determined by API concentration in the coating liquid and its spray rate. While one key consideration of obtaining lower %RSD is to maintain the maximum spray zone width with a uniform spray, longer coating time is always another top option to achieve the same goal. The coating time can be adjusted by changing the spray rate or API concentration in the coating liquid. In addition, it is noteworthy that the %RSD values on the surface labeled with "800 min" are lower with fewer changes (in the range of 1–4%) no matter how tablet velocity or ETSA varies. In contrast, a shorter coating time (e.g., the surface labeled with "200 min") leads to higher %RSD values with a wider range (3–8%) if the tablet velocity or ETSA varies. These model analyses suggest that a longer coating time is beneficial in achieving a most robust active film coating process by providing lower %RSD values with a narrower range. Any changes of other model parameters, such as tablet velocity and ETSA, are more forgivable in such a process when the coating time is longer.

Some input material attributes and process parameters are often predetermined for a particular product such as the tablet size and shape, as well as the spray nozzle design. The total spray zone width is usually maximized to cover the

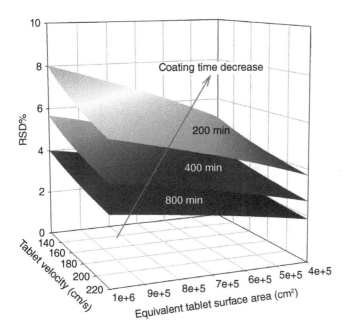

Figure 8.25 Design space for pan coating by model.

entire axial length of the tablet bed for a given pan load. Tablet velocity is also often maximized to the extent that the tablets are not damaged during pan rotation. Therefore, only the remaining process parameters from the aforementioned list such as pan load, spray rate, pan speed, and solid concentration in the coating liquid may be studied in the DOE as design space factors in order to achieve the desired content uniformity of the final product. Therefore the prior use of the RSD model to analyze the input material attributes and process parameters enables the factors that need to be studied in the DOE to be significantly reduced.

The RSD model can be used to estimate the failure limits of a theoretically conceived design space from tablet content uniformity perspective. As shown in Figure 8.26, pan load, pan speed, and spray rate, each with two levels (high or low), are the process variables chosen to be studied in a proposed design space. A total of 10 to 11 experimental runs in a full factorial DOE, including eight "corner" batches and two to three "center point" batches, would be needed in a randomized sequence of production (runs 1–11) in order to explore this design space. All of the anticipated input material attributes and process parameters that may affect the tablet content uniformity are plugged into the %RSD model (Equation 8.26) to calculate predicted tablet content uniformity %RSD values. The predicted %RSD values may be desirable (e.g., met the requirements of the uniformity of the

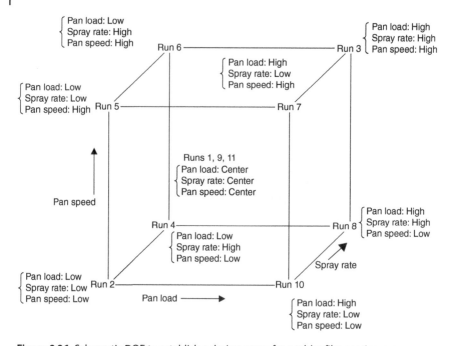

Figure 8.26 Schematic DOE to establish a design space for a tablet film coating process with the aid of the RSD model.

dosage units in the USP) or not desirable. If the results are not desirable, the model parameters or associated process variables can be adjusted until the predicted %RSD values for all of the experimental runs are acceptable in the proposed design space. This model prediction exercise is typically followed by experimental runs. The actual tablet content uniformity results are then compared with the predetermined acceptance criteria. If the results are acceptable, the process variable combinations can be considered the "points" on the "surface" of the design space. If the results are not acceptable, the "points" should be considered part of the knowledge space, which is usually larger than the design space. Other process input attributes and process parameters can be considered in a similar manner to establish the design space for the active film coating process if necessary. During the course of these evaluations, the actual tablet content uniformity results are compared with the RSD model prediction to assess the impact of the variables on the desired result or outcome. If the experimental results are within the ±95% of the predicted value, with 95% confidence interval, it is generally considered that the results are comparable.

Another way of using the RSD model to establish the design space for the active film coating process is to reduce the number of experimental runs in a full factorial DOE. It is also possible to include more than one product in a single DOE. Depending on the manufacturing scale and the cost of the material, production run time and headcount of establishing a design space for an active film coating process can be very high. With the aid of the RSD model, an effective approach for creating a design space could be the combination of film coating experimental runs and model prediction. To verify the accuracy of the model's prediction of the tablet content uniformity of the final product, some of the "corner" and "center" runs identified in the aforementioned proposed design space will need to be carried out and the products tested. If more than one product is involved in the DOE, it should be anticipated that the levels of each process variable for the different products could be different from each other and steps should be taken to enable the different impacts of the variables on the products to be isolated and measured. Products different in tablet size, shape, or dose strengths may be randomly assigned to the various locations of the proposed design space. Upon experimental verification of the RSD model for the manufactured runs, predictions for the tablet content uniformity of the other batches that have not been run in the proposed design space may be assumed based on the predictions from the model. As a result, a design space that is suitable to cover several products may be established all at once by correlating RSD model predictions with actual experimental results. In practice, this RSD model has been found suitable for tablets in a wide range of weights (from 200 to 1500 mg), shapes and sizes (from concave tablets of 4 mm in diameter to biconvex-shaped tablet with a major diameter of 25 mm), and scales (from 40 to 500 kg).

One of the fundamental principles behind the development of a robust manufacturing process is to understand the process through the establishment of a design space and avoid operations near or at the edge of failure. The establishment of a design space for a tablet active film coating process is one pathway to obtaining the flexibility to carry out the operations at different scales within this space without further extensive development work. It is also a way to show that a thorough understanding of the process has been gained. Theoretically, it is acceptable to make any changes within this established design space. If any changes are to be made, all of the anticipated input material attributes and process parameters that may affect the tablet content uniformity should be plugged into the RSD model equation to calculate the predicted tablet content uniformity value. The changes would be considered acceptable if the final product quality attributes of the batch meet the predetermined acceptance criteria. It is important to note that not every individual point around the edge of the design space has been experimentally verified. Therefore it is prudent to exercise some caution if "points" near or on the "surface" of the design space are selected. It is a good practice to conduct experimental runs using the anticipated process parameters in order to confirm the model prediction.

8.4.2 Model-Based Scale-Up

Like many other unit operations in solid dosage form manufacture, the schemes for active film coating scales are commonly categorized into lab scale (1–20 kg), pilot scale (20–150 kg), and commercial scale (150–600 kg). Generally speaking, it is more efficient and cost effective to build a design space at the lab scale or pilot scale in order to gain sufficient understanding of the process. With the knowledge and experience gained at the smaller scale, the ultimate goal is to use that information to build a design space at the commercial scale in order to gain sufficient confidence in the manufacturing process and to obtain possible regulatory relief to operate within the boundaries of the design space.

Results from ongoing research have highlighted what scale-up parameters are to be considered for a general film coating process in a perforated pan coater [42]. Pandey *et al.* [25] suggested that coating pan geometric, dynamic, and kinematic similarity should be maintained at different scales. For the same type of pan coaters of the same operating principles, equipment manufacturers need to ensure that the aspect ratio of pans across scales is constant or similar. The height, width, and shape of the baffles in the coating pan should also be proportional across different scales. Additionally, it is useful to keep the pan load fill level, that is, the ratio of the tablet volume to total pan volume, similar to each other across different scales. While these precautions are being taken, it is also important to maintain similar levels of drying capacity, spray rate, and coating time per unit weight of pan load as they are beneficial in achieving the scale-up goals.

Achieving satisfactory tablet-to-tablet content uniformity is often the focal point when scaling up a tablet active film coating process. As described in the RSD model (Equation 8.26), the content uniformity %RSD value is inversely proportional to the square root of N, total number of tablet cycles between the dry zone and spray zone. It means that whether the coating is conducted at lab, pilot, or commercial scale, the way to achieve the same level of %RSD value is to keep N the same at each scale. For example, the total coating time, t, can be adjusted by changing the API concentration in the coating liquid, provided that the tablet velocity and spray zone width are already maximized for that tablet size and shape at a known pan load level. Since the active film coating RSD model described in this chapter is a mechanistic model and scale-independent, only limited amount of work at the lab or pilot scale is necessary in order to gain preliminary understanding of the active film coating process. It is possible to avoid extensive developmental work at the lab or pilot scale and still come up with a design space at the commercial scale. In many cases, the overall cost of conducting experimental work in order to establish a design space for the commercial scale may be reduced by 20–80%.

In cases where a design space has already been established at lab or pilot scale, additional work may still be needed to establish a design space for the commercial scale. If the scale-up is to be performed using a coating equipment of the same design and operating principles, the starting point is to generate a set of process parameters by proportionally increasing the process variable values established at the smaller scale based on the pan load change. The second step is to verify, on paper, whether the thermodynamic conditions measured with the EEF are acceptable or consistent with the initial process setups. If necessary, some adjustments of the process parameters may be needed based on prior knowledge (e.g., from the film coating process for other products with the same coating equipment) and/or actual experimental runs at this scale. Next, following the same course as used in establishing a design space at other equipment scales, all of the anticipated process parameters that may affect the tablet content uniformity should be plugged into the RSD model equation to calculate a predicted tablet content uniformity %RSD value, followed by an actual experimental run at the commercial scale to validate the %RSD values. As part of the scale-up effort, it is critical to ensure that the thermodynamic conditions are suitable for a smooth film coating operation prior to addressing the content uniformity of the coating process by utilizing the aforementioned methods. An overly wet process conditions with low EEF may cause condensation of API coating liquid onto the metal parts in the coater (e.g., spray boom). The condensate may drip onto the tablet bed in the coater upon accumulation, contributing to an ununiform distribution of the API coating liquid on tablet surface. Next, following the same course as in establishing a design space at other equipment scales, all of the anticipated process parameters that may affect the tablet content uniformity should be plugged into the RSD model equation to calculate a predicted tablet content uniformity %RSD value, followed by an experimental run at commercial scale. The ideal scenario is for the experimental run to go smoothly from a general film coating standpoint and for the actual tablet content uniformity result to be satisfactory and comparable with the RSD model prediction. By this point, it should be possible to propose a design space justified based on the model prediction and verified experimentally. An example of comparing process variables of a tablet active film coating process at 50 and 400 kg scales using a perforated pan coater is shown in Table 8.2.

Whenever possible, the design space should be established at the most desired scale, often commercial production scale. If a suitable design space is established for the commercial-scale process, the need of going outside this space ought to be rare. A decision tree illustrating how to approach a tablet active film coating scale-up in order to develop a design space for a commercial scale is presented in Figure 8.27. This decision tree is based on the assumption that the film coating process runs smoothly at each equipment scale.

Table 8.2 Process variables for a tablet active film coating process at pilot and commercial scales.

Process variable	Perforated pan coater at pilot scale	Perforated pan coater at commercial scale
Pan load (kg)	35–55	300–450
Inlet air flow rate (CFM)	400–600	2500–3000
Inlet air temperature (°C)	50–55	50–55
Exhaust air temperature (°C)	36–44	36–44
API concentration in coating liquid (%)	5–15	5–15
Pan speed (rpm)	12–20	7–11
Spray rate (g/min)	50–125	300–600
Atomization air flow rate (CFM)	250–400	400–650
Pattern air (CFM)	200–450	300–800
Content uniformity (%RSD)	2.5–3.5	2.5–3.5

The equipment for a tablet active film coating process at different scales may be different in design or the coating pan is not proportionally sized sometimes. In this case, two things need to be done: the first thing to do is to ensure that coated tablets with acceptable appearance are obtained under proper thermodynamic conditions of the coating process at the larger scale, and the second is to experimentally verify the suitability of the RSD model with the coating equipment in question at the larger scale. If all results are acceptable, the next steps should follow the same format as described in previous sections in this chapter.

8.4.3 Model-Based Process Troubleshooting

Once coating process is developed and the model validated with the experimental results, the results should match the model prediction. If a deviation occurs, the tablet coating behavior will deviate from the model predictions. The deviations can be used to assist in troubleshooting the process. During this investigation, deviations to the quality of mixing and quality of spray can be used to evaluate each potential process parameter that can contribute to these deviations and can provide some insight into the system that failed and assist in assigning the root cause of the deviation. Based on the root cause, remedial measures can be explored in the next batch.

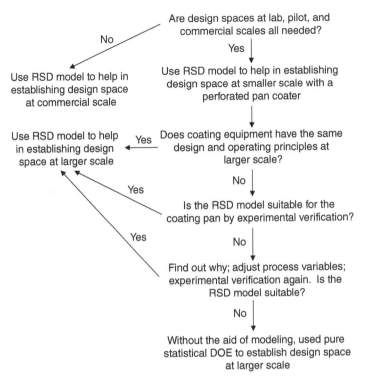

Figure 8.27 A decision tree illustrating how to scale up a tablet active film coating process using a perforated pan coater and with the aid of the RSD model.

8.5 Summary

As outlined in the introduction of this chapter, an active film coating technology is useful to address some unique formulation requirements and to develop formulations using the QbD paradigm. For traditional tablet formulations, many unit operations follow in sequence including mixing of API and excipients, dry or wet granulation, milling, mixing with a lubricant, tableting, and cosmetic coating. By incorporating the API in the coating, especially for drugs requiring low strengths, those who are compression sensitive, or those requiring handling in the containment area, handling of the drug can be reduced to only one unit operation. With mechanistic understanding and thorough control of the coating operation, all these issues can be addressed satisfactorily by focusing on one unit operation, namely, active film coating. The mechanistic model presented here will facilitate the necessary understanding and provide a tool to pharmaceutical scientists not only to develop active film coating technology for any size tablet but also to enable them to scale up the technology to commercial-scale perforated coaters.

Notations

C_{A0}	initial tracer concentration, g/ml
C_{A1}	tracer concentration after first pass, g/ml
d_p	volume equivalent diameter, cm
f	tablet flow rate through spray zone, #/s
m_i	weight gain of tablet i, g
$\bar{m}$	mean tablet weight gain, gm
M	total coated mass on the tablet at the coating time t
n_{pan}	pan speed, rpm
N	total number of cycles that tablets pass between two zones; $N = t/t_m$
N_t	total number of tablets in the total zone
R	pan radius, cm
t	coating time, s
t_m	mean residence time in drying zone, s
t_p	mean residence time in spray zone, s
$u(t - t_p)$	unit step function; $u = 0$ when $t \le t_p$ and $u = 1$ when $t > t_p$
V or v	tablet velocity, cm/s
V_m	total number of tablets in drying zone
V_p or n	total number of tablets in spray zone
V_T	peripheral velocity, cm/s
w	spray zone width, cm
ρ	tablet number density, #/cm^2
σ_m	standard deviation of tablet weight gain
$\sigma\Delta_r$	standard deviation of residence time in spray zone
σ_t	standard deviation of residence time in total zone

References

1 Desai, Divyakant; Rao, Venkatramana; Guo, Hang; Li, Danping; Stein, Daniel; Hu, Yue; and Kiesnowski, Chris; An active film-coating approach to enhance chemical stability of a potent drug molecule, Pharmaceutical Development and Technology, 2012, 17, 227–235.

2 Lipper, Robert A.; Desai, Divyakant; and Kiang, San; Case Study: Implementation of Design Space Concepts in Development of an Active-Coated Tablet, Real World Applications of PAT and QbD in Drug Process Development and Approval, Arlington, VA, September 11, 2006.

3 Kwan, Henry K. and Liebowitz, Stephen M.; Stable extended release oral dosage composition comprising loratadine and pseudoephedrine, US patent 5,314,697, May 24, 1994.

4 Bova, David J. and Dunne, Josephine; (Monitoring only) Combinations of HMG-COA reductase inhibitors and nicotinic acid compounds and methods for treating hyperlipidemia once a at night, Publication no. US2009/0226518A1, 2009.

5 Sherony, Dominic F.; A model of surface renewal with application to fluid bed coating of particles, Chemical Engineering Science, 1981, 36, 845–848.

6 Cheng, X. X. and Turton, Richard; The prediction of variability occurring in fluidized bed coating equipment. 1. The measurement of particle circulation rates in a bottom-spray fluidized bed coater, Pharmaceutical Development and Technology, 2000, 5, 311–322.

7 Turton, Richard; Challenges in the modeling and prediction of coating of pharmaceutical dosage forms, Powder Technology, 2008, 181, 186–194.

8 Joglekar, Anand; Joshi, Nitin; Song, Yongxin; and Ergun, James; Mathematical model to predict coat weight variability in a pan coating process, Pharmaceutical Development and Technology, 2007, 12, 297–306.

9 Kalbag, Arjun; Wassgren, Carl; Penumetcha, Sai Sumana; and Pérez-Ramos, José D.; Inter-tablet coating variability: Residence times in a horizontal pan coater, Chemical Engineering Science, 2008, 63, 2881–2894.

10 Kalbag, Arjun and Wassgren, Carl; Inter-tablet coating variability: Tablet residence time variability, Chemical Engineering Science, 2009, 64, 2705–2717.

11 Freireich, Ben; Litster, Jim; and Wassgren, Carl; Using the discrete element method to predict collision-scale behavior: A sensitivity analysis, Chemical Engineering Science, 2009, 64, 3407–3416.

12 Pandey, Preetanshu; Katakdaunde, Manoj; and Turton, Richard; Modeling weight variability in a pan coating process using Monte Carlo simulation, AAPS PharmSciTech, 2006, 7, E1–E10.

13 Fichana, Daniel; Marchut, Alexander J.; Ohlsson, Pernille H.; Chang, Shih-Ying; Lyngbery, Olav; Dougherty, Jeffrey; Kiang, San; Stamato, Howard; Chaudhuri, Bodhisattwa; and Muzzio, Fernando. Experimental and model-based approaches to studying mixing in coating pans, Pharmaceutical Development and Technology, 2009, 14, 173–184.

14 Denis, C.; Hemati, Mehrdji; Chulia, Dominique; Lanne, J.-Y.; Buisson, B.; Daste, Georges; and Elbaz, Frantz; A model of surface renewal with application to the coating of pharmaceutical tablets in rotary drums, Powder Technology, 2003, 130, 174–180.

15 Yamane, Kenji; Sato, Toshiharu; Tanaka, Toshitsugu; and Tsuji, Yutaka Tsuji; Computer simulation of tablet motion in a coating drum, Pharmaceutical Research, 1995, 12, 1264–1268.

16 Chen, Wei; Chang, Shih-Ying; Kiang, San; Marchut, Alexander; Lyngberg, Olav; Wang, Jennifer; Rao, Venkatramana; Desai, Divyakant; Stamato, Howard; and Early, William; Modeling of pan coating processes: Prediction of tablet content uniformity and determination of critical process parameters, Journal of Pharmaceutical Sciences, 2010, 99(7), 3213–3225.

17 Danckwerts, Peter V.; Continuous flow systems: Distribution of residence times, Chemical Engineering Science, 1953, 2, 1–13.

18 Papoulis, Athanasois; Probability, random variables and stochastic processes, 3rd ed., McGraw-Hill, New York, 1991, p. 93.

19 Levenspiel, Octave; Chemical reaction engineering, 2nd ed., John Wiley & Son, Inc., New York, 1976, pp. 253–315.

20 Alexander, Albert; Shinbrot, Troy; and Muzzio, Fernando J.; Scaling surface velocities in rotating cylinders as a function of vessel radius, rotation rate, and particle size, Powder Technology, 2002, 126, 174–190.

21 Levin, Michael; Pharmaceutical process scale-up, Drug and the pharmaceutical sciences, Volume 118, Marcel Dekker Inc., 2002, pp. 117–132.

22 Muller, Ronny and Kleinebudde, Peter; Prediction of tablet velocity in pan coaters for scale-up, Powder Technology, 2007, 173, 51–58.

23 Shukla, Atul J.; Chang, Rong-Kun; and Avis, Kenneth E.; Pharmaceutical unit operations: Coating, Drug manufacturing technology series, Volume 3, Interpharm/CRC, Boca Raton, FL, 1998, pp. 113–171.

24 Muller, Ronny and Kleinebudde, Peter; Scale-down experiments in a new type of pan coater, Die Pharmazeutische Industrie, 2005, 67, 950–957.

25 Pandey, Preetanshu; Song, Yongxin; Kayihan, Ferhan; and Turon, Richard; Simulation of particle movement in a pan coating using discrete element modeling and its comparison with video-imaging experiments, Powder Technology, 2006, 161, 79–88.

26 Chen, Wei; Chang, Shih-Ying; Kiang, San; Early, William; Paruchuri, Srinivasa; and Desai, Divyakant; The measurement of spray quality for pan coating processes, Journal of Pharmaceutical Innovation, 2008, 3, 3–14.

27 Kvasnak, William; Ahmadi, Goodarz; and Schmidt, David J.; An engineering model for the fuel spray formation of deforming droplets, Atomization and Sprays, 2004, 14, 289–339.

28 Mellmann, Jochen; The transverse motion of solids in rotating cylinders-forms of motion and transition behavior, Powder Technology, 2001, 118, 251–270.

29 ICH Q8 (R1), Pharmaceutical Development, http://www.ich.org/workProducts/ICHGuidelines/QualityGuidelines, approved November 13, 2008.

30 ICH Q9, Quality Risk Management, http://www.ich.org/workProducts/ICHGuidelines/QualityGuidelines, approved November 9, 2005.

31 ICH Q10, Pharmaceutical Quality Systems, http://www.ich.org/workProducts/ICHGuidelines/QualityGuidelines, approved June 4, 2008.

32 Prpick, Andrew; am Ende, Mary T.; Katzschner, Thomas; Lubczyk, Veronika; Weyhers, Holger; and Bernhard, Georg; Drug product modeling predictions for scale-up of tablet film coating—a quality by design approach, Computers and Chemical Engineering, 2010, 34, 1092–1097.

33 Potter, Chris; PQLI application of science- and risk-based approaches (ICH Q8, Q9, and Q10) to existing products, Journal of Pharmaceutical Innovation, 2009, 4, 4–23.

34 Nosal, Roger and Schultz, Tom; PQLI definition of criticality, Journal of Pharmaceutical Innovation, 2008, 3, 69–78.

35 Davis, Bruce; Lundsberg, Line; and Cook, Graham; PQLI control strategy model and concepts, Journal of Pharmaceutical Innovation, 2008, 3, 95–104.

36 Am Ende, David J.; Seymour, Christine B.; and Watson, Timothy J. N.; A science and risk based proposal for understanding scale and equipment dependencies of small molecule drug substance manufacturing processes, Journal of Pharmaceutical Innovation, 2010, 5(3), 72–78.

37 Lioberger, Robert A.; Lee, Sau Lawrence; Lee, LaiMing; Raw, Andre; and Yu, Lawrence X; Quality by design: Concepts for ANDAs, The AAPS Journal, 2008, 10(2), 268–276.

38 Huang, Jun; Kaul, Goldi; Cai, Chunsheng; Chatlapalli, Ramarao; Hernandez-Abad, Pedro; Ghosh, Krishnend; and Nagi, Arwinder; Quality by design case study: An integrated multivariate approach to drug product and process development, International Journal of Pharmaceutics, 2009, 382, 23–32.

39 Mueller, Ronney and Kleinebudde, Peter; Influence of scale-up on the abrasion of tablets in a pan coater, European Journal of Pharmaceutics and Biopharmaceutics, 2006, 64, 388–392.

40 Ebey, Glen C.; A thermodynamic model for aqueous film-coating, Pharmaceutical Technology, 1987, 4.

41 Strong, John C.; Psychrometric analysis of the environmental equivalency factor for aqueous tablet coating, AAPS PharmSciTech, 2009, 10(1), 303–309.

42 Pourkavoos, Nazaneen and Peck, Garnet E.; Effect of aqueous film coating conditions on water removal efficiency and physical properties of coated tablet cores containing superdisintegrants, Drug Development and Industrial Pharmacy, 1994, 20(9), 1535–1554.

43 Novit, Edward S.; Understanding the effects of process-air humidity on tablet coating, Tablets & Capsules Coating Desktop Reference, July 11–15, 2008.

9

Quality by Design: Process Trajectory Development for a Dynamic Pharmaceutical Coprecipitation Process Based on an Integrated Real-Time Process Monitoring Strategy

Huiquan Wu* and Mansoor A. Khan**

Division of Product Quality Research (DPQR, HFD-940), OTR, Office of Pharmaceutical Quality, Center for Drug Evaluation and Research, Food and Drug Administration, Silver Spring, MD, USA
* Current affiliation: Process Assessment Branch II, Division of Process Assessment 1, Office of Process and Facilities, Office of Pharmaceutical Quality, Center for Drug Evaluation and Research, FDA, Silver Spring, MD, USA
** Current affiliation: Rangel College of Pharmacy, Texas A&M University Health Science Center, College Station, TX, USA

9.1 Introduction

Increasing prevalence of poorly water-soluble drugs in pharmaceutical development provides notable risks of new products demonstrating low and erratic bioavailability. This dissolution-limited bioavailability may have consequences for safety and efficacy, particularly for drugs delivered by the oral route of administration. Several novel drug delivery technologies have been developed to improve drug solubility, dissolution rates, and bioavailability. Among those are solid dispersion, nanotechnology, supercritical fluid technology, lipid-based technology, and crystal engineering. Although these strategies are available for enhancing the bioavailability of drugs with low aqueous solubility, the success of these approaches is not yet guaranteed and is greatly dependent on the physical and chemical nature of the molecules being developed. On the other hand, crystal engineering [1] offers a number of routes such as cocrystallization [2, 3] and coprecipitation [4–6] to improve solubility and dissolution rate. Coprecipitation of poorly soluble drugs with polymers, an important technique for improving the dissolution and absorption of drugs, has been modified in recent years to prepare extended-release preparations. Previous work on coprecipitation was largely focused on formulation development and product characterization [4–6], for example, optimization of process variables for the preparation of ibuprofen coprecipitates with Eudragit S100, screening of process and formulation variables for

Comprehensive Quality by Design for Pharmaceutical Product Development and Manufacture,
First Edition. Edited by Gintaras V. Reklaitis, Christine Seymour, and Salvador García-Munoz.
© 2017 American Institute of Chemical Engineers, Inc. Published 2017 by John Wiley & Sons, Inc.

the preparation of extended-release naproxen tablets with Eudragit L100-55, and preparation and characterization of coprecipitate of ibuprofen using different acrylate polymers. It was not until recently that the process analytical technology (PAT) approach [7–11] was explored to gain insights about the coprecipitation process and process monitoring.

PAT [7] offered unprecedented opportunities for the pharmaceutical community to take advantage of the availabilities of modern process analyzers and wealth of experience from other industry sectors. Successful implementation of PAT can enable real-time process data acquisition and extract process information and knowledge from real-time data for better process understanding and better process control. One of the frequently used technologies for pharmaceutical PAT applications is near-infrared (NIR) spectroscopy (NIRS). NIRS is capable of monitoring many pharmaceutical unit operations [9, 10, 12], in addition to its well-established pharmaceutical applications in quality control (QC) and quality assurance (QA) areas [13]. It is a well-known fact that NIR spectrum of pharmaceutical material or dosage form is information rich and it embraces both the physical and chemical information of pharmaceutical material or dosage form [10].

One of the technical challenges for implementing PAT in the pharmaceutical sector is how to handle the dense flow of process data, including spectra acquired by real-time online/in-line/at-line process analyzers and process data acquired by other process measurement techniques. To extract meaningful information from real-time process data for enhanced process understanding and ultimately process control, one must use chemometrics and process modeling techniques. Depending on specific systems studied and real-time process monitoring techniques implemented, various data analysis and modeling techniques can be used. Multivariate statistical modeling techniques such as principal component analysis (PCA) [10, 14, 15], principal component regression (PCR), partial least squares (PLS) [14–16], and artificial neural network (ANN) [17] are often used to develop a predictive model for depleting raw material components and/or increasing product components, multivariate process trajectory, and process map for robust production with optimized process conditions, productivity, and yield.

Naproxen is a potent nonsteroidal anti-inflammatory drug used in the treatment of rheumatoid arthritis, osteoarthritis, and acute gout and as analgesic and antipyretic. However, its use is frequently limited due to significant gastrointestinal side effects. In this study, naproxen was selected as a model drug as it has a short half-life, has gastrointestinal effects, is soluble in alcohol, and is practically insoluble in water. The literature data for the experimental and predicted solubility of naproxen in water are 15.9 and 51.0 mg/l, respectively (http://www.drugbank.ca/drugs/DB00788, accessed on April 16, 2010). Eudragit L100 is the commercial name of an enteric polymer of methacrylic acid–methylmethacrylate that belongs to a class of reversible soluble/insoluble polymers. Eudragit L100 was used as it yields coprecipitates with naproxen without any need of additives. The dynamic coprecipitation process designed in this work was proceeded via gradually introducing water to the ternary system of naproxen–Eudragit L100–alcohol at controlled temperature of 25°C. When water is introduced into the initially transparent ternary solution sequentially, the overall composition point of naproxen–Eudragit L100–alcohol–water system will be moved accordingly within the phase diagram. Due to significant difference of solubility in alcohol and water for both naproxen and Eudragit L100, the naproxen–Eudragit L100 coprecipitates out from the solution when water is introduced. As a result of the coprecipitation phenomena, the initial transparent solution of the four-component system will become cloudy at the onset of nucleation. In this work, an integrated PAT real-time monitoring strategy was developed to follow the dynamics of the coprecipitation process. As discussed previously, the real-time in-line data acquired can help to illustrate the progress of a multiphase process and to elaborate the sequential events that take place during the process [9]. In this work, multivariate process trajectory was constructed based on the real-time process monitoring data sets, which would be helpful for achieving rational particulate process design and ultimately particulate process control.

9.2 Experimental

9.2.1 Materials

Naproxen USP was obtained from Albemarle Corporation (Orangeburg, SC). It is an odorless, white to off-white crystalline substance. Eudragit L100 was obtained from Röhm America Inc. (Somerset, NJ). It is a white powder with a faint characteristic odor. Solvent reagent alcohol (HPLC grade) was purchased from Fisher Scientific. Nonsolvent DI water was obtained from a FDA in-house facility and was kept in refrigerator at 4°C prior to use. All of these chemicals and solvents were used without any further processing or purification prior to the use for this experimental work.

9.2.2 Equipment and Instruments

Process NIR spectra were acquired with a Luminar^{TM} acousto-optic tunable filter (AOTF)-based NIR process spectrometer (Brimrose Corporation of America, Baltimore, MD), equipped with a transflectance probe. This AOTF NIR process spectrometer has the capability of in-line monitoring both the changes in the concentration of the key components and phase changes in the coprecipitation system or subsystems. The acquisition parameters of Brimrose NIRS being used in experiment included the following: the number of spectra average was 50; no background correction was applied; normal scan type; and the gain was 2. One spectrum was recorded and saved automatically every 6 s. A 1.0 cm probe extension was attached to the 679B dark-field diffuse reflectance/ transmission probe for the actual process monitoring.

Online turbidity measurements were made using a Model 2100AN Laboratory Turbidimeter (Hach Company, Loveland, Colorado) with a Masterflex® L/S® model 7518-10 digital fluid pump (Cole Parmer Instrument Company, Chicago, IL). A flexible tubing connection between the coprecipitation vessel and the sample flow cell of the turbidimeter was made, such that the fluid could be pumped from the coprecipitation vessel to the sample flow cell of the turbidimeter for online turbidity measurement and then be returned to the coprecipitation vessel in the reverse direction. Prior to its use for this work, the 2100AN turbidimeter was calibrated using formazin stock solution as recommended by the supplier. Data were reported using nephelometric turbidity units (NTU), which are specifically compliant for a 90° measurement technique.

Pure components of naproxen and Eudragit L100 were scanned by FOSS NIRsystems (Foss NIRsystems, Silver Spring, MD) offline. The schematic of the experimental setup is shown in Figure 9.1. As shown in Figure 9.1, among many important formulation and process variables, only the drug/polymer ratio was changed during the coprecipitation experiment of this study, while

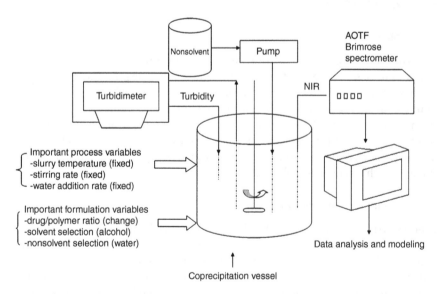

Figure 9.1 Schematic diagram of experimental setup and process flow with an integrated PAT monitoring system for API (naproxen)/polymer coprecipitation process using in-line process NIR and online turbidity measurement simultaneously.

others were fixed. For terminologies about in-line/at-line/online definitions can be found in the FDA PAT guidance [7].

9.3 Data Analysis Methods

9.3.1 PCA and Process Trajectory

PCA is a very useful tool as it can reduce the dimensionality of a data set consisting of a large number of interrelated variables while retaining as much as possible the variation present in the data set. PCA has been extensively applied in almost every discipline, such as chemistry, biology, engineering, meteorology, and pharmaceutical. It can be used in process monitoring, QC, data visualization, batch trajectories, and other areas. For pharmaceutical QbD and PAT applications where process analyzers are installed, frequently a huge amount of interrelated process data has to be handled, in a way that critical process information and knowledge can be extracted for both process and QC purposes. Therefore, PCA is a powerful chemometric technique for such applications. With the help of PCA, the following aspects could be examined: in what aspect one sample is different from another, which variables contribute most to this difference, and whether those variables contribute in the same way (i.e., are correlated) or independently of each other. It can also help to detect patterns and to quantify the amount of useful information, as opposed to noise or meaningless variation, contained in the data set. In this work, plots of a PC versus another (e.g., PC1 vs. PC2), called score plot, are used to depict the process trajectory. Process trajectory method has been used for batch process supervision [18], process monitoring, and diagnosis [19] in chemical and biotech sectors. However, relatively few applications have been reported for pharmaceuticals, especially for small molecule drug manufacturing process monitoring and process control [20]. In this work, all of the NIR spectra were acquired for the wavelength range of [1100, 2300] nm. PCA was conducted on raw NIR spectra without any preprocessing in all cases.

9.3.2 Singular Points of a Signal

It has been noticed that for a time-varying signal in a dynamic system, the information content is not homogeneously distributed throughout [21]. Some landmarks such as extreme values and shape changes in the data, termed as singular points (SPs), in the process trajectory contain more information about the dynamic behavior than others. Mathematically, a singularity is in general a point at which a given mathematical object is not defined or a point of an exceptional set where it fails to be well behaved in some particular way, such as differentiability. Therefore, for a differential equation, an SP is a point that is a singularity for at least one of the known functions appearing in the equation. Geometrically, an SP is a point (ii) on a curve at which the curve possesses no smoothly turning tangent, or crosses or touches itself, or has a cusp or isolated point, or (ii) on a surface whose coordinates, x, y, and z, depend on the parameters u and v, at which the Jacobians $D(x, y)/D(u, v)$, $D(y, z)/D(u, v)$, and $D(z, x)/D(u, v)$ all vanish. The property of singularity and SP could be very useful for many important engineering applications such as rational process design and process control. For example, SPs are used to segment the process signal into regions with homogeneous properties. Because SPs have physical meaning such as beginning or ending of a process event, they can be directly used for state identification, process monitoring, and process supervision. Examples of SPs include points of discontinuities, trend changes, and extrema. SPs were used to detect phase shifts during rifamycin B fermentation experiments [22] and to characterize microscopic flows [23]. In our previous works, SPs were used to determine powder blending process end point [15] and phase change during a coprecipitation process [9]. In this work, SPs are used to detect the onset of nucleation and crystal growth and to establish the process transition window.

In the area of multivariate data analysis and modeling under the pharmaceutical QbD/PAT framework, it is important to recognize the difference between SP and outlier. In multivariate statistics, an outlier is an observation that is numerically distant from the rest of the data. Outliers can occur by chance in any distribution, but they are often indicative either of measurement error or that the population has a heavy-tailed distribution. Outliers may be indicative of data points that belong to a different population than the rest of sample set. The commonly used method for identifying outliers in multivariate analysis is based on the squared Mahalanobis distance (MD) [24]. Points for which MD^2 value is large are identified as atypical or outliers and evaluated using the χ^2 distribution with the appropriate degrees of freedom. SP, as discussed earlier, is in general a point at which a given mathematical object is not defined or a point of an exceptional set where it fails to be well behaved in some particular way, such as differentiability. Therefore, mathematically outlier and SP are two totally different concepts; physically, they have different meanings that are related to different properties.

9.4   Results and Discussion

To investigate the utility of process trajectory for pharmaceutical process monitoring, a step-by-step approach was taken. First, we examined simple cases of binary systems for which either two dry powder components or two liquid components are involved. The NIRS was used for process assessment and process monitoring of the binary systems. Then, we examined a complicated system of four-component coprecipitation process (both solid and liquid phases are involved) for which NIRS was used for real-time in-line process monitoring.

9.4.1 Using Offline NIR Measurement to Characterize the Naproxen–Eudragit L100 Binary Powder Mixing Process

The binary systems of naproxen–Eudragit L100 with various ratios of Eudragit L100 over Naproxen (E/N ratio) were mixed and assessed by FOSS NIRS. The blending end point was determined by the STDEV method described in our previous work [15]. Care was exercised to ensure the sufficient homogeneity for the blends and the sampling representativeness. Only the NIR spectra of the final well-blended binary mixtures at each E/N ratio were used for the PCA. It was found that PC1 and PC2 can explain 98 and 2% of the total variability captured by the NIR spectra, separately. As shown in Figure 9.2, process trajectory depicted by the PCA score plot demonstrated three distinct segments or clusters that can be approximated by three straight lines:

1) The leftmost segment starts with pure component of naproxen and is a cluster of binary blends dominated with naproxen. In addition, the plot of PC2 score versus PC1 score has a positive slope. Furthermore, the higher the Eudragit L100–naproxen ratio, the larger the scores of both PC1 and PC2. However, the situation in the vicinity of the transition is complicated. As expected, the PC2 score value for point E/N = 0.257 is larger than that for point E/N = 0.206. This follows the trend displayed by other three points (E/N = 0, 0.098, 0.113).

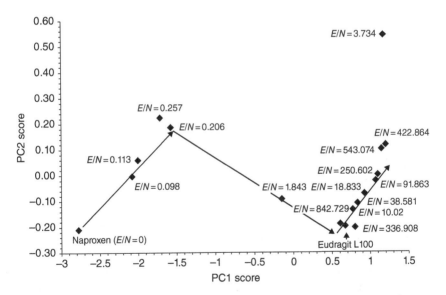

Figure 9.2 PCA-based process trajectory for binary powder blends of naproxen and Eudragit L100. The binary powder blends were made by mixing naproxen and Eudragit L100 with various weight ratios as shown in the figure by E/N values. After mixing well, the blends were transferred to vials that were then scanned by FOSS NIRS offline. The PCA was conducted on NIR spectra of the binary blends with various E/N values.

However, the absolute PC1 score value for point $E/N = 0.257$ is a bit smaller than that for point $E/N = 0.206$. This does not follow the trend exhibited by other three points. Possible factors that may attribute to this situation include (i) abnormal powder behavior in the vicinity of transition point, such as local inhomogeneity due to back-mixing, and (ii) experimental error.

2) The second segment links the leftmost segment and the rightmost segment, which is a cluster of binary blends dominated by either naproxen or Eudragit L100, respectively. In addition, the plot of PC2 score versus PC1 score has a negative slope.

3) The third segment is on the rightmost, starts with the pure component of Eudragit L100, and is a cluster of binary blends dominated by Eudragit L100. The plot of PC2 score versus PC1 score has a positive slope. It was found that apparently the sign change of the slope for the plot of PC2 score versus PC1 score might be linked to the NIR detection limit of each component in the binary mixture. That is, when the actual weight (mole) fraction of one component is decreased to such an extent that it becomes a minor component in the binary mixture, the signal of the minor component becomes so weak that the presence of the minor component cannot be evidenced solely and directly by examining the NIR raw spectra of the blends. As demonstrated in this case, the process trajectory based on PCA score plots has the capability of differentiating the information embedded. For example, it is able to distinguish when the dominant source of contributing component is switching, as demonstrated by the second segment in Figure 9.2. The second segment serves as a process transition window.

9.4.2 Using In-Line NIR Spectroscopy to Monitor the Alcohol–Water Binary Liquid Mixing Process

In-line NIRS was used to monitor the binary mixing process of alcohol and water in real time. Alcohol and water with various volume ratios were mixed and homogenized. The mixing process was monitored using in-line NIRS. The NIR spectra of the well-mixed binary mixture were complied as an NIR data set. PCA was then applied to this data set. It was found that PC1 and PC2 account for 98 and 2%, respectively, of the total variability embedded with the NIR spectra of the binary mixtures. As shown in Figure 9.3, process trajectory depicted by the PCA score plot demonstrated that there is a process transition window where the slope of the plot of PC2 score versus PC1 score changes from positive to negative. The left segment starts with pure component of alcohol and consists of a cluster of alcohol-dominant binary solutions. When water is gradually introduced into alcohol, the scores of both PC1 and PC2 increase. In contrast, the right segment starts with pure component water and consists of a cluster of water-dominant binary solutions. When alcohol is gradually introduced into water, the PC1 score is decreased, while the PC2 score is increased. Furthermore,

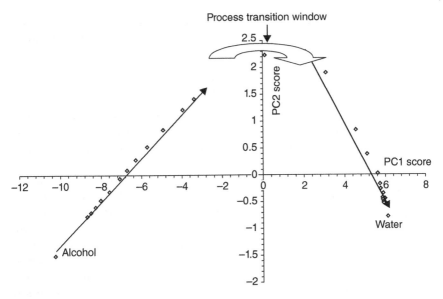

Figure 9.3 PCA-based process trajectory for binary liquid mixtures of alcohol and water with various volume ratios at room temperature. The binary liquid mixtures were made by mixing alcohol and water with various volume ratios. After mixing well, the blends were measured by real-time Brimrose in-line NIR spectroscopy. The PCA was conducted on NIR spectra of the binary blends with various alcohol/water ratios.

it was found that apparently the sign change of the slope for the plot of PC2 score versus PC1 score is linked to the NIR detection limit of each component in the binary mixture. When the actual weight (mole) fraction of one component is decreased to such an extent that it becomes a minor component in the binary mixture, the signal of the minor component becomes so weak that the presence of the minor component cannot be evidenced solely and directly by examining the NIR raw spectra of the blends. However, apparently the process trajectory based on PCA score plots helps to differentiate the information embedded. In other words, the switching of the dominant source of contributing component is demonstrated by the process transition window in Figure 9.3. As in this case, the inhomogeneous nature of the information embedded as well as the discontinuity exemplified by the sign change of the slope in the process trajectory collectively evidenced the existence of an SP.

9.4.3 Real-Time Integrated PAT Monitoring of the Dynamic Coprecipitation Process

When the number of components involved in a multicomponent system is increased, typically the process complexity is increased accordingly. In this subsection, two aspects will be explored: (i) if a similar methodology can be used to develop the process trajectory and (ii) if the PCA-based process trajectory can reveal critical process events such as nucleation and growth.

9.4.4 3D Map of NIR Absorbance–Wavelength–Process Time (or Process Sample) of the Coprecipitation Process

When in-line NIRS is applied to pharmaceutical unit operations, a lot of process-related information may be captured in real time. For the dynamic coprecipitation process evaluation, prior to the introduction of the nonsolvent (water) to the coprecipitation vessel, there was a ternary system of naproxen–Eudragit L100–alcohol already present in the vessel. When water is gradually introduced into the vessel, according to classical thermodynamics, the overall composition point in the four-component system phase diagram is moved from the liquid phase toward the solid (crystal)–liquid equilibrium line; the zone that is bordered by the solid–liquid equilibrium line (the solubility curve) and the metastable limit constitutes the metastable zone for coprecipitation. The graphical illustration of the movement of equilibrium line and metastable zone is shown in Figure 9.4. Once the solid–liquid equilibrium line is crossed, coprecipitate begins to form and grow. The overall composition point will be located within the solidus region. As discussed later, the three-dimensional (3D) process map can illustrate how the coprecipitation process progress and what process events occur during the process.

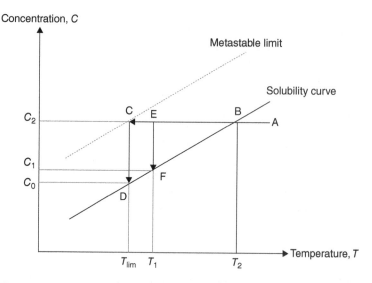

Figure 9.4 Graphical illustration of the movement of equilibrium line and metastable line zone. Concentrations C_0, C_1, and C_2 represent the solubility of solute at temperature T_{lim}, T_1, and T_2, respectively.

To enhance the resolution of the 3D process map for identifying process signatures, the addition rate of the nonsolvent into the system was kept slow, such that the transition of one process state to another state could be observed clearly. The formulation components used, the amount of the nonsolvent added and its adding sequence, and significant visual observations or indications from turbidity measurement are summarized in Table 9.1. As one example, the process 3D (NIR spectral number–wavelength–NIR absorbance) map for batch started with formulation components of 4.7617 g naproxen and 1.205 g Eudragit L100 (batch code (a) in Table 9.1) is shown in Figure 9.5.

9.4.5 Process Signature Identification

The 3D process map is able to provide visual evidence about significant process event for the ongoing process, such as addition of a new component into the existing system, depletion of an existing component, etc. The sensitivity and resolution of this kind of 3D direct and raw process map depend on a number of factors such as (i) the sensitivity and resolution of the instrumentation and the signal/noise ratio of the instrumentation used, (ii) the mass transfer rates in the vicinity of process probe tip and the bulk solution in the vessel, and (iii) the hydrodynamics of the fluid and mixing characteristics in the vicinity of water dispersion area inside the vessel. The physical distance between the process probe tip and the point where the nonsolvent water is introduced and in contact with the bulk solution presents a factor to limit the mass transfer process. In our study, despite the fact that the magnetic stirring of the solution could limit the possible concentration gradient within the vessel eventually, the time lag between the initial moment when the process event take places locally and the moment when the process probe is able to actually detect is always unavoidable due to the existence of physical distance. When a signal associated with a process event is relatively weak, especially when the process event is at its embryo stage, its appearance on the 3D process map will not be that obvious. In this case, other tools may be needed for magnifying or identifying the signature. Furthermore, there are some important questions associated with the 3D process map to be answered. Questions include but are not limited to: (1) What are the implications of the inflection points in the 3D map? (ii) Is there any new phase formation during the process? In this case, other techniques such as chemometrics may play a vital role in terms of identifying process signature associated with process event and provide essential information regarding process progression. In this work, PCA was applied to the process NIR spectra to construct process trajectory and identify SPs such that critical process information and knowledge could be extracted.

Applying PCA to the process NIR spectra of the aforementioned process batch showed that two principal components are sufficient to characterize the variability embedded with NIR spectra. PC1 and PC2 are able to account for

Table 9.1 Formulations used, process stream introduced, and subsequent process events observed via process NIR system (visual observation for the coprecipitation system during the course of adding water into the ternary system of naproxen, Eudragit L100, and alcohol to form a four-component coprecipitation system).

| Batch code | Naproxen (g) | Eudragit L100 (g) | Solvent: alcohol (ml) | Nonsolvent: water (ml) | | Process events as evidenced from integrated online process monitoring strategy | | |
				Adding sequence	Amount added (ml)	Visual observation	Real-time turbidity monitoring (NTU)	Real-time process NIR monitoring
(a)	4.7617	1.205	200	First	51	Clear	Not measured	No new phase formed
				Second	50	Slightly cloudy	Not measured	New phase formed
				Third	50	Cloudy	Not measured	New phase formed
(b)	0.132	6.09708	320	First	100	Clear	1.7–2.6	No new phase formed
				Second	100	Clear	50–70	No new phase formed
				Third	100	Slightly cloudy	140–150	No new phase formed
				Fourth	100	Cloudy	2360–2770	New phase formed
				Fifth	100	Cloudy	4000–5180	New phase formed

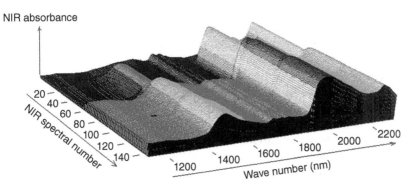

Figure 9.5 Three-dimensional map of the process state over the entire process course of adding water to the ternary system. An inflection point occurs among process spectra window between spectral numbers #60 and #67. New phase is detected between spectral numbers #60 and #67, which maybe an indication of nucleation and growth.

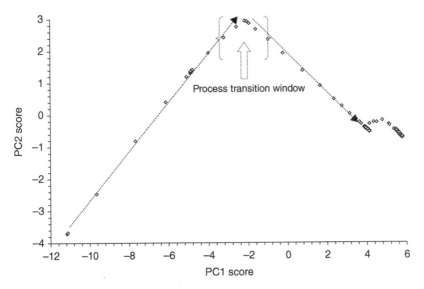

Figure 9.6 PCA score plot for process NIR spectra of the process batch started with formulation components of 4.7617 g naproxen and 1.205 g Eudragit L100.

93 and 7%, respectively, of the total variability embedded with the process NIR spectra for this batch. As shown in Figure 9.6, when PC2 score is plotted against PC1 score, there is a sharp transition on its PCA score plot as reflected by the change in the slope of the straight line of PC2 score versus PC1 sore. The slope

is dramatically changed from positive value to negative value. As a matter of fact, the two segments of the process trajectory consisting of PC2 score versus PC1 score are almost perpendicular to each other. When tracking the sample number or NIR spectral number on this PCA score plot, it was found that (i) the process trajectory that has positive slope value corresponds to the process state of transparent solution during the course of adding water into the ternary system, (ii) the sharp transition of the process trajectory corresponds to the period that solution starts to become cloudy when water is introduced to the ternary system for the third time, and (iii) the process trajectory that has negative slope value corresponds to the process state of cloudy solution during the course of adding water into the ternary system.

The process trajectory depicted by the PCA score plot demonstrated that certain process samples #61–66 can be identified, which cover the process transition window where nucleation and growth take place, as supported by both visual evidences and real-time turbidity profile of the dynamic coprecipitation process. In order to further verify this finding, similar data analysis method was extended to dynamic coprecipitation processes that have different drug/polymer ratios in the starting formulation preparation. In-line NIRS was used to monitor the dynamic coprecipitation process in real time for various batches with different drug/polymer ratios. PCA was applied to each NIR spectra data set associated with a particular drug/polymer ratio. Process trajectory for each coprecipitation process was constructed based on the PCA results. These results together with our previous work [9] collectively demonstrated that process trajectory based on online NIR real-time process monitoring and PCA can differentiate various distinguishable process events and accurately track various process stages such as incubation, nucleation, and crystal growth.

9.4.6 Online Turbidity Monitoring of the Process

For the process runs being conducted in this work, we also used online turbidity measurement to monitor the turbidity change real time during the process. The real-time process slurry turbidity profile of a dynamic coprecipitation process (batch code (b) in Table 9.1) is shown in Figure 9.7. As we can see from Figure 9.7, there are several plateaus observed during the course of introducing water into the ternary system. Before adding any water to the ternary system, the baseline turbidity data of the ternary system were around 0.33–0.45 NTU. The first plateau (1.7–2.6 NTU) and the second plateau (50–70 NTU) correspond to the first and the second addition of 100 ml of water into the ternary system, respectively. The third plateau (140–150 NTU) occurred due to third addition of 100 ml of water into the ternary system. Given that the turbidity value of first plateau of the four-component system is still very close to the baseline turbidity value of the ternary system, most likely it indicates that

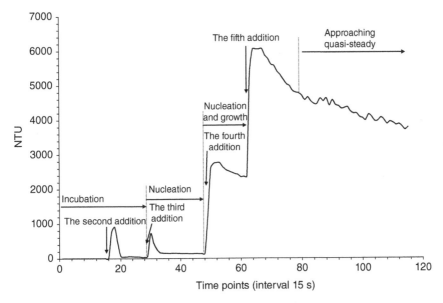

Figure 9.7 Turbidity profile during the process of sequentially adding water to the ternary system of API (0.1320 g), Eudragit L100 (6.0970 g), and alcohol (320 ml) to form the coprecipitate.

the system remains as transparent solution and no new phase is formed (as confirmed by the online process NIR monitoring). The relative low turbidity values for the second turbidity plateau suggest that the process is still at the incubation stage. At this early stage, the composition point of the system is still located within the liquid region. Therefore, no nuclei were formed and the solution was still transparent and thus had a small NTU value. When more water was introduced gradually, the third turbidity plateau (140–150 NTU) occurred, which is probably correlated with the nucleation process stage. At this stage, the system composition point reaches the solid–liquid equilibrium line, and thus nuclei were initiated and a bit higher turbidity values than the second stage were detected. Afterward, when more water was introduced, the system composition point reaches the metastable zone. Consequently, a large supersaturation was created and a crystal growth period was observed. This stage has a turbidity plateau of 2360–2770 NTU. When more and more water was introduced, the system approaches a quasi-steady state where an even higher turbidity plateau of 4000–5180 NTU was observed.

The aforementioned experimental observations could be explained from a multiphase transfer perspective. When water was introduced into the ternary system initially, there were some small nuclei formed initially in the vicinity of the water dispersion area due to local building up of supersaturation; then, the newly formed local nuclei were dissolved quickly due to the de-supersaturation phenomena via mass transfer (diffusion and convection) enhanced by the magnetic stirring of the slurry solution. However, since the existence of fine nuclei was so short and did not have sufficient time to travel to the area that was directly exposed to the process NIR probe tip, this short shelf-life nuclei was not detected by the Brimrose Process NIRS system. The relative high turbidity values for the third turbidity plateau (140–150 NTU) plus the visible slightly cloudiness of the solution suggest that the process is at the nucleation stage. At this stage, the system composition point reaches the solid–liquid equilibrium line, and thus nuclei were initiated and a bit higher turbidity values than the second stage were detected. When more water was introduced gradually, the fourth turbidity plateau (2360–2770 NTU) occurred, which is probably correlated with a stage that both nucleation and crystal growth are involved since the system composition point reaches the metastable zone. Afterward, when more water was introduced, consequentially, a large supersaturation was created and a crystal growth period was observed.

9.5 Challenges and Opportunities for PCA-Based Data Analysis and Modeling in Pharmaceutical PAT and QbD Development

PCA is a popular and powerful technique for feature extraction and dimensionality reduction and probably one of the most employed techniques of multivariate analysis. Maggio *et al.* reported a new PCA-based approach for testing "similarity" of drug dissolution profiles [25]. Comparison between the area enclosed by the confidence ellipses of the weighted scores plot and the region obtained from the bootstrap-calculated acceptable values of the corresponding f_2 tests suggested that PCA confidence region represents, in general, a more discriminated standard. Otsuka *et al.* used PCA for identifying and predicting the most important variables in the process of granulation and tableting [26]. However, there are some limitations with PCA that we should take into consideration, which include the following: (i) the absence of an associated probability density or generative model; (ii) the subspace itself is restricted to a linear mapping as PCA is a linear method; and (iii) PCA does not reveal any reliable information on time scales that are not actually sampled [27] (e.g., a short molecular dynamics trajectory does not yield an accurate covariance matrix of protein fluctuations). The high-order statistical information is discarded during the linear mapping. Basically PCA involves rotating the ellipsoid in such a way that the direction of the variance of the data comes as the first component. This works fine as long as the X/Y relation is fairly linear. For a situation where the X/Y relation is nonlinear, there is a problem. While PCA still tries to produce components by variance, it fails as the largest variance is not along a single vector, but along a nonlinear path, with the assumption that the observed data is independent. In real-world process measurement environment, this may not be a valid assumption due to the facts that (i) the measurement data at consecutive time points could be interrelated and (ii) NIR absorbance values at nearby wavelengths could be correlated too. If a system is highly nonlinear or the observed data are not independent, then the limitations of PCA-based process trajectory become a problem. In this challenging case, other methods may provide better option to handle the nonlinearity encountered, as discussed in the following text.

A number of strategies have been proposed to address the aforementioned limitations of PCA. For example, probabilistic principal component analysis (PPCA) [28], kernel principal component analysis (KPCA) [29], and probabilistic kernel principal component analysis (PKPCA) [30–32] have been developed to deal with the first two limitations of PCA. In addition, a hidden Markov model (HMM) [33] was used to obtain an optimized representation of the observed data through time. On the other hand, neural networks [17, 34] are perfectly capable of dealing with nonlinear problems and can on their own do this. Furthermore, they can do scaling directly so that the principal components can be scaled by their importance.

In the pharmaceutical manufacturing setting, it is possible to encounter nonlinear process features due to many variables (such as formulation variables, process variables, environmental variables, etc.) coexisting and possible interactions. Risk analysis and risk assessment [35] may help to rank the relative importance of those variables and thus provide a list of critical variables for further scientific investigation and design space development [17]. On the other hand, depending on the ranges of various variables selected, the impact of interaction among variables could be significant, marginal, or insignificant. This can be quantitatively assessed via design of experiments (DOE) and ANOVA, as illustrated previously [17]. For those critical variables identified via risk analysis and initial DOE study mentioned earlier, another DOE could be conducted to establish the linkage between critical variables and essential response variables (such as key quality attributes) as demonstrated in most DOE-based approaches. Or, as demonstrated in our recent work, an integrated PAT and DOE approach can be developed to establish dynamic linkages between the real-time process behaviors and the essential response variables at both transition state and steady state [17]. Both approaches are important to ensure product quality and help to achieve QbD.

The PCA-based process trajectory is useful to diagnose process healthiness and identify outliers and abnormal situations. However, the applicability of the PCA-based process trajectory for this kind of process QC and process QA really depends on the scope of the design space [36, 37] (formulation and process variables) investigated, for which the process trajectory was based on. Therefore, it is important to have a clear understanding of where and when PCA-based process trajectory can be applied for process control purpose and of how robust the PCA-based process trajectory is to the process disturbance in real applications, especially for the cases where events and disturbances appear in many different time scales. Issues surrounding PCA-based process trajectory, design space development, PAT process monitoring and control strategies, and QbD methodologies, as discussed briefly in this section, deserve much attention in future research and development during the implementation of PAT [7, 37] and QbD [7, 36–39] in the pharmaceutical sector.

9.6 Conclusions

This work provides a QbD case study that focuses on process trajectory development for a dynamic pharmaceutical coprecipitation process based on an integrated real-time PAT process monitoring strategy. The dynamic coprecipitation process is visualized via three-dimensional map of NIR absorbance–wavelength–process time. The process trajectory based on the results of applying PCA to real-time process NIR spectra data clearly demonstrated that physical meanings can be assigned to various SPs that occurred. Furthermore, those SPs in the process trajectory are directly linked to various distinguishable process events and process signatures such that incubation, nucleation, and crystal growth could be accurately tracked and differentiated. This information and knowledge are essential for developing a suitable design space and operational process space for a pharmaceutical unit operation. The challenges and opportunities of PCA-based process trajectory under the PAT framework and ICH Q8(R2) have been discussed.

Acknowledgments

Dr. San Kiang at the Pharmaceutical Research Institute of the Bristol-Myers Squibb Company (New Brunswick, NJ) and Mr. Steve Ware and Mr. Brendan Simon at the Chemglass Inc. (Vineland, NJ) are acknowledged for their technical supports on our crystallization system. Dr. Vincent Vilker and Dr. Ajaz Hussain are also acknowledged for their encouragements to our pharmaceutical crystallization research program during their tenures in the Center for Drug Evaluation and Research, FDA. The authors wish to thank the invitation from the editors of this book and their diligent efforts subsequently. The inputs from two anonymous reviewers are recognized. The literature support from the FDA Biosciences Library is greatly appreciated.

References

1 Blagden N, Matas M, and Gavan PT, York P. Crystal engineering of active pharmaceutical ingredients to improve solubility and dissolution rates. Advanced Drug Delivery Reviews. 2007;59:617–630.

2 Mcnamara DP, Childs SL, Giordano J, Iarriccio A, Cassidy J, Shet MS, Mannion R, O'Donnell E, and Park A. Use of a glutaric acid cocrystal to improve oral bioavailability of a low solubility API. Pharmaceutical Research. 2006;23:1888–1896.

3 Trask AV, Motherwell WDS, and Jones W. Pharmaceutical cocrystallization: engineering a remedy for caffeine hydration. Crystal Growth & Design. 2005;5:1013–1021.

4 Khan MA, Bolton S, and Kislalioglu MS. Optimization of process variables for the preparation of ibuprofen coprecipitates with Eudragit S100. International Journal of Pharmaceutics. 1994;102:185–192.

5 Zaghloul AA, Faltinek J, Vaithiyalingam SR, Reddy IK, and Khan MA. Naproxen-Eudragit microspheres: screening of process and formulation variables for the preparation of extended release tablets. Pharmazie. 2001;56(4):321–324.

6 Kislalioglu MS, Khan MA, Blount C, Goettsch RW, and Bolton S. Physical characterization and dissolution properties of ibuprofen: Eudragit coprecipitates. Journal of Pharmaceutical Sciences. 1991;80:799–804.

7 FDA. Guidance for Industry. PAT—A Framework for Innovative Pharmaceutical Development, Manufacturing, and Quality Assurance. 2004. Available at: http://www.fda.gov/downloads/Drugs/ GuidanceComplianceRegulatoryInformation/Guidances/UCM070305.pdf (accessed on April 20, 2017).

8 Yu L, Lionberger RA, Raw AS, D'Costa R, Wu H, and Hussain AS. Applications of process analytical technology to crystallization processes. Advanced Drug Delivery Reviews. 2004;56:349–369.

9 Wu H and Khan MA. Quality-by-design (QbD): an integrated process analytical technology (PAT) approach for real-time monitoring and mapping the state of a pharmaceutical co-precipitation process. Journal of Pharmaceutical Sciences. 2010;99:1516–1534.

10 Wu H, Lyon RC, Hussain AS, Drennen JD III, DasGupta D, Voytilla RJ, and Bejic L. Application of principal component analysis in assessing pharmaceutical formulation design: exploring the casual links between the tablet processing conditions and drug dissolution rate. In: Proceedings of Symposium on Innovations in Pharmaceutical and Biotechnology Development and Manufacturing, 2003 AIChE Annual Meeting [TE001], 2003, pp. 635–642. San Francisco, CA, November 17–21, 2003.

11 Wu H, Heilweil EJ, Hussain AS, and Khan MA. Process analytical technology (PAT): quantification approaches in terahertz spectroscopy for pharmaceutical application. Journal of Pharmaceutical Sciences. 2007;97(2):970–984.

12 Sulub Y, Wabuyele B, Gargiulo P, Pazdan J, Cheney J, Berry J, Gupta A, Shah R, Wu H, and Khan M. Real-time on-line blend uniformity monitoring using near-infrared reflectance spectroscopy: a noninvasive off-line calibration approach. Journal of Pharmaceutical and Biomedical Analysis. 2009;49:48–54.

13 Ciurczak EW and Drennen JD III. Near-infrared spectroscopy in pharmaceutical applications. In: Handbook of Near-Infrared Analysis, 2nd edition, edited by Burns DA and Ciurczak EW, Marcel Dekker, New York, 2001.

14 Xie L, Wu H, Shen M, Augsburger L, Lyon RC, Khan MA, Hussain AS, and Hoag SW. Quality-by-design (QbD): effects of testing parameters and formulation variables on the segregation tendency of pharmaceutical powder measured by the ASTM D 6940-04 segregation tester. Journal of Pharmaceutical Sciences. 2008;97(10):4485–4497.

15 Wu H and Khan MA. Quality-by-design (QbD): an integrated approach for evaluation of powder blending process kinetics and determination of powder blending end-point. Journal of Pharmaceutical Sciences. 2009;98(8):2784–2798.

16 Wu H, Tawakkul M, White M, and Khan MA. Quality-by-design (QbD): an integrated multivariate approach for the component quantification in powder blends. International Journal of Pharmaceutics. 2009;372(1–2):39–48.

17 Wu H, White M, and Khan MA. Quality-by-design (QbD): an integrated process analytical technology (PAT) approach for a dynamic pharmaceutical co-precipitation process characterization and process design space development. International Journal of Pharmaceutics. 2011;405:63–78.

18 Ündey C, Tatara E, and Çinar A. Real-time batch process supervision by integrated knowledge-based systems and multivariate statistical methods. Engineering Applications of Artificial Intelligence. 2003;16:555–566.

19 Kourti T and MacGregor JF. Process analysis, monitoring and diagnosis, using multivariate projection methods. Chemometrics and Intelligent Laboratory Systems. 1995;28:3–21.

20 Wu H, Hussain AS, and Khan MA. Process control perspective for process analytical technology: integration of chemical engineering practice into semiconductor and pharmaceutical industries. Chemical Engineering Communications. 2007;194(6):760–779.

21 Srinivasan R and Qian MS. Off-line temporal signal comparison using singular points augmented time warping. Industrial and Engineering Chemistry Research. 2005;44:4697–4716.

22 Doan X-T, Srinivasan R, Bapat PM, and Wangikar PP. Detection of phase shifts in batch fermentation via statistical analysis of the online measurements: a case study with rifamycin B fermentation. Journal of Biotechnology. 2007;132:156–166.

23 Volpe G, Volpe G, and Petrov D. Singular-point characterization in microscopic flows. Physical Review. 2008;E77:037301–037304.

24 Jackson DA and Chen Y. Robust principal component analysis and outlier detection with ecological data. Environmetrics. 2004;15:129–139.

25 Maggio RM, Castellano PM, and Kaufman TS. A new principal component analysis-based approach for testing "similarity" of drug dissolution profiles. European Journal of Pharmaceutical Sciences. 2008;34:66–77.

26 Otsuka T, Iwao Y, Miyagishima A, and Itai S. Application of principal component analysis to effectively find important physical variables for optimization of fluid bed granulation conditions. International Journal of Pharmaceutics. 2011;409:81–88.

27 Balsera MA, Wriggers W, Oono Y, and Schulten K. Principal component analysis and long time protein dynamics. Journal of Physical Chemistry. 1996;100:2567–2572.

28 Tipping M and Bishop C. Probabilistic principal component analysis. Journal of the Royal Statistical Society, Series B. 1999;21(3):611–622.

29 Schölkopf B, Smola A, and Müller K. Nonlinear component analysis as a kernel eigenvalue problem. Neural Computation. 1998;10(5): 1299–1319.

30 Tipping M. Sparse kernel principal component analysis. In: Advances in Neural Information Processing Systems, NIPS'00, vol. 13, edited by Leen TK, Dietterich TG; Tresp V, pp. 633–639, 2001.

31 Zhou C. Probabilistic analysis of kernel principal components: mixture modeling and classification. CFAR technical report, car-tr-993, University of Maryland, Department of Electrical and Computer Engineering, College Park, MD, 2003.

32 Zhang Z, Wang G, Yeung D, and Kwok J. Probabilistic kernel principal component analysis. Technical report hkust-cs04-03, The Hong Kong University of Science and Technology, Department of Computer Science, Hong Kong, 2004.

33 Alvarez M and Henao R. Probabilistic kernel principal component analysis through time. In: The 13th International Conference on Neural Information Processing, Lecture Notes in Computer Science, vol. 4232, edited by King I, Wang J, Chan LW, and Wang D, pp. 747–754, Springer-Verlag, Hong Kong, China, October 3–6, 2006.

34 Despagne F and Massart DL. Neural networks in multivariate calibration. The Analyst. 1998;123:157R–178R.

35 FDA/ICH. Guidance for Industry. Q9 Quality Risk Management. 2006. Available at: http://www.fda.gov/downloads/Drugs/ GuidanceComplianceRegulatoryInformation/Guidances/UCM073511.pdf (accessed on April 20, 2017).

36 FDA/ICH. Guidance for Industry. Q8 Pharmaceutical Development. 2006. Available at: http://www.fda.gov/downloads/Drugs/ GuidanceComplianceRegulatoryInformation/Guidances/UCM073507.pdf (accessed on June 20, 2012).

37 FDA/ICH. Guidance for Industry. Q8(R2) Pharmaceutical Development. 2009. Draft available at: http://www.fda.gov/downloads/Drugs/ GuidanceComplianceRegulatoryInformation/Guidances/ucm073507.pdf (accessed on June 20, 2012).

38 FDA/ICH. Q10. Pharmaceutical Quality System. Draft Consensus Guideline. 2007. Available at: http://www.fda.gov/downloads/Drugs/ GuidanceComplianceRegulatoryInformation/Guidances/UCM073517.pdf (accessed on April 20, 2017).

39 Wu H and Khan MA. Quality-by-design (QbD): an integrated process analytical technology (PAT) approach to determine the nucleation and growth mechanisms during a dynamic pharmaceutical Co-precipitation process. Journal of Pharmaceutical Sciences. 2011;100(5):1969–1986.

10

Application of Advanced Simulation Tools for Establishing Process Design Spaces Within the Quality by Design Framework

Siegfried Adam[1], Daniele Suzzi[1], Gregor Toschkoff[1], and Johannes G. Khinast[1,2]

[1] Research Center Pharmaceutical Engineering GmbH, Graz, Austria
[2] Institute for Process and Particle Engineering, Graz University of Technology, Graz, Austria

10.1 Introduction

Over the last several years, the pharmaceutical industry has been undergoing a significant change regarding the way products and their associated processes are developed. The goal is to move toward science- and risk-based holistic pharmaceutical development [1]. The fundamental concepts of quality by design (QbD), which is a systematic scientific and risk-based framework for pharmaceutical product and process development and life cycle management, are by now well established in the scientific community. After some initial hesitation, pharmaceutical companies have become increasingly proficient in the application of the QbD principles (e.g., Am Ende *et al.* [2], Hilden *et al.* [3], Hou *et al.* [4], Michaels *et al.* [5], Prpich *et al.* [6]). Several open-access case studies (e.g., the A-Mab study) have demonstrated skill and provided knowledge throughout the industry. As the pharmaceutical industry faces new challenges associated with increased market globalization, such as patent expiry, higher customer expectations, a need to increase profitability, tighter regulations, and the increasing demand to reduce time to market, the QbD adaption process will become vital to promoting business, efficacy, and quality goals. The traditional approach of taking a product from the laboratory scale to a pilot plant and finally to production is no longer feasible. Since process and product development are often initiated simultaneously, rapid prototyping for the development of formulations, unit operations, and associated manufacturing equipment is increasingly in demand [7]. This requires efficient development tools to assist process characterization and the employment of a QbD approach.

Comprehensive Quality by Design for Pharmaceutical Product Development and Manufacture,
First Edition. Edited by Gintaras V. Reklaitis, Christine Seymour, and Salvador García-Munoz.
© 2017 American Institute of Chemical Engineers, Inc. Published 2017 by John Wiley & Sons, Inc.

The fundamental assumption for QbD is that understanding the critical sources of variability offers control over product quality and performance via the manufacturing process that mitigates variability in the input parameter properties [8]. Within the desired state of QbD according to ICH Q8 [9], a process is regarded as well understood when all critical sources of variability are identified and explained, when the variability is managed by the process, and when the product's quality attributes can be accurately and reliably predicted over the design space. The implementation of QbD preferably begins at the early stages of the formulation and process development when a certain knowledge base has been created and continues to expand over the entire product's life cycle.

Figure 10.1 shows a systematic overview of critical elements of an overall QbD approach and overlapping components, modified according to Schousboe and Hirsh [10].

An important primer for the definition of development strategies and subsequent characterization studies is the foundation of product understanding (left part of Figure 10.1) that links the clinical profile of the product to the development goals and initiates a risk-based approach by defining critical quality attributes (CQAs). CQAs are physical, chemical, biological, or microbiological properties or characteristics that must be controlled directly or indirectly to ensure the quality of the product [11], that is, the product profile according to Figure 10.1. The defined CQAs govern the establishment of process understanding (middle left part of Figure 10.1). Here, extensive process characterization studies that examine potentially critical input parameters and their effects on CQAs are required to provide the necessary fundamental process understanding. Since characterization studies at the manufacturing scale are mostly not feasible due to high costs and the limited availability of resources, alternative approaches must be explored, for example, downscaled models or (less frequently used) simulation methods, supported by design of

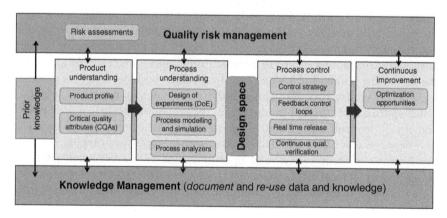

Figure 10.1 Systematic QbD approach, modified based on Schousboe and Hirsh [10].

experiments (DOE) and sophisticated process analyzers. Ultimately, the acquired process understanding is collected and documented in a design space (middle part of Figure 10.1) that links fundamental understanding of input–output parameter correlations to adequate process control strategies (middle right part of Figure 10.1). The control strategy also sets the agenda for what continuously needs to be monitored and followed up during the review cycles of continuous quality verification [10] (right part of Figure 10.1). The life cycle process of quality risk management and knowledge management supports the systematic approach according to Figure 10.1. Quality risk management drives the definition and control of patient risks associated with causal relationships, and knowledge management provides an overview of these cause-and-effect relationships [10].

In this context, the advantages of in silico experiments for gaining process understanding according to Figure 10.1 have been increasingly scrutinized in the pharmaceutical development community [1] to guide the risk assessment of potentially critical input parameters and to develop an in-depth understanding of how process variability affects the quality of the associated products. Numerical tools for "in silico" experimentation include multiphase computational fluid dynamics (CFD), lattice Boltzmann methods, or a discrete element method (DEM). For example, Suzzi *et al.* [12] recently applied multiphase CFD methods to pharmaceutical tablet coating. A combined QbD and DEM simulation approach for the analysis of powder blending was presented by Adam *et al.* [1]. Persson *et al.* [13] performed a comparative experimental study of flowability of surface-modified pharmaceutical granules. Chivilikhin *et al.* [14] introduced a CFD analysis of hydrodynamic and thermal behavior of continuous flow reactors. While these simulation approaches are widely referred to as analysis tools for various processes, applying computer simulations as distinctive QbD tools is still not straightforward and has not been adopted by regulatory authorities and industry to any appreciable extent.

There are several advantages to a simulation approach within the QbD framework. Computer simulations are typically cheaper, safer, and faster than real-life experiments. In many cases they are the only way of testing changes in the parameter space, for example, in a full-scale operation of large fungal fermentors. Moreover, they make precise in silico experiments and changing experimental parameters beyond experimentally accessible ranges possible. Additionally, sound scientific data and fundamental process understanding can be generated more efficiently to fully characterize a system at multiple locations. However, it is critical for simulation-based studies that the underlying simulation model be fully predictive and that it accurately capture the real physics of the system.

In a QbD environment, the application of data generated via computer simulation may be twofold. On the one hand, numerical results can be used in combination with risk assessment tools at an early stage of process characterization as a screening tool. Screening may be applied to map out potentially critical input factors (e.g., raw material characteristics, formulation attributes, process parameters, and equipment design) and to prioritize them for further in-depth process studies, for example, via DOE. On the other hand, simulations may be employed directly during the main phase of process characterization to generate good manufacturing practice (GMP)-relevant data to be included in the pharmaceutical dossier together with the data from experimental characterization.

Experimental studies have numerous sources of error. First, experiments may be planned poorly. For example, the temperature, pH, or concentrations reported at a sensor location may not reflect the temperature, pH, or concentration where a reaction takes place due to strong gradients not known to the experimenter. Second, measurements themselves are subject to errors, as all sensors show some level of variability or drift. Third, experimental systems may be characterized by significant variability. Examples include cell cultures, input materials from natural sources, or complex processes, such as high-shear granulation. Fourth, some probes may impact the process and lead to false results, for example, thieve probes in powder sampling, which are known to introduce segregation in the system. Lastly, operator errors, environmental conditions, or unanticipated effects (e.g., unknown catalytic effects of a reactor material) may impact the measurements.

Nevertheless, experiments are considered the gold standard in process and product development and are rarely questioned if reported properly. Similarly, numerical simulations are subject to errors and inaccuracies, and—same as for experimentation—it is important to gauge the level of accuracy of a simulation. This is prerequisite in order to replace real experimentation to guarantee a certain level of quality of prediction and reliability. Therefore, efficient validation strategies that prove the predictive capability and robustness of a simulation within a sufficient confidence interval are considered an indispensable prerequisite for the use of simulation tools in a pharmaceutical environment governed by the GMP guidelines. Moreover, the need to qualify the simulation software to be used in a GMP environment creates additional challenges. Since the accuracy and integrity of data generated by the simulation software are essential throughout the product life cycle, that is, from research and development to fundamental characterization of the final process and manufacturing of validation batches, a risk-based approach to compliant GxP computerized systems according to GAMP [15] must be taken to provide assistance in implementation, validation, and operation of the simulation tool.

Specifically, we propose the following steps to allow the use of simulations tools in a regulatory environment and to apply the data for establishing a (partially) simulation-based dossier:

a) Acquiring and retaining a qualified status of simulation tools according to GAMP 5 to assure GMP-compliant simulation hardware and software throughout the life cycle of the process/product.

b) Establishing a simulation domain: risk-based deduction of boundary conditions and relevant parameter variations (according to the systematic QbD approach).
c) Setup and review of underlying simulation models (using previous knowledge, i.e., literature, model validation studies, theoretical assessment of 1 : 1 correspondence to real-life systems, etc., within a risk-based approach to define and justify the required assumptions and simplifications).
d) Detailed documentation of model validity for the considered simulation space based on point step c.
e) Performance of simulations.
f) Comparability check to fulfill GMP requirements: simulation validation procedure to document the simulation's validity and verify the simulation's results with real-life process characterization and validation data.
g) Data evaluation and virtual design space establishment.

Detailed studies of the role of computer simulations within the overall QbD environment and their use in and applicability to the pharmaceutical development and life cycle management are required to further reveal the real capability of this approach. In the following sections we describe a combined application of QbD and computer simulation methods to two key unit operations and processes in the pharmaceutical industry. First, a blending unit operation, which is a key process step in the manufacturing of solid oral dosage forms, is characterized. The impact of potentially critical formulation attributes and process parameters on the quality of the final blend is investigated via a DEM simulation. Next, a correlation between the evaluation of the optimal blending quality and the regulatory requirements is established. Second, a coating unit operation is investigated with the main objective to provide a science-based quantitative understanding of the physicochemical parameters that influence the uniformity of the coating layer on a single tablet. This knowledge is key for process design, optimization, and rational scale-up and thus can form a basis for future studies.

10.2 Computer Simulation-Based Process Characterization of a Pharmaceutical Blending Process

10.2.1 Background

Blending is a key unit operation for solid dosage form manufacturing. It assures the blend's homogeneity that directly affects content uniformity, which is a potential CQA of the final product. According to ICH Q8 [9], a CQA is a characteristic of the final product that directly relates to safety and efficacy and, therefore, has to be controlled. It is especially true for narrow therapeutic index and high-potency drugs [16]. Content uniformity is mainly affected by poor powder flows in dies, segregation in the hopper or feed frame, agglomeration, and poor blending [1]. In particular, inefficient blending can lead to increased variability of the active component in the final dosage form, posing a threat to the patients' health [8]. For this reason, the homogeneity of pharmaceutical blends and dosage units is highly scrutinized by regulatory bodies throughout the world, and pharmacopoeias require an assessment of content uniformity to be performed on every batch of solid dosage forms [17]. In most cases, process characterization and validation mainly focus on blending efficiency with respect to the active pharmaceutical ingredient(s) (API(s)) [18]. Although excipients constitute a large portion of most drug products, they are viewed as a low-risk aspect of a drug product's safety and efficacy [19]. However, due to their specific functional properties, they contribute greatly to the success of medicinal products (e.g., in melt-extrusion drugs the matrix is inactive, yet the matrix determines release kinetics, stability, and bioavailability) and are thus considered key elements of QbD [1]. In-depth understanding of how formulation components, blending process attributes, and blending equipment design affect the blending kinetics, the blending end point, and the final blending quality is required to generate mechanistic process understanding, promoting the establishment of a robust design space, regulatory flexibility, and a reliable scale-up.

Studies published in the literature are used as "prior knowledge" to set up predictive and accurate simulations that capture the real physics of the system, including such phenomena as agglomeration in cohesive systems and segregation. Granular flows and the mixing dynamics in various blender geometries have been studied by several groups. For example, mixing mechanisms, mixing regimes, and segregation phenomena were described by Lacey [20] or Bridgwater [21]. Also, various mixer types were studied, including tumbling and convective blenders. Granular flows in these convective mixers have been experimentally studied and numerically reproduced by Stewart *et al.* [22]. Similar studies on non-cohesive powder in agitated blenders have been presented by Remy *et al.* [23, 24], highlighting the impact of fill level, particle size, and other parameters on the blending process. Recently, Radl *et al.* [25] analyzed the mixing mechanisms of wet granular flow, both numerically and experimentally.

Double-cone mixers are frequently used for batch mixing operations [8]. They impart little to moderate shear on the powder and can be implemented in large-scale systems. The slow mixing across the plane symmetry is a well-known problem for these types of blenders. For example, Brone and Muzzio [26] experimentally analyzed axial and radial mixing by means of digital camera images and a vacuuming technique. They found that the rate of radial mixing was 20–30 times faster than the rate of axial mixing across the plane of symmetry. Therefore, the performance of a double-cone blender is strongly affected by the loading protocol, for example, if an axial composition gradient is obtained during loading, inhomogeneity could last many hundreds of revolutions, compromising the entire mixed batch [1].

Today, characterization studies of powder blending processes are mainly conducted by photographing particles through transparent walls or by using thieve probes to assess mixing quality. These techniques are tedious and introduce a great deal of uncertainty since only surface particles can be analyzed [27] or probes lead to segregation phenomena, affecting the accuracy of data. Positron emission particle tracking (PEPT), a highly accurate method, is expensive and impractical for industrial equipment. In contrast, computer simulation-assisted process characterization, for example, numerical simulations based on DEM, can overcome these problems [1], as it is relatively fast and inexpensive as well as accurate. Numerical analyses of double-cone blenders have been, for example, presented by Moakher *et al.* [28] and Adam *et al.* [1].

10.2.2 Goals

Powder blending, which is one of the key unit operations for solid dosage form manufacturing, strongly affects process performance and the quality of the final product and, therefore, closely relates to safety and efficacy. Thus, resource-consuming process characterization studies of the blending process are required in order to obtain data to generate a design space within the QbD environment. (Within the QbD framework, a design space is defined as the combination of process and design variables that yields product with acceptable quality attributes. Most importantly, manufacturing changes within the design space are not subject to regulatory review.) In most cases, process characterization is carried out via experimentation, typically using small-scale models supported by appropriately designed experiments. However, experimental methods, even within the QbD environment, have some pitfalls. For instance, blending in tumbling mixers relies on the flow of a small layer of material on top of a stagnant, marginally deforming bed. The thickness of this high-shear layer is a function of the material properties. As the thickness of this layer does not change upon scale-up, the ratio of the deforming volume to the nondeforming volume decreases, changing the overall dynamics of the blending process and invalidating the scale-up approach [1, 8].

Scale-dependent problems, such as the one described earlier, may be solved by a computer simulation. Additionally, QbD development activities rely on various risk assessment approaches in order to prioritize potentially critical input parameters (e.g., material attributes, process variables, formulation parameters) for subsequent characterization studies. Thus, prior knowledge of product and process characteristics is indispensable to risk assessment. However, in many cases, the number of potentially critical input variables is high due to the lack of additional knowledge that would guide the risk assessment approach (e.g., in the very beginning of the characterization of a new manufacturing process or unit operation). Here, computer simulation-aided process characterization may contribute to screening potentially critical variables, thus providing a basis for choosing an efficient and cost-effective experimental program.

In this case study, we focus on the benefits of computer simulations within the QbD framework and their contribution to a thorough process understanding. The impact of two variables derived from a risk assessment exercise on blend homogeneity is investigated, and a three-dimensional knowledge space is generated. This knowledge space is then used to define activities for further "real" characterization studies and to create an appropriate control strategy that assures the quality of the blend at the end of the blending unit operation. In summary, the objectives of this case study were the following:

- To apply QbD concepts under ICH Q8 and Q9 to characterize a blending unit operation, including a definition of a CQA, risk assessments for prioritizing potentially critical input variables of process characterization, computer simulation-based process characterization, and a definition of an appropriate control strategy according to the established knowledge and design spaces
- To show the benefits of using DEM computer simulations for evaluating the effect of formulation parameters and process variables on the blending quality, dynamics, and end point

Please note that there are several potentially critical input variables that may affect the quality output of a blending operation. The effects of selected parameters on the blending quality are in many cases well known but not fully understood, and our primary intention was to demonstrate that simulation models can quantitatively elucidate correlations between the input factors and the quality of output. The mechanistic understanding we generated can be applied to new and modified blending processes and to the blending equipment design.

10.2.3 Material and Methods

10.2.3.1 Application of QbD Concepts

A systematic scientific and risk-based QbD approach was used to (i) establish a CQA to guide development, (ii) define potentially critical input variables whose variability may influence the established CQA, (iii) carry out computer simulation-assisted process characterization to elucidate quantitative correlations between process input and output, and (iv) map out a blending experimental space as a subset of the blending knowledge space (e.g., established by prior knowledge) that was used to develop a design space and an associated control strategy.

For a practical demonstration, the blending of a two-component system, that is, acetylsalicylic acid (ASA) as API and lactose as excipients yielding a mean tablet weight of 0.9 g, was selected. ASA is a commonly known API, and lactose is widely used as excipient in various formulations for direct compression and tableting [29].

Following the definition of CQAs, various risk assessment exercises were conducted to identify potentially critical input parameters (e.g., material attributes, process variables, and formulation parameters) to be introduced into subsequent characterization studies. The characterization step is fundamental to understanding the effect of the most important variables on the overall process performance because typically not all variables can (nor need to) be analyzed in great detail [1]. The resulting data can then be used to define the acceptable variability in input parameters [1, 30–37].

Content uniformity of the final product may be regarded as potential CQA of the finished product as it directly relates to safety and efficacy. Regulatory requirements include content uniformity in the release specification of the finished product and establish blending homogeneity as its prerequisite. Thus, as content uniformity is strongly influenced by the blending unit operation, blending homogeneity can be considered an intermediate CQA. The blending end point, that is, the point at which an adequate blend of all components is achieved, is generally established by setting acceptance criteria for the relative standard deviation (RSD) of the API content. The calculation of the RSD can be used as a quantitative surrogate to determine the blend's homogeneity [1]. Using a computational approach via DEM, measuring the RSD is straightforward since the position of all particles is available. Moreover, errors introduced by the sampling procedure, which are a major concern for experimental protocols, can be completely avoided [38–42].

As mentioned earlier, there is a whole range of input factors that may potentially affect the quality and the performance output of a blending unit operation. Adam *et al.* [1] have performed a failure mode and effects analysis (FMEA) based on risk identification using Ishikawa diagrams to identify potentially critical input parameters of a blending unit operation (Table 10.1). This risk-based preprocess characterization identifies knowledge gaps with regard to certain input factors and their effect on the blending process. Note that blending inhomogeneity is one of the potential failure modes that may lead to out-of-specification results for content uniformity of the final dosage form. Downstream processes may also influence content uniformity as they may promote segregation of the powder mixture.

The objective of the subsequent process characterization step is to evaluate the effect of factors identified in the risk assessment exercise on the CQA and the process performance and thus to define a design space in which the blending process can operate in an acceptable way with respect to product quality and process consistency [33]. Operating within the

Table 10.1 Risk assessment of the blending process step using FMEA according to Adam et al. [1].

Effect	Severity (S)	Failure mode	Detectability (D)	Cause		Probability (P)	RPN
Varying content uniformity of final dosage form	5	Inhomogeneity of final blend	4	Raw material properties	Particle-size ratio	5	100
					Weight ratio	5	100
					Particle shape	5	100
					Cohesivity	5	100
					Material density	4	80
				Process parameters	Fill volume	4	80
					Blending time/total revolutions	5	100
					Blend rpm	1	20
					Order of addition	3	60
				Equipment design	Blender type	4	80
					Blender total volume	3	60
				Environment	Relative humidity	3	60
					Temperature	2	40

Five-point scales for S, D, and P, with 5 as worst-case value and 1 as best-case value, are used. An RPN of 50 is defined as the cutoff limit.

acceptable ranges whose combination will ultimately define the process design space provides quality assurance [43].

In this case study the use of computer simulation via DEM for process characterization in the QbD environment was demonstrated by using a binary model formulation containing ASA and lactose. During the simulation, the impact of two potentially critical input factors defined by the FMEA exercise, that is, the weight ratio between the two ingredients and the number of mixer revolutions, was evaluated. The aforementioned input factors were selected as potentially critical parameters since they are known to impact blend quality. Using the weight ratio as a potentially critical parameter allowed us to understand the effect of various API concentrations (e.g., if various dosage strengths are produced) on the blend quality [1]. Furthermore, as the two components of the binary mixture differ in density, the variability in weight ratio may have different segregation effects and thus affect the outcome of the blending process.

10.2.3.2 Model and Numerical Simulation

Computer simulations were carried out using DEM. At each time step the forces acting on each particle were computed first. Then Newton's equations of motion and a rotational momentum balance were solved in order to obtain the new positions and velocities.

The forces acting on a single particle were:

- The body force $\vec{F}_{b,i}$;
- The normal and tangential contact forces $\vec{F}_{c,n,ij}$ and $\vec{F}_{c,t,ij}$ between particles i and j

Hydrodynamic forces are typically not considered for DEM simulations and thus were excluded from this case study. This assumption is valid for granular flows involving relatively large, heavy particles, when body forces are much greater than hydrodynamic forces. This is typically true for particles larger than $100\,\mu m$ in air. We assumed a mixture of free-flowing materials and excluded interparticle cohesive forces from our numerical model. However, the friction between particles was included into an appropriate model for the tangential contact forces. More details on the basic principles of DEM simulations can be found in the review paper of Zhu *et al.* [44]. The collision between particles was modeled using the so-called soft-sphere approach or time-driven method, under which collisions were represented as contacts of deforming particles lasting over a few time steps. New values of particle positions and velocities were computed at fixed time intervals. Particle deformation was modeled by overlaps of colliding spheres, and the contact forces were a function of these overlaps. The soft-sphere method allowed an accurate reproduction of particle elasticity and of damping and sliding effects due to friction [45]. The linear spring–dashpot model [46] was used, whose main advantages were that the parameters were readily determined and that analytic solutions were available for a direct verification of the numerical schemes. This model is still widely used in the area of DEM simulations since more sophisticated nonlinear interaction models provide minor benefits compared to the additional computational effort [47].

The linear spring–dashpot model calculates the normal force by adding an elastic term modeled by a spring and a second term representing dissipation modeled by a dashpot:

$$F_{c,n,ij} = k_n \delta_{n,ij} + c_n \dot{\delta}_{n,ij} \tag{10.1}$$

k_n is the normal stiffness coefficient, c_n is the normal damping coefficient, and δ_n is the normal overlap between the particles. The normal stiffness coefficient was evaluated based on material and geometrical properties of the particles and from a characteristic collision velocity [48].

The forces acting in the tangential direction were similarly represented by a linear spring and a dashpot model with a frictional slider:

$$F_{c,t,ij} = \begin{cases} k_t \delta_{t,ij} + c_t \dot{\delta}_{t,ij} & \text{for} \quad F_{c,t,ij} \leq \mu F_{c,n,ij} \\ \mu F_{c,n,ij} \dfrac{\delta_{t,ij}}{|\delta_{t,ij}|} & \text{for} \quad F_{c,t,ij} > \mu F_{c,n,ij} \end{cases} \tag{10.2}$$

Similarly to the normal force model, the term $k_t \delta_{t,ij}$ represents the force of the spring in the tangential direction, while $c_t \dot{\delta}_{t,ij}$ describes the contribution of the damping element, that is, the dashpot. The friction coefficient μ between the two materials in contact characterizes the frictional slider. Furthermore, rolling friction was accounted for by

$$M_r = \mu_r |F_{c,n,ij}| r_i \cdot \frac{\ddot{\theta}_i}{|\ddot{\theta}_i|}. \tag{10.3}$$

where $\|$ denotes the norm of a vector.

The commercial code EDEM of DEM Solutions Ltd. was used to perform the simulations. A code that can simulate many millions of particles [49] will be used for future studies for larger and more complex systems.

10.2.3.3 Process Characterization Experimental Design

The blender geometry corresponded to the experimental double-cone blender design analyzed by Brone and Muzzio [26]. A graphical representation of the double-cone geometry according to Adam *et al.* [1] is shown in Figure 10.2.

The particles were made of spherical ASA and lactose particles with a diameter of 4 mm. Recent publications have shown that the dynamics of the particle mixing process are independent of particle size once a certain

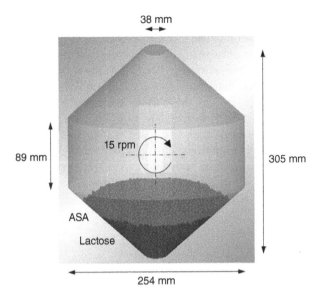

Figure 10.2 Graphical representation of the considered double-cone blender. *Source*: Adam *et al.* [1]. Reproduced with permission of Elsevier.

threshold of particles in the system has been exceeded [24]. The blender loading was performed by initializing 30 000 particles representing 40% of the blender volume in random and non-overlapping positions, allowing the particles to fall and rest before the simulation begins. The density of ASA is 1.530 kg/m³ and of lactose 1.400 kg/m³. In order to reduce the computational time, the shear module G was set to 1.3×10^9 Pa for both materials, leading to maximal particle overlaps that were under 5% and, consequently, to a minimal effect on the simulation results [22]. The simulations were conducted with coefficient of restitution (COR) values of 0.6 for lactose and 0.7 for ASA. The sliding friction coefficient was measured for both lactose and ASA in a shear cell, with the values of 0.3 and 0.4, respectively [50]. The rolling friction coefficient was set to a constant value of 0.01, based on the work of Iwashita and Oda [51].

The rotational velocity of the blender was 15 rpm, which corresponds to a Froude number of 0.038. For blending operations the Froude number was defined as

$$Fr = \frac{\omega^2 r}{g} \tag{10.4}$$

with ω and r being the rotational speed and the blender radius, respectively. As reported by Brone *et al.* [52] and Brone and Muzzio [26], the value of Fr below 0.2 corresponds to powders flowing in an inertial regime. In this regime the degree of mixing is also mostly independent of the rotation speed, which puts the simulation results in direct relation to the blender revolutions.

Based on the outcome of the quality risk assessment exercise and prior knowledge, the potentially critical formulation attribute (i.e., the weight ratio between the two components of the binary model formulation) and the potentially critical process parameter (CPP) (i.e., the number of blending revolutions) were varied between 10 and 50 wt% and 0–15 revolutions, respectively.

Based on a statistical experimental design, the input factors were varied simultaneously, and for each combination the quality and performance output of the blender was analyzed in terms of RSD of the particle mixture in order to normalize the results of cases with different initial mixtures. The RSD is defined as the ratio between the standard deviation σ and the mean mass fraction $\bar{W}$:

$$\text{RSD} = \frac{\sigma}{\bar{W}} \tag{10.5}$$

with the standard deviation σ:

$$\sigma = \sqrt{\frac{1}{n-1}\sum_{i=1}^{n}\left(W_i - \bar{W}\right)^2} \tag{10.6}$$

where n is the number of samples, W_i is the mass fraction of ASA in sample i, and $\bar{W}$ is the mean ASA mass fraction of all samples taken.

Adam et al. [1] developed a sampling procedure for simulated particulate systems, taking into account the scale of scrutiny and the overall particle number of the simulated system. We performed sampling by defining a three-dimensional grid of equal and not overlapping cubical bins. Each bin contained $M = 120$ particles. Nine samples were taken after each revolution and used in the calculation of the RSD value. The number of particles per sampling bin corresponded to the scale of scrutiny, that is, the sample size for which material homogeneity was investigated. Relying on too large sample size may mask mixing problems within a given system. According to FDA's powder blending guide [53], sample size can be defined as 1–10 times the mass of the final dosage unit. In the system under investigation, the sample size of 120 particles corresponded to 6 g sample weight, that is, almost six times the final dosage form.

RSD was then compared with the theoretical RSD limits for a two-component system with the theoretical upper limit (completely segregated mixture) of

$$\text{RSD}_s = \frac{\sigma_s}{\bar{W}} \tag{10.7}$$

where

$$\sigma_s = \sqrt{p\left(1-p\right)} \tag{10.8}$$

For a certain composition of noninteracting particulate and a given sample size, there is a minimum theoretical value of standard deviation that can be achieved by random mixing [54]. This theoretical lower limit of RSD (randomly mixed mixture) is

$$\text{RSD}_r = \frac{\sigma_r}{\overline{W}} \tag{10.9}$$

where

$$\sigma_r = \sqrt{\frac{p(1-p)}{M}} \tag{10.10}$$

Here, p is the mass fraction of one component in the mixture and M is the total number of particles in the sample [20, 38, 55, 56]. It can be seen that for random mixtures, the sample variance is inversely proportional to the sample size. The actual values of the mixture RSD must be between the two extreme values of RSD_r and RSD_s, that is, $\text{RSD}_r < \text{RSD} < \text{RSD}_s$. According to Equations 10.9 and 10.10, the RSD_r partially depends on the total number of particles in the sample. Thus, a well-mixed system with few particles has a higher "perfect" RSD than a mixture with a total number of particles that is high. This has to be taken into account when comparing systems with different numbers of particles. The limit of blend RSD $\leq 5.0\%$ suggested by the FDA [53] refers to actual mixtures with a total amount of particles in the system that is several orders of magnitude higher than in a DEM-simulated system, with the RSD_r value approximating to zero. When considering the effect of the system's total particle number on the assessment of blend homogeneity, the comparability of the mixing metric for low-particle-number systems with the FDA requirements must be assured. Thus, a modification to the FDA limit according to Lacey's index [20] was suggested, that is,

$$M_{\text{FDA}} = \frac{\sigma_s^2 - \sigma_{\text{FDA}}^2}{\sigma_s^2 - \sigma_{r\,\text{FDA}}^2} = \frac{\text{RSD}_s^2 - \text{RSD}_{\text{FDA}}^2}{\text{RSD}_s^2 - \text{RSD}_{r\,\text{FDA}}^2} \tag{10.11}$$

with RSD_s as defined earlier, $\text{RSD}_{\text{FDA}} = 5\%$, and $\text{RSD}_{r\,\text{FDA}} = 0$ for actual mixtures with a sufficient number of particles. This conversion directly links RSD as a quality attribute for high-number particle systems to M as a quality attribute for low-number particle systems. Taking into account the effect of particle numbers on the blend RSD, the blend homogeneity of low-number systems in relation to the regulatory-relevant index can be interpreted.

M_{FDA} was then compared to Lacey's index for every sampling time point in our DEM-simulated system:

$$M = \frac{\sigma_s^2 - \sigma^2}{\sigma_s^2 - \sigma_r^2} = \frac{\text{RSD}_s^2 - \text{RSD}^2}{\text{RSD}_s^2 - \text{RSD}_r^2} \tag{10.12}$$

with RSD, RSD_s, and RSD_r as defined earlier.

10.2.4 Results and Discussion

The aim of this case study was to show the benefits of a computer simulation-assisted process characterization approach within the overall QbD framework for knowledge generation and the characterization of a blending unit operation.

Following the systematic QbD approach according to ICH Q8 [9] that is described in a more practical and relevant way in Adam *et al.* [1], we began with the definition of blend homogeneity as one of potentially (intermediate) CQAs of oral solid dosage form manufacturing since it directly relates to the safety and efficacy of the final product. Subsequent quality risk assessment exercises identified "weight ratio between ASA and lactose" as a potentially critical formulation attribute and the "number of mixing revolutions" as a potential CPP of the blending unit operation for subsequent process characterization studies by means of DEM simulation.

The results of the blending process characterization have already been published in Adam *et al.* [1] and are summarized as follows.

The effects of varying blending time and different weight ratios between API and lactose on blending homogeneity were monitored. The simulations were run up to 15 complete revolutions of the blender. As described by Brone and Muzzio [26], in the case of a double-cone blender with layered loading, an acceptable homogeneity can be reached after 10–20 revolutions. The RSD evolution for three different mixtures (i.e., 10, 30, and 50 wt% ASA content) is shown in Figure 10.3. The amount of particles corresponded to 0.4% of the total number of particles, which was assumed to be an acceptable level of scrutiny [1]. The sample number was chosen according to recent reports in the literature [29].

After 15 revolutions the RSD of the binary mixture containing 10 wt% API settled at a higher value compared with the 30 and 50 wt% mixtures. The RSD values of the three mixtures after 15 revolutions were not within the FDA requirements. Obviously, there was a problem with the blending efficiency in our simulated system. The higher RSD of the 10 wt% API mixture can be explained by a segregation tendency of the API in the powder blend, which may be due to trajectory segregation based on the density differences between the two components of the mixture. In combination with the low concentration of API in the final blend, this may lead to a relatively high variance between samples, resulting in high RSD values.

As discussed earlier, RSD strongly depends on the sample size, that is, an increase in the sample weight leads to a decrease in the mixture standard deviation [54]. Therefore, the RSD inaccurately represents the blend quality in low-particle-number simulated systems [1]. In order to better assess the blend quality of low-particle-number mixture systems with varying amounts of drug content, a modified Lacey index was introduced according to Equation 10.12. In practical terms, Lacey's index is the ratio of "mixing achieved" to "mixing

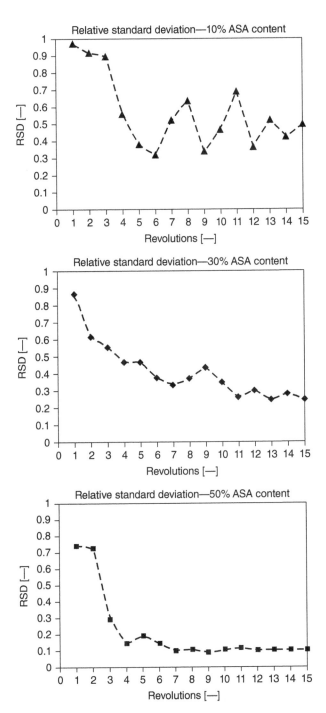

Figure 10.3 Evolution of RSD value for binary mixtures with 10, 30, and 50 wt% API content, respectively. *Source*: Adam *et al.* [1]. Reproduced with permission of Elsevier.

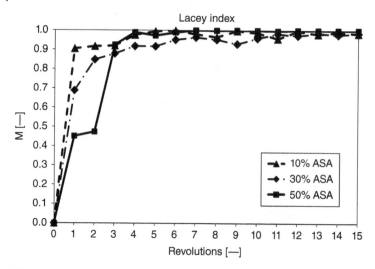

Figure 10.4 Evolution of Lacey's index for binary mixtures with 10, 30, and 50 wt% API content. *Source*: Adam *et al.* [1]. Reproduced with permission of Elsevier.

possible" [56] for the total number of particles in the system. This index is frequently utilized as a mixing metric [57] and has been discussed by several researchers [58–60]. Furthermore, the conversion of the blend RSD limit of ≤ 5.0%, as required by the FDA [53], to Lacey's mixing index M_{FDA} according to Equation 10.11 offered a regulatory-relevant assessment of the blend quality of mixtures with a relatively low number of total particles. The evolution of Lacey's index for each mixture is shown in Figure 10.4.

The illustration of the blend quality using Lacey's index reveals the actual mixing degree by considering the overall sample particle number and thus the total particle number in the system. Figure 10.4 shows that Lacey's index for all three binary mixture systems rapidly increases with the number of revolutions. Taking the M_{FDA} limit into account, the blend homogeneity of the 50 wt% mixture was well within the M_{FDA} limit after 12 revolutions. The results of the blend RSD and Lacey's index indicate that appropriate blending quality was obtained with a higher mass fraction of ASA, although an acceptable quality of the blend was not reached after 15 revolutions for the 10 and 30 wt% mixtures.

The DEM simulation-based characterization study provided data regarding the effect of various combinations of the investigated input parameters on the quality of the final blend at the end of the blending unit operation. The generated data were used to establish a blending process experimental space with the process design space as its subset, which was a multidimensional combination of the weight ratio and the blending time that assures an acceptable blend

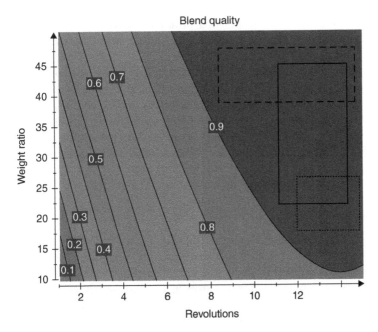

Figure 10.5 Blending unit operation experimental space, containing various possible design spaces as subset (rectangles on the right). *Source*: Adam *et al*. [1]. Reproduced with permission of Elsevier.

quality. The experimental space according to Adam *et al.* [1] that contained the blending process design space is illustrated in Figure 10.5.

The experimental space represented the effects of blending time and ASA mass content on the blending quality. The blending quality via Lacey's index was predicted over the experimentally explored space. Generally, an experimental space within the QbD approach represents a whole range of interactions between critical input parameters and their effects on related CQAs containing combinations of parameters that lead to acceptable CQA quality, as well as so-called edges of failure, which stand for the combination of inputs that results in out-of-specification CQA values. According to the blending process design space, the best mixing quality in the simulated system was achieved by a combination of a blending time between 7 and 15 revolutions and the highest API mass fraction of 50 wt%. Low API contents in the simulated system required longer blending times.

The final step in our QbD-guided process characterization was the definition of an appropriate control strategy that assured adequate quality of the CQA "blend homogeneity." According to ICH Q8 [9], a control strategy is defined as "a planned set of controls, derived from current product and process understanding that assures process performance and product quality."

Hence, the quality control of the finished product is closely related to the blending process design space (the design space documents the established process understanding).

Monitoring design space input variables according to ranges derived from process understanding and predicting values for the blending quality index via the established design space model could assure quality of the blending process. This means that all possible combinations of input variables that allow the CQA to stay within predefined specifications are accepted. Our case study demonstrated that this criterion could be met after seven revolutions for mixtures with a relatively high API content (i.e., 50 wt%). Operating within these input parameter ranges assured the blending quality as defined in the product profile.

10.2.5 Conclusion

This case study demonstrates the benefits of using a computer simulation for process characterization within the QbD environment. We showed how combined QbD and DEM approaches could be used for gaining process. The effect of two parameters on blending quality was investigated. We used the correlations between the input and the quality output that are generally known in the pharmaceutical and manufacturing community for validating the simulated system. Blend homogeneity was quantified and compared to the appropriate FDA limit by means of Lacey's index, which characterizes low-particle-number systems.

A blending process design space was set up, in which combinations of investigated input factors led to a required blend quality. This optimal region had a relatively high API content and a relatively long mixing time. Such information may be used for the development phase of a blending unit operation, to optimize the equipment design, and to select a mixing principle.

In addition to particle properties and blender geometry, other potentially critical input parameters, for example, particle loading method, fill volume, and sampling, can be tested for further process optimization. One major limitation of DEM is its intensive computational requirements. However, recent novel developments allow a simulation of many millions of particles within a reasonable amount of time [49] that will be used in future studies to further develop the combined QbD and simulation approach.

10.3 Characterization of a Tablet Coating Process via CFD Simulations

10.3.1 Introduction

Coating is the last key unit operation in many tablet production processes. Generally, the goal of coating is generating a thin solid layer around the tablet core. The function of the coating ranges from purely cosmetic improvements to sophisticated alterations of tablet's release characteristics. For example, adding color to the film material helps to distinguish between tablets and market the product by increasing brand recognition. Coating can be used to mask unpleasant odor or taste that many APIs have and thus improve patient acceptance. Targeted modification of the tablet's release properties by changing the release rate and/or the release location is performed more frequently. An enteric coating, for instance, is dissolved in the intestines and not by gastric juices, thus protecting the stomach from direct drug contact and the API from degradation in the acid environment.

Since coating originated from the confectionery industry, predominantly sugar coatings were used initially to mask taste and odor and to protect the API from the environmental impact. However, the process of sugar coating was more art than science [61], and the functionality of the film is limited. Significant progress was made by the introduction of spray coating using polymer-based coating materials that provided additional functionality. Coating processes in general became more robust, efficient, and less dependent on the operator. It should be noted that in some respects spray coating is inferior to sugar coating (e.g., bridging failures), and one could argue that, from a purely aesthetic point of view, the elegance of a smooth glossy surface of sugar-coated tablets is still unsurpassed [62]. While organic solvents offer distinct advantages, in light of ecological, economical, and health hazard issues, the general trend at the moment is toward aqueous formulations. Most commonly, aqueous polymer dispersions (e.g., cellulose ethers, acrylic polymers) are used, together with plasticizers and insoluble solid components (e.g., talcum). When the water evaporates, the distance between the disperse polymer agglomerates is reduced until they are in contact. Neighboring polymer chains begin to connect, and a continuous film is formed.

In nearly all cases, the uniformity of the film layer is of utmost importance. This is especially true when the coating is intended to protect the API at the core, to modify the release, or to contain the API itself. In these cases, it is clear that coating uniformity directly relates to safety and efficacy and can therefore be identified as an intermediate CQA according to ICH Q8 [9]. However, experimental determination of space-resolved coating thickness and thus coating uniformity remains a challenge. In this area, noninvasive technologies are of special interest as they allow further measurements of the same tablet. Recent applications include terahertz pulsed imaging [63], optical coherence tomography [64, 65], confocal laser scanning microscopy [66], or combinations thereof [67, 68].

Although the tablet coating process is based on seemingly simple principles, a constant high product quality is not easily achieved. Although a common trial-and-error approach can be applied to some degree, fundamental process understanding is necessary to optimize the processes. For this reason, numerous researches have investigated the underlying principles. The effect of a range of CPPs was investigated by Patel *et al.* [69] using a full factorial DOE.

Tobiska and Kleinebudde [70] examined the influence of the tablet size, the batch size, the pan speed, and the axis inclination on coating uniformity and efficiency by performing experimentations following a $3^{(3-1)}$ factorial DOE, including zero-level repetition. This knowledge helps to systematically set up new processes in accordance with the basic concepts of QbD from the very beginning. For example, Prpich *et al.* [6] applied a QbD approach to determine the design space for scale-up of a coating process.

The majority of the efforts to study the details of tablet coating were based on experimental and empirical investigations. In recent years, computational simulations became increasingly accepted. A commonly used technique is the DEM, which can track the movement of each single tablet particle inside the coating drum [71, 72] and, based on this, determine quality attributes, such as inter-tablet uniformity or residence time distribution. The downside is that tablet shapes other than spherical present difficulties [73]. While DEM simulations are best suited for capturing the inter-tablet uniformity, less work has been focused on intra-tablet uniformity. Examples include investigations of Suzzi *et al.* [12] and Freireich and Wassgren [74], which presented a Monte Carlo-based algorithm.

In our work, a QbD-based approach was used to investigate the influence of various coating parameters on the film development and quality in a singular coating event. The analysis was carried out by using CFD. The model described the film formation on the tablet surface by solving momentum, heat, and mass transfer equations. The simulation was set up with boundary values taken from measurements in an attempt to model the local environment of the film application and formation as realistically as possible. Subsequently, the influence of selected CPPs on the key quality attributes was investigated.

10.3.2 Background

The typical design of a coating apparatus is schematically shown in Figure 10.6. While details vary depending on the manufacturer, the general principle remains the same. Spray guns that apply the coating liquid are usually mounted on an arm inside the pan and are supplied with coating liquid and pressurized air. The bed-to-nozzle distance for production-scale spray nozzles is in the order of 20 cm. Due to the velocity profile of the tablets on the top, the spray nozzles are typically mounted in the upper third of the tablet bed [75, 76]. The rotation of the coating pan induces radial mixing of the tablets. For axial movement and to generally enhance tablet mixing, baffles of some kind are normally mounted inside the drum. In summary, tablets continuously emerge on the top of the tablet bed and move down, passing the spray zone and reentering the bed at the bottom. In this context two typical times can be defined: the residence time (i.e., the time a tablet spends inside the spray zone) and the circulation time (i.e., the time between two successive reentries into the spray zone).

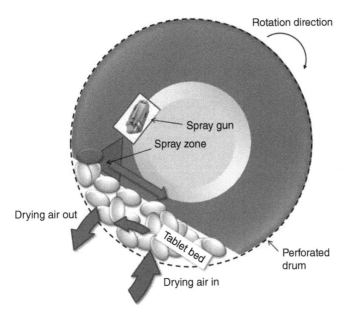

Figure 10.6 Schematic representation of a coating apparatus. A tablet emerges on the top, traverses the spray zone, and reenters the tablet bed at the bottom. In this design, drying air is introduced and extracted through the perforated drum.

During each cycle, a tablet receives a certain amount of coating liquid (solvent, film-forming substances, and additives) in the spray zone. From this point on, the amount of solvent is decreased by evaporation until the next spraying event for the same tablet. Therefore, after an initial phase, the amount of solvent introduced into the system and the amount evaporated and removed with exhaust air reach a dynamic equilibrium. The coater geometry and the process setup determine the amount of spray liquid that can be added to the system without over-wetting the tablet cores in the equilibrium state.

The focus of this work was to investigate the details of this unit operation: the coating process of one dedicated tablet as a representative of the entire tablet ensemble. Conceptually, a single coating event consists of three distinct phases according to Figure 10.7: *spraying* (application of the coating liquid to the tablet), *wetting* (spreading of the liquid on the surface), and *drying* (actual film formation). During the entire process, a single tablet passes the spray zone numerous times, receiving a partial coverage each time. Ideally, the applied film is dried sufficiently while the tablet is still on the surface of the tablet bed to avoid film damage by contact with other surfaces and the pollution of the coating drum by film transfer. In many cases, the transfer of coating solution from one tablet to another can be neglected.

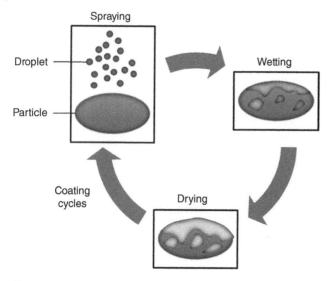

Figure 10.7 Basic steps in an aqueous coating process: application of a partial film by spraying, evolution of the film on the surface, and finally drying of the partial coating. *Source*: Suzzi *et al.* [12]. Reproduced with permission of Elsevier.

While the motion pattern of tablets inside the drum determines the distribution of mass between the tablets (inter-tablet uniformity), the distribution of coating on a single tablet (intra-tablet uniformity) depends on the properties of the spray that need to be investigated.

The quality of the spray immediately affects the quality of the generated film and thus the overall quality of the final product: the droplet size distribution determines the pattern of a coating, which is directly connected to the intra-tablet variability. In the coating system that was considered in this work, a two-component spray nozzle was used; in this type of nozzle, the main operation parameters that determined the size and velocity of the droplets were [76]:

- Atomizing air
- Pattern air
- Spray rate

Although the composition and with it the viscosity of the coating liquid play a role, the influence is less significant than that of the aforementioned parameters. In addition, for a given set of operational parameters, the results depend to some degree on the chosen spray nozzle model [77, 78].

10.3.3 Methods

The data used for our simulations were taken from industrial-scale Bohle BTC 400 (L.B. BOHLE Maschinen + Verfahren GmbH, Ennigerloh, Germany), featuring a perforated drum with an efficient introduction and removal of drying air. The coating was applied by eight spray nozzles of Schlick Model 930 (Düsen-Schlick GmbH, Untersiemau/Coburg, Germany).

The liquid in the simulation consisted of three components: water, glycerol, and talcum. Physical properties of the components were a function of temperature and were calculated based on the available literature data and manufacturer's specifications [79] using linear interpolation. As each component had different evaporation rates, the composition of the droplets was allowed to change over time.

10.3.3.1 Model and Numerical Simulation

CFD simulations were performed using the software package AVL FIRE v2009. For simulation purposes, the whole process was divided into a gas phase (the air around the tablet) and a liquid phase (the spray droplets in the airstream), plus a model for the deposition and evolution of a liquid film on the surface. For the modeling of the multiphase airflow around the tablet, a Reynolds-averaged Navier–Stokes (RANS) approach with a k-epsilon turbulence model was used. For the description of the coating spray and of the coating film that grows on the tablet surface, special simulation models, which are described in the following sections, were employed.

Discrete Droplet Method

Because the number of droplets in the spray was very high, it was neither feasible nor necessary to track each single particle. Instead, a statistical approach was taken, under which each so-called parcel represents a number of identical particles. The spray was assumed to consist of various parcels, with the number of parcels lower than the number of individual droplets but still high enough not to oversimplify the system. The time evolution of the parcels, that is, the change of mass, momentum, and energy with time, was calculated using the Lagrangian discrete droplet method (DDM) approach [80]. For the calculation of the momentum, Newton's second law was directly integrated over time for each spray parcel:

$$m_{id} \frac{du_{id}}{dt} = F_{i,D} + F_{i,G} + F_{i,P} + F_{i,Ext} \tag{10.13}$$

with $F_{i,D}$ being the drag force, $F_{i,G}$ the combined gravity and buoyancy force, $F_{i,P}$ the pressure force, and $F_{i,Ext}$ possible additional external forces. As the size of the droplets in this work was taken directly from measurement at the point of the tablet bed, the effects that alter the droplet size (i.e., secondary atomization, collision, and coalescence) were neglected.

Equation 10.13 describes the movement of a body under the influence of different forces. Solving the equation for each time step gives the trajectories of the parcels. To describe spray coating phenomena that are not directly connected with the droplet movement in the airstream, such as particle/wall interaction and evaporation, additional models were included. For each parcel, equations for mass and energy conservation were solved to obtain the mass and temperature change of the spray. The coating liquid is a mixture of materials (at the very least, solvent and solute were present), and, therefore, the equations included more than one species. Additional models described the interaction between the tablet's surface and the spray. Finally, the formation of the film on the tablets was accounted for via wall-film models, which describe the evolution of the film. For both droplets and wall film, the evaporation was governed by the continuity equation. For the droplets (represented as parcels), the continuity equation can be written as

$$\frac{dm_{id}}{dt} = -\dot{m}_{iE} \tag{10.14}$$

where m_{id} is the mass of parcel i and $\dot{m}_{iE}$ is the corresponding rate of mass change caused by evaporation. As stated earlier, the terms for the breakup and collision/coalescence processes were neglected.

Droplet–Wall Interaction

After traveling from the nozzle outlet to the tablet bed, the droplets arrives at the tablet surface. Depending on parameters such as the droplet size and velocity and the surface conditions, a number of outcomes is possible, ranging from complete deposition on the tablet to partial splashing and re-bouncing of the entire droplet. Various models can determine the outcome for each single parcel based on the parameters that are available for a simulation. The model that was used in this work to describe the droplet–wall interaction was a derivation proposed by Mundo *et al.* [81] In this model, the impact may have two results: splashing (the breaking into smaller droplets) or deposition (the entire droplet mass stays on the surface and forms a film there or adds to the existing film). The outcome is determined based on two commonly used dimensionless numbers: the Reynolds number Re and Ohnesorge number Oh:

$$\mathrm{Re} = \frac{\rho_L v_{d\perp} D_d}{\mu_L} \tag{10.15}$$

$$\mathrm{Oh} = \frac{\mu_L}{\sqrt{\rho_L \sigma_L D_d}} \tag{10.16}$$

Experimental investigations have shown that the Re–Oh parameter space, that is, the range of all possible combinations of Re and Oh numbers, is divided into two regimes. In one regime splashing predominantly occurs. In the other one deposition occurs. The regions are separated by an empirically determined curve given as

$$\mathrm{Oh}_{crit} = 57.7 \mathrm{Re}^{-1.25} \tag{10.17}$$

Therefore, whenever a parcel hits a surface, depending of the local properties of the parcel and the impact location, two outcomes are possible: either the total liquid mass is transferred to the wall film (deposition), or only a part is transferred and from the rest new particles are generated (splashing regime). The droplets generated by the splashing are new parcels themselves, that is, they are treated by the software as equal to the original spray parcels, which implies that they may impact the tablet surface a second time.

Coating Film Treatment

The behavior of the coating solution once it has been deposited on the tablet surface is a central aspect of the present tablet coating analysis. The quality of the final coating layer depends to a great extent on the distribution of the tablet coating after a singular coating event, which in turn depends on

1) The pattern of coating that is applied by the spray
2) The change of this pattern by the tablet film movement
3) The evaporation of solvent from the coating solution

Point 1 is governed by the attributes of the generated spray and was discussed earlier. Points 2 and 3 are captured by the wall-film model of the simulation software as described in the following paragraphs.

The treatment of a liquid flowing on a surface can include complex phenomena and interactions and can therefore be computationally intensive. In order to arrive at a compromise between accuracy and computational effort, model assumptions were made to simplify a problem. For example, it can be assumed that the flow velocity of the liquid film on the tablet surface is low compared with the typical air velocity, partially due to the high viscosity of the coating solution. In addition, drying of the film, that is, the evaporation of the solvent, is known to happen relatively fast in most modern tablet coating processes, which further limits the film flow. Another assumption that is valid for organic and aqueous coatings is "thin film," meaning that the film thickness is much lesser than the characteristic length scale of the simulated domain. Even the typical thickness of the final coating layer of about 100 μm is well below the threshold for this assumption. Based on the "thin film" approximation, the volume of the film can be neglected in the sense that the wall film is described as a two-dimensional layer. During a simulation, the film layer is made of discrete computational cells. The third dimension, that is, the thickness of the film, is still calculated, but for each cell the liquid surface is assumed to be parallel to the wall surface. Therefore, the film thickness is uniform in each cell and can be stored as a single value per cell.

Under these assumptions, the predominant influences on the film are interfacial shear stresses and wall friction. The inertial forces and tangential shear are negligible and therefore omitted in the force balance [82]. External forces are limited to the forces originating from the surrounding air, plus the gravity force. Effectively, this means that the liquid layer is assumed to be in a steady state at each time step. A change in surrounding conditions is in this sense mirrored "instantaneously" by the velocity profile.

The main effect in connection with film formation is the evaporation of the various components from the liquid film at various rates, changing the liquid composition. The corresponding numerical treatment is therefore as detailed as possible, without major simplifications. Similarly, the enthalpy equation is solved as well to calculate the changing temperature of the liquid. For more details, see Ref. [83].

10.3.3.2 Simulation Design and Characterization

The simulations were set up to capture the environment of the real coating process as closely as possible. Boundary conditions for the simulation were taken from measurements of an industrial aqueous coating process.

The simulation region was a rectangular box made of tetrahedral and hexahedral elements (Figure 10.8). A biconvex tablet was stationary in the center of the box. The boundaries of the box were treated as open by using a "constant pressure" boundary condition and setting the pressure to that encountered inside the coater. The size of the box was chosen such that the influence of the boundary on the simulation region could be considered negligible. The coating spray was initialized a short distance from the tablet. In the boundary above the spray generation region, an inlet region was defined, in which an airstream entered with a velocity equal to the mean droplet velocity to account for the airflow generated by the atomizing air and by the droplets. It was assumed that

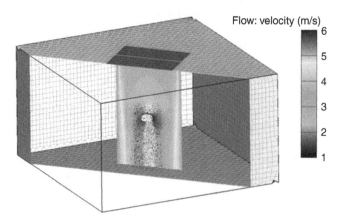

Figure 10.8 Graphical representation of the simulation setup. The rectangular box marks the simulation region; part of the front walls is set transparent. The dark gray region in the middle of the top wall denotes the air inlet. In the center, the tablet is white and the spray droplets are black. The velocity distribution of the airstream is shown in the rectangular cut surface surrounding the tablet.

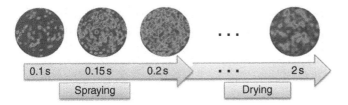

Figure 10.9 Evolution of the coating film on the top surface of the tablet. The coverage increases during the spray phase and decreases again during drying until the shown final state.

due to air friction, the droplets and the surrounding air matched each other's velocity. In any case, the drying air had little influence during the short time it took the droplets to travel from generation to tablet surface. The properties of the spray at initialization, that is, droplet diameter and velocity distribution, were taken from phase Doppler anemometry (PDA) measurements for the initialization position (details are given in Section 10.3.4). In other words, as far as the tablet was concerned, the immediate environment was equivalent to that inside the spray zone of the coating apparatus.

A simulated coating event consisted of three phases of different duration (Figure 10.9):

- 0–0.1 s: Initialization of the airflow without spray addition to arrive at a stable converged state.
- 0.1–0.2 s: Spray phase—the spray is generated, with properties according to the measurements.
- 0.2–2 s: Drying in the airflow without further generation of spray droplets; existing droplets either deposit or leave the simulation region.

In this context, an important quantity was the amount of film liquid on the tablet. Due to the geometry of the calculation mesh, the tablet surface was made of triangular cells, with each cell having an assigned thickness value of coating film. Therefore, an area-averaged mean film thickness was used:

$$\bar{d} = \frac{1}{A_{\text{tot}}} \sum_i A_i d_i \tag{10.18}$$

Equally important to the *amount* of liquid but often more difficult to control is the *distribution* of liquid on the tablet, and one of the most important intermediate CQAs is the uniformity of the film on the tablet. The film quality at the end of the coating unit operation impacts important CQAs of the final product, for example, dissolution rate and appearance. Therefore, coating uniformity can be regarded as a surrogate of these CQAs and as the main quality output of the coating unit operation. There is a range of characteristics for the qualification of coating uniformity, and a common choice is to use a standard deviation. For our work, a standard deviation of the coating thickness in each cell weighted by the cell area was used:

$$\sigma_{\text{Intra}} = \sqrt{\frac{\sum_i A_i \left(d_i - \bar{d} \right)^2}{A_{\text{tot}}}} \qquad (10.19)$$

To allow a more meaningful and comparable characterization of intra-tablet uniformity, the RSD was calculated:

$$\text{RSD}_{\text{Intra}} = \frac{\sigma}{\bar{d}} \qquad (10.20)$$

Notably, the standard deviation itself increased monotonically during a coating process, while RSD decreased. RSD could therefore be regarded as a quantitative surrogate for the quality output of the coating process.

10.3.3.3 Potentially Critical Input Parameters

It is known that several parameters influence the quality of the product [61, 69]. To get an overview of the vast parameter space, the main influences can be divided into the following classes that could be further introduced into a risk assessment exercise:

1) Coating equipment
2) Process conditions
3) State of the tablet core
4) Shape of the tablet
5) Coating liquid

A common scenario is that a tablet coating process is already in place and working, but optimal performance and quality output have not yet been reached. This means that not all classes of parameters can be adjusted. In this case, changes are often limited to process conditions (class 2). This is considered in the present work.

In addition, a distinction must be made between independent and dependent variables, that is, process parameters that can in principle be set to a certain value and variables whose values are determined directly or indirectly by other chosen parameter values. Examples of dependent variables are the outlet air temperature, the outlet air humidity, and the tablet core temperature. For independent variables, a range of parameters is theoretically available. A closer investigation showed that the following parameter subset had a high impact on the process quality and could be controlled [61]:

- Spray rate
- Inlet drying air volume
- Inlet air temperature

Table 10.2 Design levels for the potentially critical process parameters.

Level	−1	0	1
Atomizing air pressure p_A (bar)	1.2	2	3
Drying air temperature T_{In} (K)	333	343	353

- Spray atomizing pressure
- Drum rotation speed

In the special case of a coating apparatus with a perforated drum, the most of the drying air is not in direct contact with the spray. Therefore, the inlet airflow rate is of limited importance, while the inlet air temperature governs the equilibrium temperature inside the coater. In the actual coating process, the drum rotation speed is to a great extent determined by the coating equipment, and the spray rate is governed by the target process time.

Based on the evaluation of the prior knowledge mentioned earlier, we identified spray atomizing pressure p_A and inlet air temperature T_{In} as potential CPPs. A three-level full factorial DOE of the two chosen input parameters was set up (Table 10.2).

Changes in the inlet air temperature were introduced into the simulation directly by setting the air temperature at the inlet as described earlier, while the atomizing air pressure influenced the process indirectly by changing the sizes of the spray droplets. To determine the droplet size distribution that was connected with a certain atomizing air pressure, PDA measurements [84] were performed. As a coating material, an aqueous suspension of Eudragit L30D-55 (Evonik Industries AG, Essen, Germany) was prepared according to the manufacturer's description. The PDA measurements were performed using the same spray nozzle and spray rate as in the industrial process. The results for the three spray pressure levels are shown in Figure 10.10. As expected, higher pressure shifted the distribution to lower droplet diameters.

10.3.4 Results and Discussion

As mentioned, a full factorial investigation was performed, that is, a complete simulation run was performed for each of the nine possible combinations of atomizing air pressure level and air temperature level. For experimental investigations, it is good "DOE practice" to repeat measurements. For a computational simulation, as we use deterministic models, this would lead to the exact same results and was therefore omitted.

To characterize the performance of a set of parameters, the RSD as defined in Equation 10.18 and the increase in average thickness, Equation 10.20, were investigated. Both were identified as intermediate CQAs that are directly

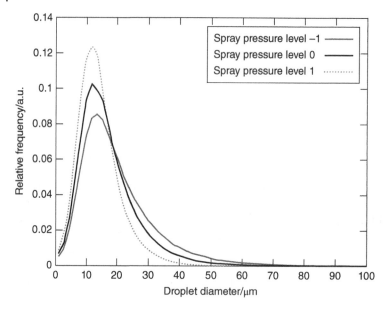

Figure 10.10 Number-weighted diameter distribution of spray droplets generated with three atomizing air pressure settings.

related to final product qualities (e.g., dissolution rate and appearance), as well as to global process characteristics (e.g., total process time).

10.3.4.1 Time Development of Mean Thickness and RSD

Figures 10.11 and 10.12 show the time evolution of the mean film thickness for constant atomizing air pressure and temperature, respectively. Beginning at the onset of spraying at 0.1 s, the thickness increased during the spray phase and decreased again during the drying phase beginning at 0.2 s. In Figure 10.11, the spray level is set to zero and the results for the various temperature levels (−1, 0, 1) are plotted. During the spray phase, for all cases, the coating was applied at the same rate, leading to roughly the same mean film thickness at 0.2 s, with only the lowest temperature showing a slightly increased value. During drying, the influence of the temperature was more pronounced and the mean thickness decreased at distinctly different rates. Consequently, for temperature levels 0 and 1, the film reached a constant thickness value after 1.6 s, while level −1 still showed some decrease. Similarly, Figure 10.12 shows the mean film thickness for various spray levels (−1, 0, 1), with the temperature level set to zero. In contrast to the temperature that had no distinct influence on the amount of coating introduced during the spray phase, the droplet size distribution had a clear effect, as can be seen by comparing the mean values at 0.2 s. Note that the total amount of spray liquid was the same in all cases but was distributed to droplets of

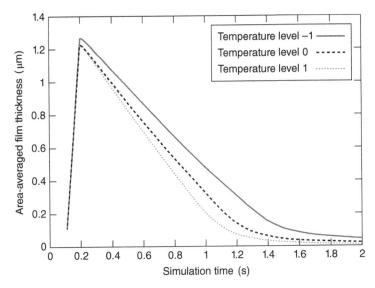

Figure 10.11 Time evolution of mean film thickness over the coating event for constant spray pressure level 0.

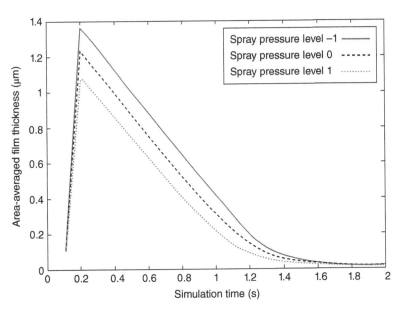

Figure 10.12 Time evolution of mean film thickness over the coating event for constant temperature level 0.

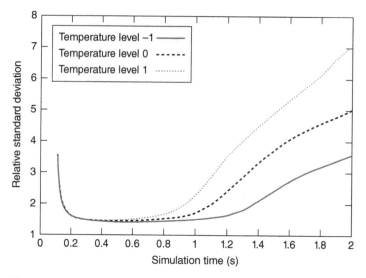

Figure 10.13 Time evolution of RSD over the coating event for constant spray pressure level 0.

different sizes. In the drying phase, the thickness decreased linearly at first and at the same rate for all spray levels. After about 1.6 s, the three curves all had zero slopes and reached a similar level.

The corresponding plots for RSD of the film thickness over time are shown in Figures 10.13 and 10.14. Similarly to the mean values, the difference in temperature had little effect during spraying but led to significant differences of the states after drying. As the plots show, the lower the temperature, the lower is the RSD and therefore the higher the uniformity of the coating after the spray event.

10.3.4.2 Knowledge Space

So far, the results of film thickness and uniformity were shown as a function of time, where one parameter was varied and the others were maintained constant. In this section, to illustrate the full influence and interdependence of the process parameters, contour plots were generated at a distinct time, either right after the spray phase or at the end of the drying phase. As described earlier, a full factorial investigation was performed, meaning that a data point was analyzed for each combination of the CPPs' spray pressure and air temperature. The corresponding intermediate CQA values were film thickness and RSD. The goal was now to establish a design space, that is, a region in the parameter space that would provide optimal results. In this case, the parameter space was two-dimensional.

In Figure 10.15, the mean coating mass applied by the spray is depicted. It can be seen that the drying air temperature had little influence, with slightly

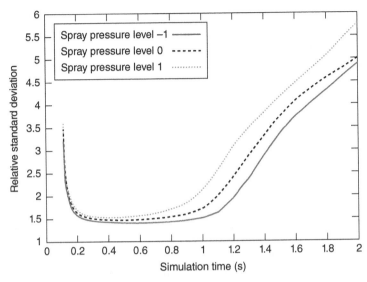

Figure 10.14 Time evolution of RSD over the coating event for constant temperature level 0.

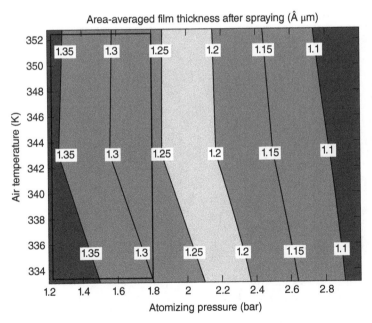

Figure 10.15 Contour plot of the mean film thickness, averaged over the entire tablet after the spray phase, as a function of atomizing air pressure and air temperature. The rectangle shows a possible choice of the design space.

increased film masses at low temperature. The influence of atomizing air pressure, on the other hand, can be clearly seen with lower pressures leading to higher deposited film mass, since for a lower pressure the size distribution of the generated droplets was shifted to larger values. Near the tablet surface, the airstream is parallel to the surface. Thus, larger droplets follow their paths and hit the tablet surface, while smaller particles leave with the exhaust air or hit nearby tablets.

In sum, considering the thickness alone, a design space could be defined as shown in Figure 10.15. There was no restriction on the air temperature and high atomizing air pressures were excluded. If the only goal was to maximize the thickness, one would restrict the design space to only include the lowest pressure. On the other hand, choosing a wide interval for the pressure equals a more flexible process that may be adapted in the future. A good compromise would be to include an interval of 1.2–1.8 bar in the design space, as the gain in flexibility still outweighs the penalty of the reduced film thickness at 1.8 bar.

For many applications, the uniformity of the finished coating is more critical than variations in the film mass. Figure 10.16 shows the RSD directly after the spray phase. The lowest and, therefore, best results were achieved for low

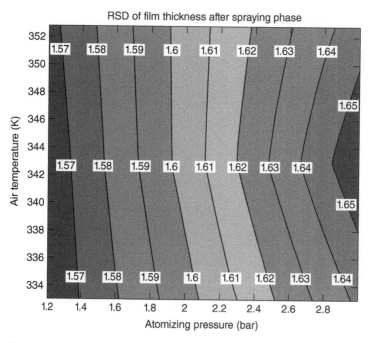

Figure 10.16 RSD of the film thickness over the whole tablet after the spraying phase as a function of atomizing air pressure and air temperature.

atomizing pressure. The influence of temperature was limited and more pronounced for higher pressure. After drying (Figure 10.17), the general trend for the atomizing air pressure remained unchanged, with low values providing a good uniformity. However, compared with Figure 10.16, air temperature is becoming an important factor as well. Higher temperature values lead to a decrease in the inter-tablet uniformity (an increase in RSD) because, due to faster drying, the wall film generated by an impinging droplet had less time to move about the tablet and could not unite with the film sections generated by neighboring droplets.

Overall, low atomizing air pressure and low temperature were desirable. However, the decrease in temperature was limited by the tablets' sensitivity to humidity. Therefore, there was a lower limit that allowed enough drying to reach an acceptable dynamic equilibrium for the humidity, that is, the cores were not over-wetted. In Figure 10.17, the design space that was derived earlier from film mass investigations is indicated by a dashed line. Based on the new knowledge of the film RSD, this design space was further refined: combinations of high temperature and pressure that result in high nonuniformity were excluded, and a limit for extremely low temperatures was introduced to reduce the risk of over-wetting the product.

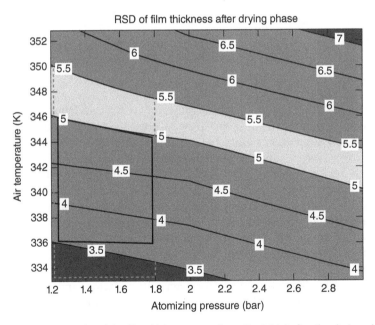

Figure 10.17 RSD of the film thickness over the entire tablet after the drying phase as a function of atomizing air pressure and air temperature. The dashed region is the design space in the case when only coating thickness is considered, and the solid region is a possible final design space with RSD optimized and over-wetting considered.

10.3.5 Summary

This case study applies QbD principles to develop a design space for a coating process via CFD. The simulation was set up in an effort to realistically mirror the local environment of a tablet during coating. To this end, boundary values of the process for the simulation were taken from an industrial coating process.

A systematic investigation of the parameters that influence the outcome identified two CPPs of particular importance: atomizing air pressure and inlet air temperature. The mean film thickness and RSD of film thickness were identified as intermediate CQAs. Based on this risk assessment, a full factorial three-level experimental design was set up. For each defined point in a parameter space, a complete simulation run for a singular coating event was performed.

The results were presented as plots of intermediate CQA evolution over time and as contour plots of the intermediate CQA at defined times. The analysis of the results shows how a design space could be developed in two steps. The resulting example of a design space is represented by a solid line in Figure 10.17 [1]. As long as the process parameters are within the designated region, the film mass increase was maximized within the given limits of the designated parameter space, thus minimizing the total coating time. Simultaneously, the inter-tablet coating uniformity was maximized and the risk of over-wetting was controlled.

10.4 Overall Conclusions

Modeling and simulation may contribute significantly to the development of a fundamental process understanding that may be extended to new and more complex processes. Simulation approaches are mainly used as design tools for equipment performance testing in combination with various pharmaceutical material attributes and process parameters. However, the proposed novel approach showed that numerical simulations can address knowledge gaps and generate basic process understanding to guide "real" experimental process characterization studies, for example, by defining ranges for variables under investigation. In this context, computer simulations may be used during pre-characterization risk assessment activities, for example, as an integrated FMEA screening tool.

We investigated in silico correlations between critical factors and quality/performance outputs that are well known but not completely understood. Our simulation approach proved to be capable of reproducing these correlations, and we validated our simulations (data not shown) by comparing them to the data provided in the literature. However, this is only the first step toward the simulation of more complex systems.

The use of combined computer simulation and process characterization approaches will evolve in the future, as computational power, and thus the ability to simulate even more complex systems, will increase significantly. In the future, computer simulation strategies will be applied to generate GMP-relevant data that can be used along with data from a "real" experimental process characterization and validation studies to be included in the regulatory dossiers. However, efficient validation strategies, which are an indispensable prerequisite in a GMP environment to prove the predictive capability and robustness of a simulation, must be available. Moreover, the need for qualified simulation software as well as qualified simulation personnel must be adequately addressed. This will be the topic of future research.

References

1 Adam S, Suzzi D, Radeke C, Khinast JG. An integrated Quality by Design (QbD) approach towards design space definition of a blending unit operation by Discrete Element Method (DEM) simulation. Eur. J. Pharm. Sci. 2011;42:106–115.

2 Am Ende DJ, Seymour CB, Watson TJN. A science and risk based proposal for understanding scale and equipment dependencies of small molecule drug substance manufacturing processes. J. Pharm. Innov. 2010;5:72–78.

3 Hilden J, Schrad M, Sloan JT, Kuehne-Willmore J, Kramer T. Model-based control of drug substance particle size to ensure drug product uniformity. Paper 108b, AIChE 2010 Annual Meeting, Salt Lake City, Utah, November 7–12, 2010.

4 Hou DS, MacPhail N, Smith EJ, Ho J, Cleary A. Implementation of a normalization risk assessment tool to link design space inputs. Paper 424b, AIChE 2010 Annual Meeting, Salt Lake City, Utah, November 7–12, 2010.

5 Michaels JN, Bonsignore H, Hudson-Curtis BL, Laurenz S, Lin H, Mathai T, Pande G, Sheth A, Sprockel O. Can a design space be built on material attributes alone? Paper 621f, AIChE 2010 Annual Meeting, Salt Lake City, Utah, November 7–12, 2010.

6 Prpich A, Doshi P, Topper L. A quality by design approach to scale-up of the melt-spray-congeal multiparticulate process. Paper 21b, AIChE 2010 Annual Meeting, Salt Lake City, Utah, November 7–12, 2010.

7 Pordal HS, Matice CJ, Fry TJ. The role of computational fluid dynamics in the pharmaceutical industry. Pharm. Technol. 2002;26:72–79.

8 Muzzio FJ, Ierapetritou M, Portillo P, Llusa M, Levin M, Morris KM, Soh JLP, McCann RJ, Alexander A. A forward-looking approach to process scale-up for solid dose manufacturing. In: Augsburger LL, Hoag SW (eds.). Pharmaceutical Dosage Forms: Tablets—Volume 3: Manufacture and Process Control. New York: Informa Health Care USA, 2008:119–152.

9 International Conference on Harmonisation of Technical Requirements for Registration of Pharmaceuticals for Human Use. Pharmaceutical Development Q8(R2). 2009.

10 Schousboe M, Hirsh I. Change management associated with QbD implementation. In: Schmitt S (ed). Quality by Design: Putting Theory into Practice. Bethesda: PDA, DHI, 2011:89–114.

11 Yu LX. Pharmaceutical quality by design: product and process development, understanding, and control. Pharm. Res. 2007;25:781–791.

12 Suzzi D, Radl S, Khinast JG. Local analysis of the tablet coating process: impact of operation conditions on film quality. Chem. Eng. Sci. 2010;65:5699–5715.

13 Persson AS, Alderborn G, Frenning G. Flowability of surface modified pharmaceutical granules: a comparative experimental and numerical study. Eur. J. Pharm. Sci. 2011;42:199–209.

14 Chivilikhin MS, Soboleva V, Kuandykov L, Woehl P, Lavric ED. CFD analysis of hydrodynamic and thermal behaviour of Advanced-Flow™ reactors. Chem. Eng. Trans. 2010;21:1099–1104.

15 ISPE. GAMP 5—A Risk Based Approach to Compliant GxP Computerized Systems. Tampa: ISPE, 2008.

16 Wu H, Tawakkul M, White M, Khan MA. Quality-by-Design (QbD): an integrated multivariate approach for the component quantification in powder blends. Int. J. Pharm. 2009;372:39–48.

17 Garcia TP, Prescott JK. Blending and blend uniformity. In: Augsburger LL, Hoag SW (eds.). Pharmaceutical Dosage Forms: Tablets, Third Edition—Volume 1: Unit Operations and Mechanical Properties. New York: Informa Healthcare USA, 2008:111–174.

18 Wu H, Khan MA. Quality-by-Design (QbD): an integrated approach for evaluation of powder blending process kinetics and determination of powder blending end-point. J. Pharm. Sci. 2009;98:2784–2798.

19 Wu H, Heilweil E, Hussain AS, Khan M. Process Analytical Technology (PAT): quantification approaches in terahertz spectroscopy for pharmaceutical application. J. Pharm. Sci. 2007;97:970–984.

20 Lacey PMC. Developments in the theory of particle mixing. J. Appl. Chem. 1954;4:257–268.

21 Bridgwater J. Fundamental powder mixing mechanisms. Powder Technol. 1976;15:215–236.

22 Stewart RL, Bridgwater J, Zhou YC, Yu AB. Simulated and measured flow of granules in a bladed mixer—a detailed comparison. Chem. Eng. Sci. 2001;56: 5457–5471.

23 Remy B, Khinast JG, Glaser BJ. Discrete element simulation of free flowing grains in a four-bladed mixer. AIChE J. 2009;55:2035–2048.

24 Remy B, Glasser BJ, Khinast JG. The effect of mixer properties and fill level on granular flow in a bladed mixer. AIChE J. 2010;56:336–353.

25 Radl S, Kolvoda E, Glasser BJ, Khinast JG. Mixing characteristics of wet granular matter in a bladed mixer. Powder Technol. 2010;200:171–189.

26 Brone D, Muzzio FJ. Enhanced mixing in double-cone blenders. Powder Technol. 2000;110:179–189.

27 Zhou YC, Yu AB, Bridgwater J. Segregation of binary mixture of particles in a bladed mixer. J. Chem. Technol. Biotechnol. 2003;78:187–193.

28 Moakher M, Shinbrot T, Muzzio FJ. Experimentally validated computations of flow, mixing and segregation of non-cohesive grains in 3D tumbling blenders. Powder Technol. 2000;109:58–71.

29 Edge S, Kibbe A, Kussendrager K. Lactose, monohydrate. In: Rowe RC, Sheskey PJ, Siân CO (eds.). Handbook of Pharmaceutical Excipients, Fifth Edition. London/Chicago: APhA, Pharmaceutical Press, 2005:389–395.

30 Seely J. Process characterization. In: Rathore AS, Sofer G (eds.). Process Validation in Manufacturing of Biopharmaceuticals. Boca Raton: Taylor & Francis, 2005: 13–30.

31 Rathore AS, Brenning RCD, Cecchini D. Design space for biotech products. Biopharm. Int. 2007;20:36–40.

32 Garcia T, Cook G, Nosal R. PQLI key topics—criticality, design space and control strategy. J. Pharm. Innov. 2008;3:60–68.

33 Harms J, Wang X, Kim T, Yang X, Rathore AS. Defining process design space for biotech products: case study of *Pichia pastoris* fermentation. Biotechnol. Prog. 2008;24:655–662.

34 Lepore J, Spavins J. PQLI design space. J. Pharm. Innov. 2008;3:79–87.

35 Nosal R, Schultz T. PQLI definition of criticality. J. Pharm. Innov. 2008;3:69–78.

36 Rathore AS. Roadmap for implementation of quality by design (QbD) for biotechnology products. Trends Biotechnol. 2009;27:546–553.

37 Van Hoek P, Harms J, Wang X, Rathore AS. Case study on definition of process design space for a microbial fermentation step. In: Rathore AS, Mhatre R (eds.). Quality by Design for Biopharmaceuticals—Principles and Case Studies. Hoboken: John Wiley & Sons, Inc., 2009:85–109.

38 Lemieux M, Léonard G, Doucet J, Leclaire LA, Viens F, Chaouki J, Bertrand F. Large-scale numerical investigation of solids mixing in a V-blender using the discrete element method. Powder Technol. 2008;181:205–216.

39 Muzzio FJ, Goodridge CL, Alexander A, Arratia P, Yang H, Sudah O, Mergen G. Sampling and characterization of pharmaceutical powders and granular blends. Int. Pharm. 2003;250:51–64.

40 Venables HJ, Wells JI. Powder sampling. Drug Devel. Ind. Pharm. 2002;28:107–117.

41 Muzzio FJ, Robinson P, Wightman C, Brone D. Sampling practices in powder blending. Int. J. Pharm. 1997;155:153–178.

42 Berman J, Planchard JA. Blend uniformity and unit dose sampling. Drug Devel. Ind. Pharm. 1995;21:1257–1283.

43 Rathore AS, Winkle H. Quality by design for biopharmaceuticals. Nat. Biotechnol. 2009;27:26–34.

44 Zhu HP, Zhou ZY, Yang RY, Yu AB. Discrete particle simulation of particulate systems: a review of major applications and findings. Chem. Eng. Sci. 2008;63: 5728–5770.

45 Dziugys A, Peters B. An approach to simulate the motion of spherical and non-spherical fuel particles in combustion chambers. Gran. Mater. 2001;3:213–265.

46 Cundall PA, Strack ODL. Discrete numerical-model for granular assemblies. Geotechnology 1997;29:47–65.

47 Di Renzo A, Di Maio FP. Comparison of contact-force models for the simulation of collisions in DEM-based granular flow codes. Chem. Eng. Sci. 2004;59:525–541.

48 DEM Solutions Ltd. EDEM 2.0 User Guide. Edinburgh: DEM Solutions Ltd., 2009.

49 Radeke CA, Glasser BJ, Khinast JG. Large-scale powder mixer simulations using massively parallel GPU architectures. Chem. Eng. Sci. 2010;65:6435–6442.

50 Koller DM, Posch A, Hörl G, Voura C, Radl S, Urbanetz N, Fraser SD, Tritthart W, Reiter F, Schlingmann M, Khinast JG. Continuous quantitative monitoring of powder mixing dynamics by near-infrared spectroscopy. Powder Technol. 2011;205:87–96.

51 Iwashita K, Oda M. Rolling resistance at contacts in simulation of shear bond developed by DEM. J. Eng. Mech. 1998;124(3):285–292.

52 Brone D, Alexander A, Muzzio FJ. Quantitative characterization of mixing of dry powders in V-blenders. AIChE J. 1998;44:271–278.

53 FDA. Guidance for Industry. Powder Blends and Finished Dosage Units—Stratified In-Process Dosage Unit Sampling and Assessment. Draft Guidance. Pharmaceutical cGMPs, 2003.

54 Muzzio FJ, Alexander A, Goodridge C, Shen E, Shinbrot T. Solids Mixing. In: Paul L, Atiemo-Obeng VA, Kresta SM (eds.). Handbook of Industrial Mixing. Hoboken: Wiley Interscience, 2004.

55 Poux M, Fayolle P, Bertrand J, Bridoux D, Bousquet J. Powder mixing: some practical rules applied to agitated systems. Powder Technol. 1991;68:213–234.

56 Rhodes M. Introduction to Particle Technology, 2nd Edition. Chichester: John Wiley & Sons, Ltd, 2008.

57 Asmar BN, Langston PA, Matchett AJ. A generalized mixing index in distinct element method simulation of vibrated particulate beds. Gran. Matter 2002;4:129–138.

58 Williams JC. The mixing of dry powders. Powder Technol. 1968;2:13–20.

59 Williams JC. The properties of non-random mixtures of solid particles. Powder Technol. 1969;3:189–194.

60 Rhodes MJ, Wang XS, Nguyen M, Stewart P, Liffman K. Study of mixing in gas-fluidized beds using a DEM model. Chem. Eng. Sci. 2001;56:2859–2866.

61 Cole G, Hogan J, Aulton M. Pharmaceutical Coating Technology. London: Informa Healthcare, 1995.

62 Muliadi AR, Sojka PE. A review of pharmaceutical tablet spray coating. Atomization and Sprays 2010;20(7):611–638.

63 Ho L, Muller R, Romer M, Gordon KC, Heinamaki J, Kleinebudde P, Pepper M, Rades T, Shen YC, Strachan CJ, Taday PF, Zeitler JA. Analysis of sustained-release tablet film coats using terahertz pulsed imaging. J. Controlled Release 2007;119:253–261.

64 Mauritz J, Morrisby R, Hutton R, Legge C, Kaminski C. Imaging pharmaceutical tablets with optical coherence tomography. J. Pharm. Sci. 2010;99:385–391.

65 Koller DM, Hannesschläger G, Leitner M, Khinast JG. Non-destructive analysis of tablet coatings with optical coherence tomography. Eur. J. Pharm. Sci. 2011;44(1–2):142–148.

66 Ruotsalainen M, Heinämäki J, Guo H, Laitinen N, Yliruusi J. A novel technique for imaging film coating defects in the film-core interface and surface of coated tablets. Eur. J. Pharm. Biopharm. 2003;56:381–388.

67 Zhong S, Shen YC, Ho L, May RK, Zeitler JA, Evans M, Taday PF, Pepper M, Rades T, Gordon KC, Müller R, Kleinebudde, P. Non-destructive quantification of pharmaceutical tablet coatings using terahertz pulsed imaging and optical coherence tomography. Opt. Lasers Eng. 2011;49:361–365.

68 Cahyadi C, Karande A, Chan L, Heng P. Comparative study of non-destructive methods to quantify thickness of tablet coatings. Int. J. Pharm. 2010;398:39–49.

69 Patel J, Shah A, Sheth N. Aqueous-based film coating of tablets: study the effect of critical process parameters. Int. J. Pharm. Technol. Res. 2009;1:235–240.

70 Tobiska S, Kleinebudde P. Coating uniformity and coating efficiency in a Bohle Lab-Coater using oval tablets. Eur. J. Pharm. Biopharm. 2003;56:3–9.

71 Yamane K, Sato T, Tanaka T, Tsuji Y. Computer-simulation of tablet motion in coating drum. Pharm. Res. 1995;12:1264–1268.

72 Ketterhagen WR, am Ende MT, Hancock BC. Process modeling in the pharmaceutical industry using the discrete element method. J. Pharm. Sci. 2009;98:442–470.

73 Song Y, Turton R, Kayihan F. Contact detection algorithms for DEM simulations of tablet-shaped particles. Powder Technol. 2006;161:32–40.

74 Freireich B, Wassgren C. Intra-particle coating variability: analysis and Monte-Carlo simulations. Chem. Eng. Sci. 2010;65:1117–1124.

75 Alexander A, Shinbrot T, Muzzio FJ. Scaling surface velocities in rotating cylinders as a function of vessel radius, rotation rate, and particle size. Powder Technol. 2002;126:174–190.

76 Müller R, Kleinebudde P. Comparison of a laboratory and a production coating spray gun with respect to scale-up. AAPS Pharm. Sci. Technol. 2007;8:425–433.

77 Aulton ME, Twitchell AM, Hogan JE. Proceedings of the Fourth International Conference on Pharmaceutical Technology 1986. APGI, Paris, France, V, pp. 133–140.

78 Aliseda A, Hopfinger E, Lasheras J, Kremer D, Berchielli A, Connolly E. Atomization of viscous and non-Newtonian liquids by a coaxial, high-speed gas jet. Experiments and droplet size modeling. Int. J. Multiphase Flow. 2008;34:161–175.

79 The Dow Chemical Company. Handbook of Pharmaceutical Excipients, 7th Edition. Rowe RC (editor), London: Pharmaceutical Press, 2012.

80 Dukowicz JK. A particle fluid numerical model for liquid sprays. J. Comp. Phys. 1980;35:229–253.

81 Mundo C, Sommerfeld M, Tropea C. Droplet-wall collisions: experimental studies of the deformation and breakup process. Int. J. Multiphase Flow 1995;21:151–173.

82 Cebeci T, Bradshaw P. Momentum Transfer in Boundary Layers. Washington/London: Hemi-sphere Publishing Corporation/McGraw-Hill Book Company, 1977.

83 AVL GmbH. FIRE v2008 User Manual, Graz, 2008.

84 Tropea C, Xu TH, Onofri F, Géhan G, Haugen P, Stieglmeier M. Dual-mode phase-Doppler anemometer. Part. Part. Syst. Charact. 1996;13:165–170.

11

Design Space Definition: A Case Study—Small Molecule Lyophilized Parenteral

Linas Mockus[1], David LeBlond[2], Gintaras V. Reklaitis[1], Prabir K. Basu[3], Tim Paul[4], Nathan Pease[4], Steven L. Nail[4], and Mansoor A. Khan[5,6]

[1] *Davidson School of Chemical Engineering, Purdue University, West Lafayette, IN, USA*
[2] *Wadsworth, IL, USA*
[3] *Mt Prospect, IL, USA*
[4] *Pharmaceutical Development, Baxter Pharmaceutical Solutions, LLC, Bloomington, IN, USA*
[5] *Rangel College of Pharmacy, Texas A&M University Health Science Center, College Station, TX, USA*
[6] *Division of Product Quality Research (DPQR, HFD-940), OTR, Office of Pharmaceutical Quality, Center for Drug Evaluation and Research, Food and Drug Administration, Silver Spring, MD, USA*

11.1 Introduction

As stipulated by ICH Q8 R2 [1], the prediction of critical process parameters based on process modeling is a part of an enhanced Quality by Design approach to product development. It defines design space as the "multidimensional combination and interaction of input variables (e.g., material variables) and process parameters that have been demonstrated to provide assurance of quality."

The process to establish critical quality variables to be used in the construction of design space is quite well established on the basis of recent discussions at industry conferences [2]. However, the procedure for constructing a design space from data composed of such variables is still in its infancy.

The response surface determined using traditional statistical tools often represents a "mean" response surface and does not quantify the assurance (probability) that product critical quality attributes (CQAs) will be met. Moreover, traditional design space boundaries based on single-point estimates provide insufficient information to a formulation scientist because they are not probabilistic. That is, they do not consider the uncertainties present in data or estimated parameters, do not take into account correlations among multiple responses, and convey no information about the risk of exceeding the design space boundary. Bayesian treatments, on the other hand, can readily incorporate all these considerations.

Comprehensive Quality by Design for Pharmaceutical Product Development and Manufacture,
First Edition. Edited by Gintaras V. Reklaitis, Christine Seymour, and Salvador García-Munoz.
© 2017 American Institute of Chemical Engineers, Inc. Published 2017 by John Wiley & Sons, Inc.

In addition, the Bayesian paradigm provides an effective algorithm for incorporating prior knowledge from theory or experience and naturally leads to a structured process of knowledge building. Bayesian approaches thus fully support the spirit of FDA's goal of reducing "uncertainty about product performance throughout the product life cycle through scientific research" [3].

Bayesian approaches using linear models have been applied successfully to solid oral dosage forms [4–7]. The present work extends the Bayesian treatment to a lyophilized parenteral drug and additionally illustrates the use of a nonlinear, accelerated degradation (Arrhenius) model. The application of Bayesian methodology to predict primary drying duration for a lyophilization cycle may be found in Mockus *et al.* [8].

11.2 Case Study: Bayesian Treatment of Design Space for a Lyophilized Small Molecule Parenteral

A lyophilized small molecule parenteral was selected as a case study to illustrate the Bayesian treatment of design space. There are several examples in the literature that use Bayesian approach to treat solid oral dosage forms [7]. Lyophilized (freeze-dried) parenteral was selected to extend the application of the Bayesian approach to a different type of dosage form.

Lyophilization is the most common method for manufacturing parenterals when aqueous solution stability is an issue. It is central to the protection of materials, which are temperature sensitive, require low moisture content (<1%) in order to ensure stability, and require a sterile and gentle preservation process. Sodium ethacrynate, a small molecule, was chosen as the active pharmaceutical ingredient. The commercially available drug product is a strong diuretic used to treat edema. Sodium ethacrynate was formed in aqueous solution (pre-lyo solution) by titration of ethacrynic acid with 1 N sodium hydroxide solution. Immediately after completion of titration in order to prevent degradation in the liquid state, the pre-lyo solution was freeze-dried. After completion of freeze-drying, stoppered vials containing freeze-dried cake were stored under constant conditions of 30 and 40°C to collect stability data. The reader may be referred to Mockus *et al.* [10]. for further details.

11.2.1 Arrhenius Accelerated Stability Model with Covariates for a Pseudo-Zero-Order Degradation Process

For each batch the data consisted of estimated degradation rate (k), Temperature (T), API concentration of pre-lyo solution (API), pH of pre-lyo solution (pH), and buffer concentration of pre-lyo solution (buffer). The data collected were affected by measurement and modeling errors, which were assumed to be independent and lognormally distributed, with a mean of zero and variance σ^2 (on the log scale).

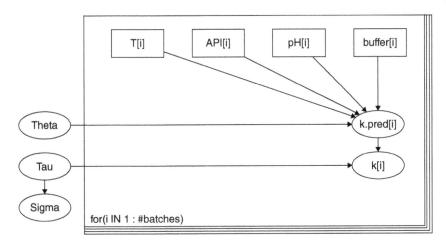

Figure 11.1 Degradation model.

The model is symbolically illustrated in Figure 11.1. It depicts a directed graph of the statistical model describing dependencies between groups of variables: k.pred, degradation rate predicted by the model; k, measured degradation rate; tau, precision of measurement and modeling error; σ, standard deviation of measurement and modeling error, $\tau = 1/\sigma^2$; theta, vector of degradation model parameters; API, API concentration of pre-lyo solution for each batch; pH, pH of pre-lyo solution for each batch; buffer, buffer concentration of pre-lyo solution for each batch; and T, temperature at which stability samples were maintained for a particular batch. Two types of nodes and arrows are featured in the figure: boxes represent variables for which values are constant and known, such as T or API; ovals represent values with unknown stochastic values, such as sigma and theta; double arrows represent direct deterministic functional relationships (e.g., $\sigma = 1/\tau^2$); and single arrows represent direct stochastic functional relationships (e.g., $k[i] \sim N$ (k.pred$[i],\sigma^2$).

The degradation process was assumed to be of zero order as suggested in the literature [9]. We define the rate of change in potency as the negative of rate constant, that is,

$$\text{Rate of change in potency} = -k, \tag{11.1}$$

where k is the pseudo-zero-order rate constant in %/day.

To assess nonlinearity (i.e., nonzero-order kinetics) in the API stability profiles, a quadratic term for storage period (days) was included for 11 randomly selected formulations stored at 30°C. The stability data at 40°C includes only two time points, and the test for nonlinearity is possible only if more than three time points were collected. This term was only statistically significant (p-value < 0.05) in one of the 11 cases. When testing 11 independent hypotheses, the

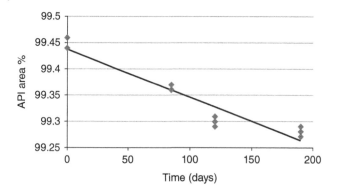

Figure 11.2 Representative potency time profile.

probability of one or more hypotheses showing significance (p-value < 0.05) by random chance is high $(1 - 0.95^{11} = 0.43)$. Therefore this one case of significance could reflect random chance. Furthermore, when the entire data set was fit as a whole to a larger model that included main effects and all estimable interactions plus a quadratic term for storage period, the quadratic term for storage period was statistically not significant (p-value of 0.33). Therefore a zero-order kinetic model was considered appropriate for the present analysis. Figure 11.2 depicts typical potency profile for a representative formulation. Diamonds indicate API, expressed as a percent of the expected HPLC peak area, at various time points during stability study, while a straight black line shows a linear regression line representing zero-order kinetics. See Section 11.A.2 for a discussion of data.

Based on [10], we use the negative of the simple linear regression slope of the stability profile (potency vs. storage time in days) as an estimate of k. According to the Arrhenius kinetic model, k is related to temperature T (in °C) as follows:

$$\ln(k) = -\frac{E_a}{R^*(T + 273.15)} + \ln(A),\tag{11.2}$$

where E_a is the true reaction activation energy in kJ/mol, $R = 0.008315$ kJ/mol/°K, T is temperature in °C, and A is the true Arrhenius "acceleration" constant.

Equation 11.2 represents a simple linear relationship between $\ln(k)$ and $1/(R^*(T+273.15))$ with $\ln(A)$ and $-E_a$ taking on the roles of "intercept" and "slope," respectively. To extend this model to include the effects of the covariates API, pH, and buffer, we might consider a linear response surface model such as is commonly used to approximate the design spaces of manufacturing processes [6]. However, the model terms that can be supported will be limited by the experimental design. The present design was essentially a partially balanced factorial that eliminated quadratic effects and some interactions from consideration. In addition, initial analyses using stepwise regression suggested that interactions between $1/(R^*(T + 273.15))$ and other covariates were absent. The final statistical model used for the response surface definition was essentially an expansion of Equation (11.2) that included six linear coefficients θ_i:

$$\ln(k) = \theta_1 + \theta_2 \cdot T_c + \theta_3 \cdot pH_c + \theta_4 \cdot API_c + \theta_5 \cdot buffer_c + \theta_6 \cdot pH_c \cdot API_c + \varepsilon,$$
(11.3)

in which ε represents constant normal measurement error with a mean of 0 and standard deviation σ. T_c, pH_c, API_c, and buffer$_c$ are transformations of the original temperature and covariables that have been centered and standardized to facilitate interpretation and improve convergence of the Gibbs sampling simulation algorithm of WinBUGS [11]:

$$T_c = \left(\frac{1}{T + 273.15} - \left(\frac{1}{30 + 273.15} + \frac{1}{40 + 273.15} \right) \Big/ 2 \right) \Big/ R$$
$$pH_c = (pH - 7)\big/1$$
(11.4)
$$API_c = (API - 20)\big/10$$
$$buffer_c = (buffer - 30)\big/25$$

Thus $\theta_1 = \ln(A)$ and $\theta_2 = -E_a$ when the covariables are set to their central values. Then θ_3 to θ_6 each represents the predicted additive effect on $\ln(k)$ of changing the covariable value from its centered to its high level. The respective effect on reaction rate (k) itself would be a multiplicative change of $\exp(\theta_i)$ fold.

We let θ represent the vector of coefficients. We used a Bayesian approach to estimate the parameters θ and τ in model [2]. The Bayesian approach requires prespecification of the prior distributions of these model parameters. Conjugate normal priors are often used for linear model coefficients such as θ, and gamma conjugate priors are often used for inverse variances (i.e., precisions) such as τ^{-1} (see Gelman *et al.* [12]). This was the approach taken here. We assumed prior independence of all seven parameters.

Typical values of activation energy for these types of applications are in the range from 60 to 100 kJ/mol, so a moderately informative prior normally distributed with mean of 80 and very large standard deviation was used for θ_1. The same rationale was employed for the prior of θ_2: it was assumed to be normally distributed with mean of -10 and a very large standard deviation.

Since the model for degradation rate was semiempirical, there was little prior information available for θ_3–θ_6. Consequently, a mean of zero was employed for these parameters. In specifying the variances of the prior normal distributions of the elements of θ, a large, non-informative variance (10^6) was used.

Similarly, a very non-informative gamma distribution (with shape and scale parameters of 0.001) was used for τ. This precision reflects the standard deviation of potency measurement, taking into consideration the number of measurements and their spacing over time. Typically, the potency measurement relative standard deviation of an HPLC assay is about 1–2%. Batch-to-batch variance is also included here as well as some content non-uniformity, which could potentially increase the standard deviation. A more detailed examination of this could lead to a more informative prior for the degradation rate data, but it will not be pursued in this study.

From Bayes' theorem, the joint posterior distribution of the parameters to estimate $P(\theta, \sigma^2 \,|\, k, T, \text{API}, \text{pH}, \text{buffer})$ is proportional to the likelihood of the data multiplied by parameter priors

$$P(\theta, \sigma^2 \,|\, k, T, \text{API}, \text{pH}, \text{buffer}) \sim P(k \,|\, \theta, \sigma^2, T, \text{API}, \text{pH}, \text{buffer}) * P(\theta) * P(\sigma^2),$$

where

$P(\theta, \sigma^2 \,|\, k, T, \text{API}, \text{pH}, \text{buffer})$ is the joint posterior distribution

$P(k \,|\, \theta, \sigma^2, T, \text{API}, \text{pH}, \text{buffer})$ is the likelihood of the data

$P(\theta)$ and $P(\sigma^2)$ are the prior distributions (as specified previously) for θ and σ^2, respectively.

The likelihood term was given by the lognormal measurement model

$$\ln(k) \sim N(\ln(\text{pred}.k(\theta, T, \text{API}, \text{pH}, \text{buffer}), \sigma^2)$$

Current standard practice in Bayesian statistics is to characterize a complicated high-dimensional posterior distribution using pseudorandom draws from the conditional distribution of each individual model parameter, given the other parameters [13]. This can be shown to be equivalent, at equilibrium, to random draws from their joint posterior distribution [12]. Based on conditional independence, the conditional distributions of individual model components are simplified to

$$P(\theta \,|\, \sigma^2, k, T, \text{API}, \text{pH}, \text{buffer}) \sim P(k \,|\, \theta, \sigma^2, T, \text{API}, \text{pH}, \text{buffer}) * P(\theta)$$

$$P(\sigma^2 \,|\, \theta, k, T, \text{API}, \text{pH}, \text{buffer}) \sim P(k \,|\, \theta, \sigma^2, T, \text{API}, \text{pH}, \text{buffer}) * P(\sigma^2)$$

Each term on the right-hand side can be numerically evaluated, given the distributional assumptions listed previously. WinBUGS uses a Gibbs sampler to generate an instance (or "draw") from the distribution of each parameter in turn, conditional on the current values of the other parameters. Once convergence is achieved and a posterior sample of parameter values is obtained, further simulation can then be performed to compute posterior distributions of quantities of interest (i.e., shelf life, frequency factor, activation energy, or probability that product specifications are met within a shelf life period).

We estimated shelf life for one specific temperature (30°C) and at the middle level of the covariates. Note that at these covariate levels, the model reduces to a simple Arrhenius interpretation. For a given posterior draw, shelf life is taken as the number of days over which potency is reduced from its starting value by 5%. Under the assumption of a zero-order process,

$$S = \frac{0.05 * C_0}{k} = \frac{5}{k},$$

where S is shelf life and C_0 is initial potency taken as 100% of API.

Shelf life at other temperatures or covariate levels can be obtained by performing the simple calculations for each draw using appropriate utilities in R [14] or JMP.

11.2.2 Design Space Definition

The design space was defined using the reliability limit concept introduced by Peterson [7]. The reliability limit is defined as the edge of the feasible operating region (i.e., design space) for a process. The design space is the set of process parameter settings for which the joint posterior predictive probability that all process responses will be acceptable is at least at the level of χ:

$$\{\theta : \Pr(Y \in Q \mid \theta, \Xi) \geq \chi\},$$

where θ is the vector of uncertain process parameters and incoming material attributes, Y is the vector of responses, Q is the acceptance region (i.e., specification boundaries) associated with the responses, and Ξ is the experimental data. The specification region was defined so that potency of API does not fall more than 5% during a shelf life of 2 years, or, in other words, degradation rate k is not greater than 2.5%/year.

11.3 Results

The WinBUGS Gibbs sampling algorithm converged rapidly to the target posterior distribution. To verify convergence, three chains, each of length 40 000 draws but with over-dispersed starting points, were obtained. History plots indicated good mixing. Figure 11.3 depicts the mixing of parameter θ_1. The mixing is good as indicated by the vast majority of draws being closely packed together.

Draws 10 001–40 000 from all three chains were taken as the posterior estimate, and the marginal distributions of each statistic, including shelf life at 30°C and centered covariable levels, is given in Table 11.1.

The 95% credible intervals for all model parameters except θ_5 exclude zero, indicating that the effects of the associated factors can be considered statistically significant. On that basis, the effect of initial buffer concentration might be considered borderline statistically significant.

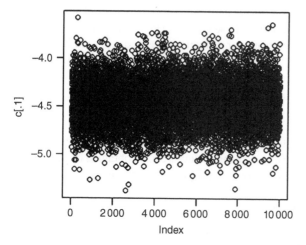

Figure 11.3 Mixing of parameter θ_1.

Table 11.1 Marginal distributions of shelf life at 30°C and centered covariable levels.

Parameter	Mean	Standard deviation	MC error	Marginal posterior distribution		
				2.5th percentile	Median	97.5th percentile
Shelf life	1165.0	280.6	3.078	709.3	1135.0	1800.0
θ_1	−4.454	0.2028	0.001923	−4.847	−4.455	−4.046
θ_2	−153.1	29.33	0.3192	−210.6	−153.2	−94.86
θ_3	−0.7884	0.2007	0.001842	−1.18	−0.7872	−0.3961
θ_4	−0.9948	0.2381	0.00239	−1.467	−0.9907	−0.5228
θ_5	0.3097	0.1638	0.001804	−0.01744	0.3096	0.6364
θ_6	0.6103	0.2638	0.00263	0.08018	0.6094	1.129
σ	0.9514	0.1447	0.002019	0.7164	0.934	1.286

The posterior distribution of shelf life (product is maintained at 30°C) and central levels of the covariables (API concentration, pH, and buffer concentration) is depicted in Figure 11.4. The thin black line indicates the highest probability density region (0.95). From a Bayesian point of view, this distribution summarizes all available knowledge about likely values of shelf life. The representation in terms of a distribution gives insight to the formulation scientist about the probability that product specifications will still be met at the end of shelf life. It is worth noting that the product is designed to be maintained at

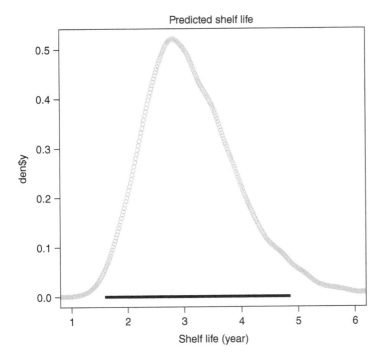

Figure 11.4 Distribution of predicted shelf life.

room temperature during its shelf life, and, therefore, the accelerated stability data supports this claim.

The design space is represented in more detail in Figures 11.5 and 11.6. The surface in Figure 11.5 depicts the probability that the product will meet specifications at the end of shelf life (assuming that the product is maintained at 30°C). The contour plot in Figure 11.6 depicts the probability that the product will meet specifications at the end of shelf life (product is maintained at 30°C). Both Figures 11.5 and 11.6 have buffer concentrations of 5 mM in pre-lyo solution shown on the left and 50 mM on the right. It is evident that with increasing buffer concentration, the probability of satisfying shelf life constraint decreases when all other factors remain constant. Low pH and API concentrations in pre-lyo solution have a detrimental effect on stability. This representation of design space is consistent with ICH Q8 definition and conveys much more information to formulation scientist than the traditional single-point estimate represented via response surface or contour plot [7]. Moreover, in the presence of multiple CQAs, the design space constructed using traditional overlapping mean response (OMR) techniques may lead to incorrect design space [7]. The reason for this is that identification of the joint overlap region using the marginal distributions ignores potential correlations among responses and is therefore not an adequate representation of the joint distribution.

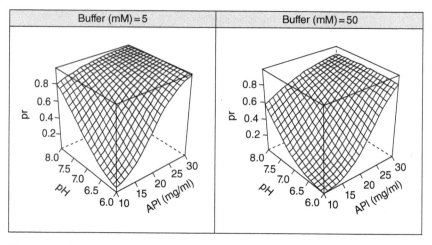

Figure 11.5 Wireframe representation of design space.

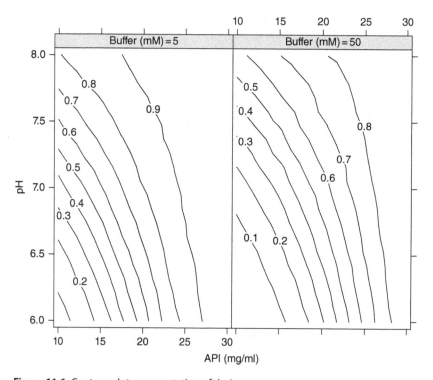

Figure 11.6 Contour plot representation of design space.

11.4 Conclusions

This chapter outlines Bayesian treatment of design space for lyophilized parenteral. The approach presented naturally aligns with the ICH Q8 definition of design space. Moreover, it is more intuitive than traditional design space representation developed using single-point estimates. It may be naturally extended to multiple unit operations (compounding, filling, and freeze-drying) by incorporating appropriate degradation models. For example, the key degradation pathways for ethacrynate sodium are different in the liquid and solid states [10], and this will reflect in the corresponding kinetics. In addition to design space definition, the proposed Bayesian approach provides estimates of key product attributes such as shelf life and activation energy that have a mechanistic interpretation. We have also demonstrated how prior knowledge such as typical activation energy or assay measurement error may be seamlessly incorporated into Bayesian decision framework. Traditional statistical approaches permit incorporation of prior information only in an ad hoc manner and therefore do not offer a framework for continuous knowledge building about a process.

The case study illustrates the Bayesian treatment of a single quality attribute, namely, the stability of the final product. Without loss of generality, multiple quality attributes may be considered in an analogous fashion (i.e., reconstitution time and stability). While the traditional OMR approach may lead to incorrect design space as shown by Peterson [7], the design space constructed using Bayesian methodology ensures that the threshold probability of final product meeting specifications is always met. Unlike traditional statistical probabilities that relate to repeated sampling performance of the statistical method, Bayesian probabilities are directly useful for risk assessment because they refer to the likely levels of underlying process parameters.

Appendix 11.A Implementation Using WinBUGS and R

While the theory and application of Bayesian approaches is straightforward, their implementation may be unfamiliar. The purpose of this Appendix is to provide interested readers with a working example as a template to facilitate their own work and encourage the use of these methods. The example is illustrated using the freely available softwares WinBUGS (the current version of WinBUGS can be downloaded from http://www.mrc-bsu.cam.ac.uk/bugs/winbugs/contents) and R (the current version of R can be downloaded from http://cran.r-project.org/bin/windows/base/; the associated library R2WinBUGS can be downloaded from http://cran.r-project.org/web/packages/R2WinBUGS/index.html). Interested readers should familiarize themselves with the underlying concepts, assumptions, and considerations of Bayesian and MCMC approaches as well as the importance of convergence assessment [12].

11.A.1 WinBUGS Model

```
model {
for (i in 1:N) {
Y[i] <- log(-Slope[i])}
# Priors
tau ~ dgamma(1.0E-3, 1.0E-3)
c[1] ~ dnorm(-10, 1.0E-6)
c[2] ~ dnorm(80, 1.0E-6)
c[3] ~ dnorm(0, 1.0E-6)
c[4] ~ dnorm(0, 1.0E-6)
c[5] ~ dnorm(0, 1.0E-6)
c[6] ~ dnorm(0, 1.0E-6)
# Likelihood
# Apply Model with Covariates
for (i in 1:N) {
# first center covariables and temperature
Tc[i] <- (1 / (Temperature[i] + 273.15) - InvTmean) / R
phc[i] <- pH[i] - 7
APIc[i] <- (API[i] - 20) / 10
Buffc[i] <- (buffer[i] - 30) / 25
eta[i] <- c[1] + c[2] * Tc[i] + c[3] * phc[i]
+ c[4] * APIc[i] + c[5] * Buffc[i]
+ c[6] * phc[i] * APIc[i]
Y[i] ~ dnorm(eta[i], tau)
}
}
```

11.A.2 Data Used for Analysis

The data (given in the succeeding text) used for analysis came from stability studies on 33 batches of lyophilized powder. Duration of stability studies were as long as 190 and 37 days for batches stored at 30 and 40°C, respectively. Formulation factor levels (API, pH, buffer) were varied for each batch as indicated in the table. The experimental design was a partially balanced factorial in all four factors, and each row in the table represents an independent manufacturing operation. The rate of change in potency (expressed as a percent of the initial level) was taken as the simple linear regression slope. Data from only 31 batches could be used because two batches yielded positive slopes. Summary of the data used for analysis is provided in Table 11.2.

Table 11.2 Summary of the data used for analysis.

API conc. (mg/ml)	pH	Buffer conc. (mM)	Temperature (°C)	Slope (%/day)
10	6	5	30	−0.024305
10	7	55	30	−0.13215
20	7	55	30	−0.005596
30	7	55	30	0.0002784[a]
10	8	5	30	−0.003072
30	8	5	30	0.0005065[a]
10	6	5	30	−0.009408
30	6	5	30	−0.000894
10	8	5	30	−0.00153
30	8	5	30	−0.000608
20	6	5	30	−0.012959
10	6	55	30	−0.059675
30	6	55	30	−0.002655
30	6	5	30	−0.001073
10	6	55	30	−0.087016
20	6	55	30	−0.031165
30	6	55	30	−0.003827
10	6	105	30	−0.129132
20	6	105	30	−0.005938
30	6	5	40	−0.005938
20	6	50	40	−0.017034
10	7	5	40	−0.182413
10	6	50	40	−0.402849
20	7	5	40	−0.020981
30	7	5	40	−0.011424
10	8	50	40	−0.019808
10	6	5	40	−0.411558
30	6	50	40	−0.052258
10	8	5	40	−0.012152
20	6	5	40	−0.007821
30	7	5	40	−0.009353
10	7	5	40	−0.103589
15	7	5	40	−0.049626

[a] Batch excluded from analysis because of a nonnegative slope. The absolute value of the slope is an order of magnitude less compared with remaining batches, which indicates that the degradation rate is not within experimental error. In order to simplify the presentation, the batch is not considered. In general, a more rigorous statistical analysis that incorporates Bayesian hierarchical modeling techniques [15] and/or errors in variables [16] could be performed.

11.A.3 Calling WinBUGS from R

```
# Get Data
data <- read.csv (" path to data file ", header =
TRUE, sep = ",",
quote = "\"", dec = ".", fill = TRUE, comment.
char = "")
data <- data [data$Slope < 0, ] # Remove observations
in which Slope is positive
API <- data$API
pH <- data$pH
buffer <- data$buffer
Slope <- data$Slope
Temperature <- data$Temperature
N <- length (API)
R <- 0.008315
InvTmean <- (1 / (30 + 273.15) + 1 /
(40 + 273.15)) / 2
data = list("N", "API", "pH", "buffer", "Temperature",
"Slope", "R", "InvTmean")
# Prepare inputs for WinBUGS
parameters <- c("c", "tau")
inits = function() {list (c = c (-10, 80, 0, 0, 0, 0),
tau = 0.25)}
chains = 1
iter = 20000
burnin = 10000
thin = 1
bugs.dir <- " path to WinBUGS.exe "
library (R2WinBUGS)
# Call WinBUGS
sim <- bugs (data = data, inits = inits, parameters.
to.save = parameters,
model.file = " path to model file ", n.chains =
chains,
n.iter = iter, n.burnin = burnin, n.thin = thin,
bugs.directory = bugs.dir, debug = TRUE)
# Save Posterior from WinBUGS
CODA output
tau <- sim$sims.list$tau #holds the draws of
measurement precision
c <- sim$sims.list$c #holds the draws of linear
coefficients
```

11.A.4 Calculating the Predictive Posterior Probability of Meeting Shelf Life

```
# Calculates posterior of shelf life (not used further)
T.pred <- 30
pH.pred <- 7
API.pred <- 20
buffer.pred <- 30
shelf.pred <- 5 / exp (model.log.minus.slope (c,
T.pred, API.pred, pH.pred,buffer.pred, InvTmean, R))
# Set up limits for design space grid
K_API = 20 # number of grid points in each dimension
K_pH = 20
K_buffer = 2
lb_API = 10 # lower bounds
lb_pH = 6
lb_buffer = 5
ub_API = 30 # upper bounds
ub_pH = 8
ub_buffer = 50
API.seq <- seq (lb_API, ub_API, (ub_API - lb_API) /
(K_API - 1))
pH.seq <- seq (lb_pH, ub_pH, (ub_pH - lb_pH) /
(K_pH - 1))
buffer.seq <- seq (lb_buffer, ub_buffer, (ub_buffer -
lb_buffer) / (K_buffer - 1))
# Calculate the predictive posterior for k for each
grid point
minus.slope.pred.post <- array (0, dim = c (length
(tau), K_API, K_pH, K_buffer))
for (i in 1 : K_API) {
for (j in 1 : K_pH) {
for (l in 1 : K_buffer) {
minus.slope.pred.post [, i, j, l] <-
exp (rnorm (length (tau),
mean = model.log.minus.slope (c, T.pred,
API.seq [i], pH.seq [j], buffer.seq [l], InvTmean, R),
sd = 1 / sqrt(tau)))
}
}
}
# If predictive posterior rate of change is below 5%
over 2 years indicator =1 else 0
```

```
minus.slope.ind.pred.post <- ifelse (minus.slope.pred.
post < 5 / 730, 1, 0)
# Set up wireframe and level plot data (probability of
passing shelf life)
API.f <- array (0, dim = c (K_API * K_pH * K_buffer))
pH.f <- array (0, dim = c (K_API * K_pH * K_buffer))
data.f <- array (0, dim = c (K_API * K_pH * K_buffer))
k = 1
for (i in 1 : K_API) {
for (j in 1 : K_pH) {
for (l in 1 : K_buffer) {
API.f [k] <- API.seq [i]
pH.f [k] <- pH.seq [j]
data.f [k] = mean (minus.slope.ind.pred.post
[, i, j, l])
k <- k + 1
}
}
}
buffer.f <- factor (buffer.seq)
wf <- wireframe (data.f ~ API.f * pH.f | buffer.f,
xlab = "API (mg/mL)", ylab = "pH",
strip = strip.custom (strip.names = c(TRUE, FALSE),
strip.levels = c (TRUE, FALSE), var.name =
"buffer (mM)",
sep = " = "),
scales = list (x = list (arrows = FALSE, tck =
1, cex = 0.4),
y = list (arrows = FALSE, tck = 1, cex = 0.4),
z = list (arrows = FALSE, tck = 1, cex = 0.4)),
zlab = list ("pr", rot = 90))
lp <- levelplot (data.f ~ API.f * pH.f | buffer.f,
xlab = "API (mg/mL)", ylab = "pH",
strip = strip.custom (strip.names = c (TRUE, FALSE),
strip.levels = c (TRUE, FALSE), var.name =
"buffer (mM)",
sep = " = "))
```

Notation

χ threshold of joint posterior predictive probability that all process responses will be acceptable

ε constant normal measurement error with mean 0 and standard deviation σ

θ (theta)	vector of degradation model parameters
Ξ	experimental data
σ (sigma)	standard deviation of measurement and modeling error, $\tau = 1/\sigma^2$
τ (tau)	precision of measurement and modeling error
A	Arrhenius "acceleration" constant
API	API concentration of pre-lyo solution for each batch
API_c	centered factor API
buffer	buffer concentration of pre-lyo solution for each batch
$buffer_c$	centered factor buffer
C_0	initial potency in % of API
E_a	activation energy (kJ/mol)
k	measured degradation rate
k.pred	degradation rate predicted by the model
pH	pH of pre-lyo solution for each batch
pH_c	centered factor pH
Q	acceptance region (i.e., specification boundaries) associated with the responses
R	universal gas constant (0.008315 kJ/mol/°K)
S	shelf life
T	temperature at which stability samples were maintained for a particular batch (°C)
T_c	centered factor T
Y	vector of responses

Acknowledgments

We are grateful to the National Institute for Pharmaceutical Technology and Education (NIPTE) and the US Food and Drug Administration (FDA) for providing funds for this research. This study was funded by the FDA-sponsored contract "Development of Quality by Design (QbD) Guidance Elements on Design Specifications across Scales with Stability Considerations" (contract number HHSF223200819929C).

References

1 ICH. ICH Harmonised Tripartite Guideline, Pharmaceutical Development: Q8(R2), ICH, Geneva, 2009.

2 Moore, C. Design Space—An FDA Perspective. Paper presented at 43rd Annual Meeting of the Drug Information Association, Atlanta, 2007.

3 Woodcock J., Woosley R. The FDA Critical Path Initiative and Its Influence on New Drug Development. Annual Review of Medicine. 2008; 59: 1–12.

4 Miró-Quesada G, del Castillo E, Peterson J. A Bayesian Approach for Multiple Response Surface Optimization in the Presence of Noise Variables. Journal of Applied Statistics. 2004; 31: 251–270.

5 Peterson JJ. A Posterior Predictive Approach to Multiple Response Surface Optimization. Journal of Quality Technology. 2004; 36: 139–153.

6 Peterson JJ, Yahyah, M. A Bayesian Design Space Approach to Robustness and System Suitability for Pharmaceutical Assays and Other Processes. Statistics in Biopharmaceutical Research. 2009; 1(4): 441–449.

7 Peterson JJ. A Bayesian Approach to the ICH Q8 Definition of Design Space. Journal of Biopharmaceutical Statistics. 2008; 18(5): 959–975.

8 Mockus L, LeBlond, D, Basu, PK, Shah, RB, Khan, MA. A QbD Case Study: Bayesian Prediction of Lyophilization Cycle Parameters. AAPS PharmSciTech. 2011; 12(1): 442–448.

9 Shamblin SL, Hancock, BC, Pikal, MJ. Coupling Between Chemical Reactivity and Structural Relaxation. Pharmaceutical Research. 2006; 23(10): 2254–2268.

10 Mockus L, Paul, TW, Pease, NA, *et al.* Quality by Design in Formulation and Process Development for a Freeze-dried, Small Molecule Parenteral Product: A Case Study. Pharmaceutical Development and Technology. 2011; 16(6): 549–576.

11 Cowles M. Review of WinBUGS 1.4. The American Statistician. 2004; 58(4): 330–336.

12 Gelman A, Carlin JB, Stern HS, Rubin DB. In: Bayesian Data Analysis, 3rd edition. London: Chapman & Hall/CRC Texts in Statistical Science; 2014.

13 Smith AFM, Roberts GO. Bayesian Computation via the Gibbs Sampler and Related Markov Chain Monte Carlo Methods. Journal of the Royal Statistical Society. 1993; 55(1): 3–23.

14 R: A Language and Environment for Statistical Computing. Vienna: R Foundation for Statistical Computing; 2009 (Manual).

15 Gelman A, Hill, J. Data Analysis Using Regression and Multilevel/Hierarchical Models. Cambridge: Cambridge University Press; 2007.

16 Reilly PM, Patino-Leal H. A Bayesian Study of the Error-in-Variables Model. Technometrics. 1981; 23(3): 221–231.

12

Enhanced Process Design and Control of a Multiple-Input Multiple-Output Granulation Process

Rohit Ramachandran

Department of Chemical and Biochemical Engineering, Rutgers, The State University of New Jersey, Piscataway, NJ, USA

12.1 Introduction and Objectives

Granulation is a particle design process of converting fine powder into larger free-flowing agglomerates. It finds application in a wide range of industries (e.g., pharmaceuticals, fertilizers, and minerals). Granulated products often have notable improvements compared with fine powders, and these include increased bulk density, improved flow properties, and uniformity in the distribution of multiple solid components. Granulation processes have been ubiquitous in the industry for many years with significant research undertaken to gain further insight into the underlying phenomena occurring during the process. However, industrial granulation processes are operated inefficiently, and the resultant wide distribution of granule properties is often the cause of large recycle rates in continuous processes and high rejection rates in batch processes. Therefore, an integrated systems approach to design and control the process will be a crucial aid to mitigate this situation [1–4].

In a continuous granulation process (which this study deals with), feed material is continuously introduced into the granulator as granulation occurs. The granulator is fitted with several spray nozzles in different positions through which the liquid binder is introduced into the granule bed. The granules formed are then dried and classified based on product specification(s). Granules that do not conform to product specification(s) are recycled and reprocessed. Prior to the actual design of the controller for the granulation process, it is important to know how well the process can be controlled and what factors may hinder the control-loop performance that may be achieved in reality. It is also imperative that appropriate plant inputs and outputs are

Comprehensive Quality by Design for Pharmaceutical Product Development and Manufacture,
First Edition. Edited by Gintaras V. Reklaitis, Christine Seymour, and Salvador García-Munoz.

selected for control purposes, and furthermore they are paired correctly as incorrect pairings may limit and hinder control-loop performance. Therefore, a plant is said to be controllable if there exists a controller that can, in principle, be able to achieve a certain output state via certain admissible input changes [5]. Controllability is an intrinsic plant property and is determined by the plant design such as sensor locations, recycle loops, internal couplings, and sizing of equipment. Controllability conditions for granulation processes are also made more challenging given the presence of internal variables [6].

Previous work has addressed the sensitivity of inputs such as binder flow rates on granule properties for a single-compartment model [7]. The aim of the present work is to systematically ascertain the sensitivity/controllability of the multi-compartment granulation process by examining the effect of control handles such as binder and feed flow rates on controlled variables such as average size, distribution width, bulk density, and moisture content. The study also investigates the best possible control-loop pairings for the purpose, assesses the overall control-loop performance, and suggests rectifications in process design, in the event of suboptimal control-loop performance.

12.2 Population Balance Model

The granulation process is an example of a multi-scale problem wherein the microscale information such as wetting kinetics, energy dissipation effects, and contact forces, in combination with process parameters, affect the mesoscale phenomena (nucleation, aggregation, breakage, consolidation), which in turn affect macroscopic properties such as granule size and bulk density [8]. The microscale properties themselves are influenced by fundamental material properties, both of the solid and the liquid. Taking the aforementioned into consideration, a continuous granulation process can be adequately modeled by a three-dimensional population balance equation as shown in Equation 12.1:

$$
\frac{\partial F\left(s,l,g,t\right)}{\partial t} + \frac{\partial}{\partial g}\left(F\left(s,l,g,t\right)\frac{dg}{dt}\right) = \mathfrak{R}_{nuc} + \mathfrak{R}_{agg} + \mathfrak{R}_{break}
$$
$$
+ R_f X_0\left(s,l,g,t\right) - R_f X\left(s,l,g,t\right)
$$

$$(12.1)$$

where $F(s,l,g,t)$ represents the population density function such that $F(s,l,g,t)ds\,dl\,dg$ is the moles of granules (adopting a number-based population balance instead of mass or volume based) with solid volume between s and $s+ds$, liquid volume between l and $l+dl$, and gas volume between g and $g+dg$. The partial derivative with respect to g accounts for consolidation that, due to compaction of the granules, results in an increase of pore saturation and decrease in porosity. R_f is the input flow rate (number density per second) of particles, and X is the molar concentration of particles and is defined as

$X(s,l,g,t) = \dfrac{F(s,l,g,t)}{\int F(s,l,g,t)\,ds\,dl\,dg}$. X_0 is the input molar concentration of the particles. The inflow and outflow terms are added to facilitate long-term running of the process from which suitable transfer functions (based on step response data) can be identified. $\mathcal{R}_{\text{nuc}}$, $\mathcal{R}_{\text{agg}}$, and $\mathcal{R}_{\text{break}}$ are the nucleation, aggregation, and breakage rates, respectively, and their model forms have been well documented in previous works [9–11] and are reproduced here for clarity.

The formation and depletion terms associated with the aggregation phenomenon ($\mathcal{R}_{\text{agg}}$) are defined in Equations 12.2–12.4 [12, 13]. In these equations, s_{nuc} is the solid volume of nuclei (assumed fixed), and $\beta(s',s-s',l',l-l',g',g-g')$ is the size-dependent aggregation kernel that signifies the rate constant for aggregation of two granules of internal coordinates (s',l',g') and $(s-s',l-l',g-g')$. β is essentially a measure of how successful collisions between two particles resulting in a larger granule are:

$$\mathfrak{R}_{\text{agg}}(s,l,g,t) = \mathfrak{R}_{\text{agg}}^{\text{formation}} - \mathfrak{R}_{\text{agg}}^{\text{depletion}} \tag{12.2}$$

$$\mathfrak{R}_{\text{agg}}^{\text{formation}} = \frac{1}{2}\int_{s'=s_{\text{nuc}}}^{s-s_{\text{nuc}}} \int_{l'=0}^{l} \int_{g'=0}^{g} \beta(s',s-s',l',l-l',g',g-g')$$
$$\times F(s',l',g',t)F(s-s',l-l',g-g',t)\,ds'\,dl'\,dg' \tag{12.3}$$

$$\mathfrak{R}_{\text{agg}}^{\text{depletion}} = F(s,l,g,t)\int_{s'=s_{\text{nuc}}}^{s_{\text{max}}} \int_{l'=0}^{l_{\text{max}}} \int_{g'=0}^{g_{\text{max}}} \beta(s',s,l',l,g',g)$$
$$\times F(s',l',g',t)\,ds'\,dl'\,dg' \tag{12.4}$$

The breakage term ($\mathcal{R}_{\text{break}}$) is mathematically described in two parts: the breakage kernel (k_{break}) and the breakage function (b). The former describes the rate at which a particle of size s, l, and g breaks into a fragments of size s_1, l_1, and g_1. The latter describes the sizes of these fragments formed. Details can be found in Refs. [14] and [15]. Therefore, $\mathcal{R}_{\text{break}}$ is defined in Equations 12.5–12.7:

$$\mathfrak{R}_{\text{break}}(s,l,g,t) = \mathfrak{R}_{\text{break}}^{\text{formation}} - \mathfrak{R}_{\text{break}}^{\text{depletion}} \tag{12.5}$$

$$\mathfrak{R}_{\text{break}}^{\text{formation}} = \int_{s}^{\infty} \int_{l}^{\infty} \int_{g}^{\infty} k_{\text{break}}(s',l',g')b(s,l,g,s',l',g')$$
$$\times F(s',l',g',t)\,ds'\,dl'\,dg' \tag{12.6}$$

$$\mathfrak{R}_{\text{break}}^{\text{depletion}} = k_{\text{break}}(s,l,g)F(s,l,g,t) \tag{12.7}$$

Similar to the formation and depletion terms associated with $\mathcal{R}_{\text{agg}}$ and $\mathcal{R}_{\text{break}}$, the nucleation is described by the rate of formation of nuclei and the rate of depletion of primary particles and is defined in Equations 12.8–12.10:

$$\mathfrak{R}_{\text{nuc}}(s,l,g,t) = \mathfrak{R}_{\text{nuc}}^{\text{formation}} - \mathfrak{R}_{\text{nuc}}^{\text{depletion}} \tag{12.8}$$

$$\mathfrak{R}_{nuc}^{formation} = \int_{s'=s_{min}}^{s_{max}} \int_{l'=l_{min}}^{l_{max}} \int_{g'=g_{min}}^{g_{max}} k_{nuc}\left(s',l',g'\right)F\left(s',l',g',t\right)ds',dl',dg'$$

(12.9)

$$\mathfrak{R}_{nuc}^{depletion} = k_{nuc}\left(s',l',g'\right)F\left(s',l',g',t\right)Nx\left(s',l',g'\right)$$

(12.10)

where N is the number of primary particles in the nucleus and $x(s',l',g')$ is the fraction of primary particles with characteristic s',l',g' that are present in the nucleus [10].

12.2.1 Compartmentalized Population Balance Model

The previously described continuous PBM is further compartmentalized to model a pilot-plant drum granulation process located at the University of Queensland, Australia, which was the basis of a previous study that focused on deriving a model predictive controller for a single-input single-output process [16]. The schematic of the drum granulator can be seen in Figure 12.1 and consists of three sections (labeled 1, 2, and 3) wherein each compartment consists of a binder spray nozzle (labeled u_1, u_2, and u_3). The solid feed inflow is denoted by Rf_{in} and the corresponding outflow as Rf_{out}. In the compartmentalized granulator (3 compartments), the population balance equation is effectively solved thrice at each time step. Assuming that u_1 is turned on and u_2 and u_3 are turned off, this means that as particles enter the first compartment, binder addition takes place, leading to net granule growth (defined to be the resultant effects of nucleation, aggregation, breakage, and consolidation), and as the particles proceed to the second and third compartments, binder addition has ceased. However, net growth still takes place due to the presence of the residual binder among the particles. Compartmentalizing the granulator in such a manner will also allow us to realistically examine the effect of spray nozzle position that is an important design parameter in the granulation process.

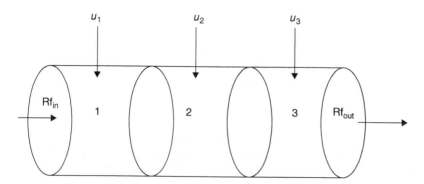

Figure 12.1 Compartmentalized model of the granulator.

In solving compartmentalized model, a hierarchical solution strategy is employed [13, 17, 18], which explicitly recasts the population balances of discretized subpopulations in terms of the underlying rate processes (i.e., nucleation, aggregation, and consolidation). This technique is based on the discretization of the three-dimensional population into finite volumes ("bins") and the derivation of the population balance in terms of these subpopulations. Thus, the partial differential integral equation given in Equation 12.1 is now solved as a set of ordinary differential equations. Further, a decomposition solution framework is employed, wherein the contributions of each of the subprocesses of nucleation, aggregation, and consolidation to the population redistribution are accounted separately. Further details of the numerical algorithm used in this study may be found in the original reference [17].

12.3 Simulation and Controllability Studies

Using the compartmentalized model, Figure 12.2a–d depicts the effect of the position of the spray nozzle on critical granule properties simulated over a period of 80 min. The properties considered in this study are (i) volume mean diameter of the granule (d_{30}); (ii) width of the granule distribution ($d_{84/16}$) (which is defined as $\sqrt{\dfrac{d_{84}}{d_{16}}}$ where d_{84} and d_{16} are the 84th and 16th percentile, respectively, of the granule size distribution); (iii) bulk density (ρ_b) of the granules, which is defined as (M_s/V_t), where M_s is the total solid mass and V_t is the total particle volume; and (iv) moisture content (m_c) of the granules (which is defined to be (V_l/V_t), where V_l is the liquid volume). It can be seen that d_{30}, m_c, and ρ_b are dependent on the spray nozzle position. The nonlinearity of the granulation phenomena is also illustrated as seen by the larger changes in granule attributes from compartment 2 and 3 and then from compartment 1 and 2. $d_{84/16}$ on the other hand is independent of the spray nozzle position. It must be noted that all simulations were performed on a model that has been previously validated against experimental data [9, 11] to ensure that all input parameters were representative of a typical granulation process. All model simulations were carried out on a 2 GHz Intel dual-core single-processor desktop computer with 1 GB RAM using the Intel Fortran compiler.

A simulation-based controllability analysis was performed on the multi-compartment process model as depicted in Figure 12.3.

Open-loop step tests were performed on the model to ascertain the controllability of the process outputs. A step change was made to the nozzle and solid feed flow rates one at a time while holding all other step inputs constant.

Although the granulation process is nonlinear, an approximate linear model that relates the four outputs to the four manipulated inputs can be considered appropriate for the purpose of identification of control-loop pairings. The model variables are defined in terms of the deviation of actual process variables from their respective nominal operating regions as seen in Equations 12.11–12.16:

$$n_i = u_i - u_i^0, \quad i = 1, 2, 3 \tag{12.11}$$

(a)

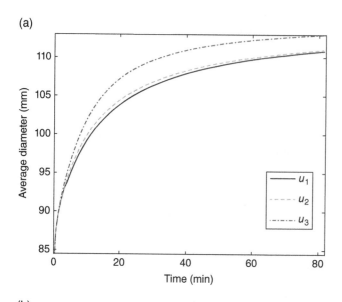

(b)

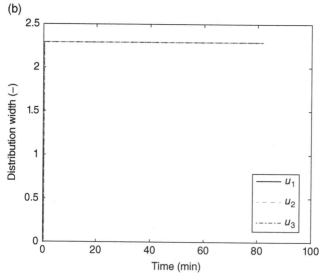

Figure 12.2 Time profiles for (a) average diameter, (b) distribution width, (c) average moisture content, and (d) average bulk density for different spray nozzle positions under start-up conditions.

(c)

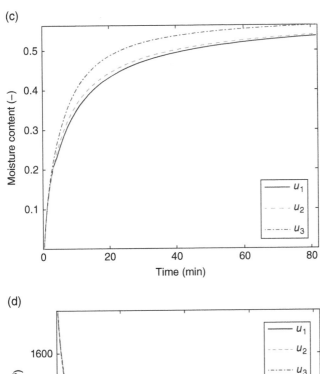

(d)

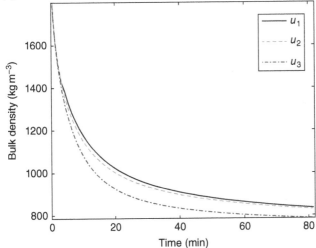

Figure 12.2 (Continued)

$$n_4 = \text{Rf}_{\text{in}} - \text{Rf}_{\text{in}}^0 \tag{12.12}$$

$$y_1 = d_{30} - d_{30}^0 \tag{12.13}$$

$$y_2 = d_{84/16} - d_{84/16}^0 \tag{12.14}$$

$$y_3 = m_{\text{c}} - m_{\text{c}}^0 \tag{12.15}$$

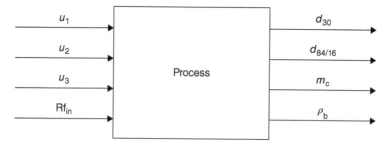

Figure 12.3 A block diagram of the process and its control objectives.

$$y_4 = \rho_b - \rho_b^0 \tag{12.16}$$

A linear continuous-time model of the form

$$\mathbf{y}(s) = \mathbf{G}_{ij}\mathbf{n}(s) \tag{12.17}$$

is obtained from the process data. The individual step response data indicate that a first-order plus time delay transfer function of the form

$$G_{ij} = \frac{k_{ij}e^{-\theta_{ij}s}}{\tau_{ij}s+1} \tag{12.18}$$

where k is the process gain, τ is the process time constant, and θ is the process time delay is adequate to represent the behavior observed in each of the responses. From the open-loop step test data, the corresponding transfer functions were determined as

$$\mathbf{y}(s) = \begin{bmatrix} \dfrac{0.9}{12s+1} & \dfrac{0.91}{11s+1} & \dfrac{0.93}{10s+1} & \dfrac{-0.27}{3s+1} \\ 0 & 0 & 0 & 0 \\ \dfrac{0.05}{7s+1} & \dfrac{0.06}{6.5s+1} & \dfrac{0.07}{6s+1} & \dfrac{-0.01}{4s+1} \\ \dfrac{-10.6}{8s+1} & \dfrac{-11}{9.15s+1} & \dfrac{-11.3}{9.10s+1} & \dfrac{0.5}{4s+1} \end{bmatrix} \mathbf{n}(s) \tag{12.19}$$

It must be noted that various other step changes (within the vicinity of the nominal operating conditions) were made to the manipulated variables. It was hypothesized that due to the nonlinearity of the process, the process dynamics could be different, resulting in different transfer functions that could then go on to result in different control-loop pairings. However, results confirmed that the transfer functions obtained were almost identical. Therefore, the process was deemed linear over the vicinity of the current operating conditions.

Results from (12.19) show that the average diameter, moisture content, and bulk density are controllable through the manipulations of u_1, u_2, u_3, and Rf_{in}. As seen from (12.19) a positive step change in the nozzle flow rates result in a corresponding increase in both the average granule diameter and moisture content and a decrease in the bulk density. In contrast, a positive step change in the solid feed flow rate results in a decrease in the average granule diameter due to a higher proportion of smaller fine particles and a decrease in the moisture content due to a higher proportion of dry particles. Bulk density on the other hand shows an increase. Results also show that step changes in the manipulations have no effect on the size distribution width. This implies that $d_{84/16}$ is not controllable and cannot be selected as a controlled variable. This observation has also been reported in [16] who observed that both their experimental and simulated $d_{84/16}$ were not influenced by any admissible inputs and therefore deemed not controllable. As a result only d_{30}, ρ_b, and m_c can be selected as controlled variables with any three of u_1, u_2, u_3, and Rf_{in} as the potential choices as control handles. In many scenarios, distribution width is an important variable that needs to be controlled. This is because differing size distributions (albeit having the same mean size) may spuriously affect downstream operations such as tableting. In such instances other existing manipulated inputs may have to be selected or the process has to be redesigned to facilitate enhanced controllability.

12.4 Identification of Existing "Optimal" Control-Loop Pairings

In the previous section, the controllability of the granulation process incorporating its key mechanisms was examined. In this section, given the possible choices for controlled and manipulated variables, an optimal combination of control-loop pairings was selected to facilitate efficient control of the process.

In a multiple-input multiple-output system where there are more manipulations available than outputs, the manipulations to use for pairing with the corresponding outputs need to be decided. Pairings must be made such that the interaction within the control loops is minimized. As there are three controlled variables and four manipulated variables, one of the manipulated variables must be discarded, while the other variables are paired optimally. This is done by means of the relative gain array (Λ). The relative gain array can be useful for establishing the best input–output pairings for use in the control of a multivariable system. The array (Λ) is defined as the element-by-element product (the Hadamard or Schur product) of the transfer function matrix G(s) and the transpose of the inverse of this matrix as seen in Equation 12.20:

$$\Lambda = G \times \left(G^{-1} \right)^{T} \tag{12.20}$$

Typically, the relative gain array is evaluated at zero frequency (i.e., at steady state). Relative gains in the range 0 to 1 indicate moderate interaction, with values of 0.5 being the worst. To minimize interaction, variables with relative gains closest to 1 should be paired as this implies that the gains are largely unaffected by the closing of other loops. Variables with negative gains should not be paired for control as (i) the overall closed-loop system may be rendered unstable, (ii) the loop with the negative relative gain is unstable by itself, or (iii) the closed-loop system is unstable if the loop with the relative negative gain is opened. Relative gains of greater than 5 usually imply severe loop interaction. More details on RGA can be found in [19].

12.4.1 Discarding n_1

Assuming that n_1 is discarded and not considered as one of the manipulations, the 3×4 system reduces to a 3×3 system, and the calculated Λ is

$$\Lambda = \begin{bmatrix} 3.31 & -3.47 & 1.15 \\ -7.24 & 8.21 & 0.02 \\ 4.93 & -3.74 & -0.18 \end{bmatrix} \qquad (12.21)$$

For a 3×3 system, there would be six possible theoretical pairings. From Equation 12.21, it can be seen that there is only one possible pairing with all relative gain array elements positive (i.e., $n_2 \leftrightarrow y_4$, $n_3 \leftrightarrow y_3$, $n_4 \leftrightarrow y_1$).

12.4.2 Discarding n_2

Assuming that n_2 is discarded and not considered as one of the manipulations, the 3×4 system reduces to a 3×3 system and the calculated Λ is

$$\Lambda = \begin{bmatrix} 1.64 & -1.76 & 1.12 \\ -3.03 & 3.95 & 0.07 \\ 2.38 & -1.19 & -0.19 \end{bmatrix} \qquad (12.22)$$

From Equation 12.22, it can be seen that there is only one possible pairing with all relative gain array elements positive (i.e., $n_1 \leftrightarrow y_4$, $n_3 \leftrightarrow y_3$, $n_4 \leftrightarrow y_1$).

12.4.3 Discarding n_3

Assuming that n_3 is discarded and not considered as one of the manipulations, the 3×4 system reduces to a 3×3 system and the calculated Λ is

$$\Lambda = \begin{bmatrix} 3.34 & -3.42 & 1.07 \\ -5.84 & 6.72 & 0.11 \\ 3.49 & -2.30 & -0.19 \end{bmatrix} \qquad (12.23)$$

From Equation 12.23, it can be seen that there is only one possible pairing with all relative gain array elements positive (i.e., $n_1 \leftrightarrow y_4$, $n_2 \leftrightarrow y_3$, $n_4 \leftrightarrow y_1$).

12.4.4 Discarding n_4

Assuming that n_4 is discarded and not considered as one of the manipulations, the 3×4 system reduces to a 3×3 system and the calculated Λ is

$$\Lambda = \begin{bmatrix} 48.42 & -94.19 & 46.77 \\ 1.54 & -10.94 & 10.39 \\ -48.97 & 106.14 & -56.16 \end{bmatrix} \tag{12.24}$$

From Equation 12.24, it can be seen that there is only one possible pairing with all relative gain array elements positive (i.e., $n_1 \leftrightarrow y_1$, $n_2 \leftrightarrow y_4$, $n_3 \leftrightarrow y_3$).

12.4.5 Discussion

Given that we are dealing with a non-square 3×4 system, there exist in total four possible options for control-loop pairings. With the first and third options (i.e., Section 12.4.1 and 12.4.3), it can be seen that λ_{22} and λ_{31} are $\gg 1$, implying that when these loops are closed, they result in large interactions with the other loops, causing potential difficulties in overall control. With the second option (i.e., Section 12.4.2), it can be seen that λ_{22} and λ_{31} are relatively closer to unity, thus resulting in better control. With the fourth option (i.e., Section 12.4.4), it can be seen that λ_{11}, λ_{42}, and λ_{42} have very high relative gains, implying that when each loop is closed, it results in severe interactions with the other loops. This would be very undesirable for control purposes. Intuitively, this can be explained as follows. Selecting the three nozzle spray rates as the manipulations is undesirable as their effect on the controlled variables is very similar (small differences in the gains and time constants). Hence, a manipulation in one of the inputs (in a closed -loop configuration) would have an almost equal impact on the all the three controlled variables. As a result, the ensuing interactions would make control and regulation of the variables very difficult. Furthermore, the dynamics of the feed flow rate are greater than that of the nozzle spray rates. This shows that a change in the feed flow rate would have a faster effect on any of the controlled variables. Therefore, the feed flow rate (n_4) should be one of the manipulations considered. Overall, the results show that the optimal control-loop pairing is that of $n_1 \leftrightarrow y_1$, $n_3 \leftrightarrow y_3$, and $n_4 \leftrightarrow y_4$ (i.e., first compartment nozzle with average diameter, third compartment nozzle with moisture content, and feed flow rate with bulk density). These pairings are valid over the vicinity of the current operating regime and therefore provide valuable insight pertaining to this localized operating regime. Deviating from this operating regime could potentially result in different control-loop pairings.

12.5 Novel Process Design

In previous work, the effect of several important formulation properties on granulation mechanisms was investigated for high-shear granulation [2]. A more detailed experimental study was performed by [8], whereby the effects of solid properties (e.g., powder density), liquid properties (e.g., binder surface tension and viscosity), and solid–liquid properties (e.g., contact angle) on granulation mechanisms were investigated. Based on the study, an integrated systems representation of the various formulation properties and operating and/or design variables that could influence both the granulation process and the resultant granule attributes was formulated (see Figure 12.4). Powder feed rate, nozzle position, multiple nozzles, binder flow rate, and the mixing rate are all important manipulations that may be adjusted to ensure target attributes are attained. In addition, the liquid binder properties can also be utilized for feedback control purposes, subject to practicality of implementation. Both the viscosity and surface tension effects have also been mentioned as playing a role

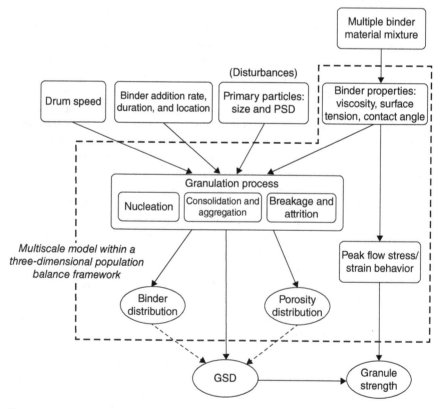

Figure 12.4 A comprehensive systems representation of the granulation process.

in some respects toward the strength of granules. Thus, in addition to manipulating the flow rates individually through each nozzle, the viscosity and surface tension of the binder could also be a manipulated variable for GSD control (through combinations of binders) while simultaneously paying attention to their effect on the granule strength. This manipulation of the viscosity and surface tension could be carried out through the mixing of different concentrations of binder solutions.

12.5.1 Identification of Kernels

A primary challenge in the development of the population balance models is the identification of appropriate kernels (rate constants) that describe the individual mechanisms. While the development of a multidimensional population balance model is motivated by the physics of the problem, it is a tougher task to obtain three-dimensional kernels that account for the dependence of the rates on the particle traits (i.e., size, binder content, and porosity). Many of the kernels (for nucleation, aggregation, and breakage) that appear in the literature are empirical or semiempirical in nature, and a comprehensive summary of them can be found in [20–22]. The disadvantages of these kernels are that (i) they have no physical basis, (ii) they have several empirical parameters that need to be tuned to provide the best fit to experimental data, (iii) they are unable to account for the effects of process parameters and material properties on the end granule properties/distributions, and (iv) they are not predictive and for each new formulation/operating condition, a new experiment has to be carried out to tune the empirical kernel/model.

12.5.2 Proposed Design and Control Configuration

The kernels used in this study are mechanistic and are based on fundamental physics and chemistry of the granulation mechanisms. The nucleation kernel (see Refs. [9, 10]) explicitly accounts for the wetting kinetics and nucleation thermodynamics. The aggregation kernel (see Ref. [13]) in turn accounts for the energy dissipation effects, balancing the net attractive potential between the kinetic/elastic energy and the viscous forces. Similarly, the breakage kernel (see Ref. [11]) relates the external stresses experienced by the particles in the granulator with the intrinsic particle strength brought about by viscous, capillary, and frictional forces. These kernels have been validated against experimental data for certain operating conditions and formulations [9, 11]. The novelty of these kernel formulations lies in (i) accounting for fundamental material properties and process parameters, (ii) capturing sensitivities in critical process variables as observed experimentally, and (iii) being able to account for important granule properties without having to perform additional experiments. Table 12.1 in [11] and Table 12.2 in [9] show all the relevant material properties and process parameters featured in the kernel derivations.

Table 12.1 Nominal values of material properties and process/design parameters.

Property/parameter	Value	Unit
γ_s	4.48×10^{-2}	$\mathrm{N\,m^{-2}}$
γ_l	7.28×10^{-2}	$\mathrm{N\,m^{-2}}$
μ_l	1.0×10^{-3}	Pa s
φ	7.54×10^{-1}	rad
ρ_s	2.72×10^{3}	$\mathrm{kg\,m^{-3}}$
ρ_l	1.0×10^{3}	$\mathrm{kg\,m^{-3}}$
μ_a	1.85×10^{-5}	Pa s
E_p	4.50×10^{10}	Pa
E_w	7.0×10^{10}	Pa
E_i	2.0×10^{11}	Pa
σ_p	0.22	—
σ_w	0.17	—
σ_i	0.30	—
e_p	0.60	—
e_w	0.50	—
e_i	0.80	—
circ	1.0	—
C_d	0.47	—
θ	0.50	—
TSA	2.02×10^{-1}	$\mathrm{m^2}$
WA	6.13×10^{-1}	$\mathrm{m^2}$
IA	3.06×10^{-1}	$\mathrm{m^2}$
D_{lower}	246	µm
A	1.5×10^{-8}	$\mathrm{s^{-1}}$
$\dfrac{da}{dt}$	5.0×10^{-2}	$\mathrm{m\,s^{-1}}$

TSA refers to the total surface area of the initial seed particles.
TSA will change as the particle density changes during the course of granulation.

Combining the experimental and modeling work reported previously, the objective in this section is to study the dynamic responses of the critical granule properties (described in Section 12.3) as a function of certain important material properties of the powder particles. The kernels that have been developed can elegantly capture many of such material properties, and in this study, two important ones are investigated, which are binder viscosity and binder

Table 12.2 Fundamental material properties and less sensitive adjustable parameters.

No	Property	Value
1	μ	5.20×10^{-3} Pa s
2	ρ_s	2.74×10^{3} kg m^{-3}
3	Y_d	1.591×10^{-4} Pa
4	E	8.20×10^{6} Pa
5	h_a	5.0×10^{-5} m
6	l_{min}	0.99 (dimensionless)
7	d_{50}	$130\,\mu$m
8	γ_{LV}	5.2×10^{-2} Nm^{-1}
9	γ_{SV}	4.0×10^{-2} Nm^{-1}
10	θ	$43.2°$
11	ε_{eff}	0.20 (dimensionless)
12	u_0	5.0×10^{-3} ms^{-1}
13	V_{up}	$0\,$m^3

surface tension. It must be noted that due to the mechanistic kernel derivations, it is possible to conduct this study. It would not be possible to evaluate the effects of material properties or even certain process parameters using empirical kernels.

The viscosity and surface tension variables are defined in terms of the deviation of actual process variables from their respective nominal operating regions as seen in Equations 12.25 and 12.26. Step changes were made to both binder viscosity (μ) and binder surface tension (γ) and their effects on granule properties were analyzed via a first-order transfer function:

$$n_5 = \mu - \mu^0 \tag{12.25}$$

$$n_6 = \gamma - \gamma^0 \tag{12.26}$$

The corresponding transfer function matrix is shown in Equation 12.27:

$$\mathbf{y}(s) = \begin{bmatrix} \dfrac{15}{3s+1} & \dfrac{11}{5s+1} \\[2ex] \dfrac{0.8}{8s+1} & \dfrac{0.3}{10s+1} \\[2ex] \dfrac{0.12}{7s+1} & \dfrac{0.09}{8.5s+1} \\[2ex] \dfrac{-14}{2s+1} & \dfrac{-12}{3s+1} \end{bmatrix} \mathbf{n}(s) \tag{12.27}$$

Combining the material properties with previous process manipulations (see Equation 12.28), an overall 4 by 4 transfer function matrix is put together and shown in Equation 12.37 based on the following deviation variables (see Equations 12.29–12.36). Thereafter, the corresponding 4×4 is calculated and shown in Equation 12.37:

$$\mathbf{y}(s) = \begin{bmatrix} \dfrac{0.9}{12s+1} & \dfrac{-0.27}{3s+1} & \dfrac{15}{3s+1} & \dfrac{11}{5s+1} \\ 0 & 0 & \dfrac{0.8}{8s+1} & \dfrac{0.3}{10s+1} \\ \dfrac{0.05}{7s+1} & \dfrac{-0.01}{4s+1} & \dfrac{0.12}{7s+1} & \dfrac{0.09}{8.5s+1} \\ \dfrac{-10.6}{8s+1} & \dfrac{0.5}{4s+1} & \dfrac{-14}{2s+1} & \dfrac{-12}{3s+1} \end{bmatrix} \mathbf{n}(s) \tag{12.28}$$

$$n_5 = \mu - \mu^0 \tag{12.29}$$

$$n_6 = \gamma - \gamma^0 \tag{12.30}$$

$$n_5 = \mu - \mu^0 \tag{12.31}$$

$$n_6 = \gamma - \gamma^0 \tag{12.32}$$

$$y_1 = d_{30} - d_{30}^0 \tag{12.33}$$

$$y_2 = d_{84/16} - d_{84/16}^0 \tag{12.34}$$

$$y_3 = m_c - m_c^0 \tag{12.35}$$

$$y_4 = \rho_b - \rho_b^0 \tag{12.36}$$

$$\Lambda = \begin{bmatrix} -0.11 & 0 & -0.12 & 1.23 \\ -0.10 & 0 & 1.42 & -0.31 \\ -1.37 & 2.03 & 0.30 & -0.06 \\ 2.49 & -1.03 & -0.60 & 0.15 \end{bmatrix} \tag{12.37}$$

Based on Equation 12.37, it can be seen that the 1 control-loop pairing is identified and the schematic is shown in Figure 12.5. The bold lines represent the pairings and the dashed lines represent the interactions. PID represents the regulatory control used for feedback control. Furthermore, based on the RGA results, (i) $d_{84/16}$ is controllable using μ as an admissible input, and (ii) indirect interactions as a result of closing all control loops is minimal compared with the previous design configuration.

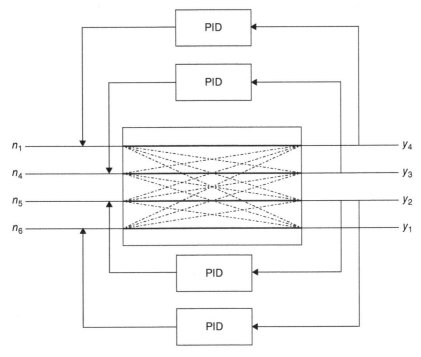

Figure 12.5 Proposed design and control configuration.

12.6 Conclusions

In this study, a continuous and compartmentalized population balance model of a granulation process was presented, which depicted a realistic model of a pilot-plant granulation process. The model considered was based on granulation mechanisms that have been previously derived from first principles and validated at the batch scale for different operating conditions and formulations. Model simulations elegantly captured the effect of nozzle positions on the granule output. As a direct consequence of the compartmentalization, the controllability of the granulation process could be examined taking into account all the important granule inputs and outputs, providing valuable insight into the control and operation of the granulation process. Results show that the average diameter, moisture content, and bulk density are realistic choices of controlled variables, with the three nozzle spray rates and the solid feed rate as possible choices of manipulated variables. The size distribution width was not sensitive to any of the manipulated variables and therefore deemed uncontrollable, thus corroborating the work of [16]. This is also an important finding as size distribution may be an important variable that needs to be tightly regulated. By means of step tests and a relative gain array analysis, the best possible control-loop pairing was identified. However, results confirmed the existence of high control-loop interactions under closed-loop conditions that would limit overall control-loop performance. As a result, this paved the way for a new process design to be proposed whereby via experiments and model simulations, it was seen that (i) granule size distribution was now controllable and (ii) less severe indirect interactions resulted under closed-loop conditions. These results are promising toward enhanced process design and control of a MIMO granulation process both in terms of regulating critical granule attributes that affect both the granulation process itself and subsequent downstream unit operations.

References

1 Mort P. R. Scaleup of binder agglomeration processes. Powder Technology. 2005;150:86–103.
2 Bardin M., Knight P. C., Seville J. P. K. On control of particle size distribution in granulation using high shear mixers. Powder Technology. 2004;140:169–175.
3 Knight P. C., Instone T., Pearson J. M. K., Hounslow M. J. An investigation into the kinetics of liquid distribution and growth in high shear mixer agglomeration. Powder Technology. 1998;97:246–257.
4 Litster J. D. Scaleup of wet granulation processes: science not art. Powder Technology. 2003;130:35–40.
5 Eek R. A., Bosgra O. H. Controllability of particulate processes in relation to sensor characteristics. Powder Technology. 2000;108:137–146.
6 Semino D., Ray W. H. Control of systems described by population balance equations-I. Controllability analysis. Chemical Engineering Science. 1995;50:1805–1824.
7 Fevotte F., Doyle III F. J. Sensitivity analysis of multi-regime population balance model for control of multiple particulate properties in granulation, in *American Control Conference*, Portland, OR, 2005.
8 Ramachandran R., Poon J., Sanders C., *et al.* Experimental studies on distributions of granule size, binder content and porosity in batch drum granulation: inferences on process modelling requirements and sensitivities. Powder Technology. 2008;188:89–101.
9 Poon J. M. -H., Ramachandran R., Sanders C. F. W., *et al.* Experimental validation studies on a multi-scale and multi-dimensional population balance model of batch granulation. Chemical Engineering Science. 2009;64:775–786.
10 Poon J. M. -H., Immanuel C. D., Doyle III F. J., Litster J. D. A three-dimensional population balance model of granulation with a mechanistic representation of the nucleation and aggregation phenomena. Chemical Engineering Science. 2008;63:1315–1329.

11 Ramachandran R., Immanuel C. D., Stepanek F., Litster J. D., Doyle III F. J. A mechanistic model for granule breakage in population balances of granulation: theoretical kernel development and experimental validation. Chemical Engineering Research and Design. 2009;87:598–614.

12 Ramkrishna D. Population Balances. Academic Press, San Diego, 2000.

13 Immanuel C. D., Doyle III F. J. Solution technique for a multi-dimensional population balance model describing granulation processes. Powder Technology. 2005;156:213–225.

14 Tan H. S., Salman A. D., Hounslow M. J. Kinetics of fluidised bed melt granulation IV: selecting the breakage model. Powder Technology. 2004;143–144:65–83.

15 Pinto M. A., Immanuel C. D., Doyle III F. J. A two-level discretisation algorithm for the efficient solution of higher-dimensional population balance models. Chemical Engineering Science. 2008;63:1304–1314.

16 Glaser T., Sanders C. F. W., Wang F. Y., *et al.* Model predictive control of drum granulation. Journal of Process Control. 2009;19:615–622.

17 Immanuel C. D., Doyle III F. J. Computationally efficient solution of population balance models incorporating nucleation, growth and coagulation: application to emulsion polymerization. Chemical Engineering Science. 2003;52:3681–3698.

18 Pinto M. A., Immanuel C. D., Doyle III F. J. A feasible solution technique for higher-dimensional population balance models. Computers and Chemical Engineering. 2007;31:1242–1256.

19 Ogunnaike B. A., Ray W. H. Process Dynamics, Modeling and Control. New York: Oxford University Press, 1994.

20 Smit D. J., Hounslow M. J., Newman R., Paterson W. R. Aggregation and gelation. 3. Numerical classification of kernels and case studies of aggregation and growth. Chemical Engineering Science. 1995;50:1025–1035.

21 Adetayo A. A., Litster J. D., Desai M. The effect of process parameters on drum granulation of fertilizers with broad size distributions. Chemical Engineering Science. 1993;48:3951–3961.

22 Adetayo A. A., Ennis B. J. A unifying approach to modeling coalescence mechanisms. AIChE Journal. 1997;43:927–934.

13

A Perspective on the Implementation of QbD on Manufacturing through Control System: The Fluidized Bed Dryer Control with MPC and NIR Spectroscopy Case

Leonel Quiñones, Luis Obregón, and Carlos Velázquez

Pharmaceutical Engineering Research Laboratory, Chemical Engineering Department, University of Puerto Rico at Mayaguez, Mayaguez, Puerto Rico

13.1 Introduction

The Quality by Design (QbD) initiative ultimately seeks to change the manufacturing mind-set. The essence of the idea is to design the operation of a process in such a way that its product always has the desired characteristics or attributes. It is a proactive approach to manufacturing where the allowable range of operating conditions to obtain the desired quality are determined beforehand so that the process can be adjusted in real time in the presence of expected disturbances to ensure the desired quality. This different thinking of manufacturing could be implemented, for example, in drying processes.

Drying is an ancient process that has been employed for ages, even with wet clothes. In the pharmaceutical and food industries, drying is often used to reduce moisture content in powder or solid products in order to inhibit microbial growth, ensure flowability, and avoid agglomeration during storage. However, it has not been until recently that drying has been looked at for control and optimization.

During the drying process, a wetting agent is removed from a solid material applying thermal and/or mechanical energy through combinations of gas flows, temperature changes, or vacuum ambient (low pressures). Despite the phenomenon used, the objective is to obtain a dry solid material. Many papers studying the drying phenomenon and its control have been published; only several are mentioned here for the sake of space. Liu and Bakker-Arkema [1] applied a model predictive controller to the drying of grains; Davidson *et al.* [2] studied the drying of ginseng roots in a controlled three-stage drying system; Nagaya *et al.* [3] controlled the airflow and temperature to dry at low

Comprehensive Quality by Design for Pharmaceutical Product Development and Manufacture, First Edition. Edited by Gintaras V. Reklaitis, Christine Seymour, and Salvador García-Munoz. © 2017 American Institute of Chemical Engineers, Inc. Published 2017 by John Wiley & Sons, Inc.

temperature a food ingredient; Yuzgeç *et al.* [4] used nonlinear predictive control to dry baker's yeast adjusting temperature and inlet humidity applying some constraints; De Temmerman *et al.* [5] used the model predictive control (MPC) strategy to control the drying of pasta; Regina-Luz *et al.* [6] applied the well-known PID control law to dry soybean meal by adjusting the inlet soybean meal rate and the inlet temperature of the drying air; and Zacour *et al.* [7] used a hybrid control for the drying of a pharmaceutical blend in a fluid bed processor including a cooling stage of the blend.

Other works have focused specifically on fluidized bed dryer (FBD) (described later) due to the widely used of these units in different industries: Temple *et al.* [8] controlled the fluid bed drying of tea using PID-based control; Atthajariyakul and Leephakpreeda [9] used fuzzy logic for optimal control of rice paddies adjusting temperature and air recycle ratio; Tasirin *et al.* [10] optimized the operating parameters (bed depth, air velocities, operating temperatures, and drying time) for the drying rate; Wang *et al.* [11] manipulated the air velocity using also PID; and Köni *et al.* [12] used another control strategy, known as genetic algorithm (GA), to adjust air temperature and flow rate for the dry baker's yeast.

The main contributions of the mentioned works are the demonstration that the drying process can be controlled, that the control can be achieved by different control schemes and by manipulating more than one variable, and that the desired moisture content or, in other words, the desired quality can be obtained. Moreover, description of how to implement the control scheme is provided in some cases. However, the works are not discussed from the point of view that the control scheme should operate inside a prescribed feasible space of the manipulating variables. This feasible space must be determined beforehand to achieve the desired quality without violating any limitation or safety of the process. This feasible space constitutes the design space that will govern the actions of the controller. Therefore, this paper focuses on presenting, from the QbD perspective, an example of the control of a fluidized bed drying of a pharmaceutical blend with the MPC strategy, using a near-infrared (NIR) spectrometer as moisture sensor and an industrial control system.

13.2 Theory

13.2.1 Fluidized Bed Dryers (FBDs)

FBDs [13] are very common in the manufacturing of a large number of products. FBD units can be operated either batchwise or continuously, utilize a continuous flow of gas to disperse the powders throughout a confined space, and remove any unbound solvent from the surface or pores of the powders. The dispersion frees all the bulk surface area of the powders increasing the drying rate obtaining in the end a more efficient drying process.

Drying involves two major phenomena: constant rate evaporation and diffusion-limited regime [13]. In the constant rate period, unbound solvent or water at the external surface of the particles is removed by evaporation at a constant temperature determined by the equilibrium temperature corresponding to the pressure of operation and thus producing a constant reduction in moisture content of the powders. This evaporation rate is heat transfer driven and thus proportional to the temperature difference and velocity of the passing gas [14]; therefore the phenomenon can be governed by manipulating the air temperature and gas velocity (independent variables).

In the diffusion-controlled regime, water entrapped in pores within particles or intra-particles diffuses to the external surface where it gets removed by the passing gas. Diffusion is a much slower process than the former [14] and happens primarily after the constant evaporation rate period is over, causing a change in the dynamics of the entire drying. By contrast to the former phenomena, the diffusion rate can only be increased by increasing the temperature [14]. However, the increase in diffusion is not linear with temperature and since the water is entrapped in the confined space of the pores, the drying gas flow rate has little impact and only has an effect after the molecule has reached the outer surface of the particles [11, 13, 14].

Many papers ([8, 15–17], to mention just a few representative type of models) have presented empirical and mechanistic models based on the phenomena discussed to relate the drying to the independent variables. These models can be employed to compute a set of values of the independent variables that when implemented will achieve specific levels of the drying state (dependent variable). Therefore, they could be used to establish the operating space that ensures the desired quality, establishing the QbD space. Moreover, the same models could then be used within industrial control systems to provide the capacity to perform similar computations during manufacturing that would ensure the achievement of the desired quality in the product, that is, the desired moisture content in the FBD case. As such, the control of the manufacturing operation would be based on developed process knowledge, gathered within the process model, rather than based on a limited experiential scheme.

13.2.2 Process Control

Control algorithms are designed to maintain at or drive the process to a target state. In general, there are many ways to achieve this goal. In the batch case, for example, the control strategies could employ one of two typical approaches: (i) controller tracks a preferred profile to achieve the desired end point or (i) the profile is only limited by operational constraints and control is purely to achieve the target end point. Independently of the approach, some control strategies, such as the classical PID, manipulate one independent variable, without considering constraints at all or using a model of the process, to drive one dependent variable to a desired target. Other strategies, such as MPC and GAs, use a process model to manipulate one or more independent variables at the same time to drive one or more dependent variables to their corresponding desired values. These strategies can be implemented so as to include constraints, typically physical, to ensure they are not violated during the operation. In the event of an external disturbance on the dependent or controlled variable, the control strategy will reject it by compensating with a suitable change in the independent variable or variables so that the target value is reached.

13.2.2.1 Proportional Integral Derivative (PID) Control

The PID control law, described by Equation 13.1, uses the difference, $e(t)$, between the desired value of the dependent variable (process or controlled variables) and its measured value to compute the changes in the independent variable (manipulated or input variables):

$$CO(t) = K_c \left\{ e(t) + \frac{\int_0^\infty e(t)\,dt}{\tau_I} + \tau_D \frac{de(t)}{dt} \right\} \tag{13.1}$$

PID is a control strategy heavily used in the industry because of its demonstrated versatility and ease of implementation. The first term, known as proportional, computes changes in the manipulated variable proportionally to the difference between the desired value and the measured value causing a rapid corrective action. The second term, known as integral, keeps modifying, using an integral, the changes in the manipulated variable as long as the error is different than zero. The third term, known as derivative, is used to accelerate the response, but it is avoided when high variability is present in the measured value.

In spite of the proven capabilities, it has limitations in handling constraints and in considering all the relationships in processes with multiple inputs and multiple outputs. The further major drawback is the inability to provide an avenue for optimizing the process, although it can achieve the desired product quality set point.

13.2.2.2 Model Predictive Control (MPC)

Details about the MPC strategy, which is based on Figure 13.1, can be found in most process control textbooks [18, 19]. MPC uses a representative model of the process to compute optimal changes of the inputs (manipulated) variables, $u(k)$ and beyond k, to drive the controlled variable $\{\hat{y}(k+1/k)\}$ toward the specified desired value (target) in an optimal manner to achieve the desired quality. During the computation of the optimal changes, it could use physical constraints to ensure a feasible operation.

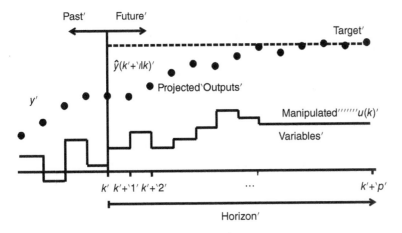

Figure 13.1 MPC computed trajectories of process variables.

The moves or changes in the inputs are based on the control performance index given in Equation 13.2:

$$J = E(k+1)^{\mathrm{T}} \mathbf{Q}E(k+1) + \Delta U(k)^{\mathrm{T}} R\Delta U(k) \tag{13.2}$$

where $E(k+1)$ would be in this case the difference between the desired value of the dependent variable (moisture content in a FBD) and its model trajectory ($\hat{y}(k+I/k)$), projected outputs), ΔU is the changes or moves of the inputs, in this case gas speed and temperature, and $\mathbf{Q}$ and R are weight or tuning matrices.

The minimization of J without constraints would produce the analytical solution given below:

$$\Delta U(k) = \left(K^{\mathrm{T}}\mathbf{Q}K + R\right)^{-1} K^{\mathrm{T}}\mathbf{Q}\hat{E}^\circ(k+1) = K_c\hat{E}^\circ(k+1) \tag{13.3}$$

where $\hat{E}^\circ(k+1)$ is the error between the target value and the projected outputs and K is a matrix that incorporates the dynamics of the process, which can be represented, for example, by Equation 13.4. The errors are obtained using the process model and previous values of the inputs. It is in Equation 13.3 that can be designed to compute the optimal values of the inputs so that the values ensure no violation of any physical constraint or requirement, and the achievement of the specified desired state or in other words the desired quality. The design of Equation 13.3 includes the implementation of the process model, the specification of the tuning parameters $\mathbf{Q}$ and R [20], and the inclusion of constraints. These degrees of freedom are the advantage of MPC over PID where information of the process behavior and constraints cannot be explicitly included in the strategy.

The process model for MPC can be of many different forms; it can be completely data driven or theoretical. The advantage of the data-driven models is that they are based on the actual behavior versus the theoretical models that, due to the many assumptions, miss the actual process behavior in some situations. Despite of the advantage of the data-driven models, caution must be exercised when applying it due to the possible large prediction errors (i) when employed outside the range of the experimental data and (ii) when predicting underlying nonlinear dynamics. Independently of the nature of the model, it should include explicitly all the inputs of the process to be useful for control and optimization and, as such, implementation of QbD for the product.

This chapter demonstrates how the QbD framework can be implemented in real time by using a process control strategy to insure that the desired quality of the product is maintained [21]. The work used a FBD, the MPC strategy, and a NIR spectroscopy-based sensor for the moisture content to demonstrate the enforcement of QbD during real time operation. The quality attribute of interest was the moisture content. It was demonstrated that MPC was able to achieve the desired quality and characteristics of the product by changing the airflow and its temperature without violating the physical constraints imposed. Therefore, the design and implementation of a control strategy such as MPC would insure that the product would reach the desired quality maintaining the operation inside the feasible design space. A perspective on the implementation of QbD in a continuous FBD is also provided.

13.3 Materials and Methods

13.3.1 Materials

The material to be dried was granulated lactose anhydrous, NF direct tableting, prepared using distilled water with a pH range of 5.5–6.5 as a binder. This was done by first adding the lactose anhydrous to a container where it was mixed with the binding solution (water) manually. Sufficient solution was added to reach the initial target moisture content of 11%. Next, the premixed materials (2 kg in total) were placed in the high shear granulator, Erweka Gmbh-type SW1/S, which was operated at 400 rpm for 5 min.

13.3.2 Equipment

Figure 13.2 depicts the P&ID of the FBD unit used for the experiments: model Aeromatic AG STREA 1. The system consists of a blower with a variable speed motor operating in the range from 55 to 120 m^3/h. The air enters the bowl from the bottom and leaves through an HEPA filter at the top. It houses an electrical

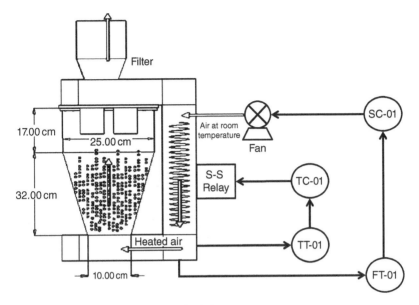

Figure 13.2 Schematic of a fluidized bed dryer.

resistance heater at the blower exit that enabled the heating of the air. The inlet air temperature was maintained from 25 to 100°C, although the actual heating capacity of the resistance heater exceeds the limit of 100°C. The heat produced by the resistance heater was manipulated by changing the power to the resistance with a solid-state rectifier as the actuator.

The FBD has a 16 l bowl to contain the material to be dried; however only a maximum of 2 kg of powders can be handled at a time. The airflow rate must be set so that the fluidized powders do not overpass the conical section of the bowl so as to minimize the amount of powders that reach the internal filters. This is necessary to avoid clogging of the filters. The bowl has a port where the sensor probe was placed for inline measurement. At the bottom of the bowl, a perforated plate serves to distribute the airflow so that potential fluidization problems, such as dead zones, are avoided. The dryer was accessorized with two sensors, one for temperature and one for airflow, to monitor the manipulated variables.

The distributed control system Delta V from Emerson Process Management was the platform used to implement the MPC strategy on the FBD. A PC was used to execute some of the calculations and to receive the signal from the NIR detector. The latter was an NIR model 128 Element T.E. cooled InGaAs Array from Control Development Incorporated (South Bend, IN). It was configured to scan on a wavelength range from 1107 to 2218 nm to monitor the actual value of the moisture content.

13.3.3 MPC Implementation

The MPC algorithm configured in the DCS used the current process data of the air inlet flow rate and temperature and the current predicted value of moisture content (received through OPC connection from the NIR system) to compute the new optimal values of the air inlet temperature, $T_{air,i}$, and flow rate, $F_{air,i}$, which were implemented and maintained until the next sampling time. The new values are implemented only if they are inside the defined design space (Figure 13.3) to achieve the desired final moisture content. If not, an appropriate mathematical logic assigns the corresponding limits of the design space or clamps the values as the new "optimal" values for the particular variable(s), and then these clamped values are the ones implemented. Figure 13.3 depicts the operating capabilities (gray) of the FBD and then the specific design space (white), including the physical constraints required, would ensure the desired quality of the dried material. The design space was established for a typical moisture content or quality attribute range from 2 to 20% (w/w).

This design space was implemented through the MPC algorithm applying the following constraints:

- $T_{air,i} \leq 85°C$
- $55 \leq F_{air,i} \leq 120\,m^3/h$

Once the optimal values were limited to feasible ones using the instructions mentioned previously, the clamped values were sent, as desired set points for air inlet temperature and flow rate, to two local PID loops (TC-01 and SC-01 in Figure 13.2), which serve to implement process variable changes to achieve

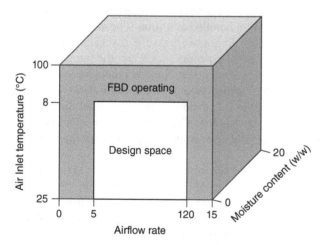

Figure 13.3 Design space of the air inlet temperature and flow rate for proper operation of the FBD and desired quality of the dried material.

those set points. The advantage of using the two local PID loops is that both ensure that the set points received are maintained despite any disturbances that could affect the air inlet temperature and speed. For example, if the temperature of the air entering the heating device decreases, the air inlet temperature entering the FBD will then decrease proportionally if the PID loop is not available. That decrease will impact the drying rate by reducing it and thus increasing the drying time. The local temperature-PID would reject the temperature disturbance, and therefore, the drying rate will be maintained. A variation of the clamping option could be the direct incorporation of the constraints, if the DCS allows it, into the algorithm to compute the optimal values of the inputs, making the clamping unnecessary.

The actual implementation of the MPC with NIR spectroscopy as sensor can present some technological challenges. In this application, the NIR scanning rate was limited to 3 s; therefore the control cycle was executed each 3 s. The NIR value was computed in the separate computer using the integration softwares specified in Figure 13.4: Spec32, Instep, and MATLAB®. This integration had a technological-based limitation to handle large quantities of data and on the speed to handle it, which imposed the 3-s scanning rate. However, this

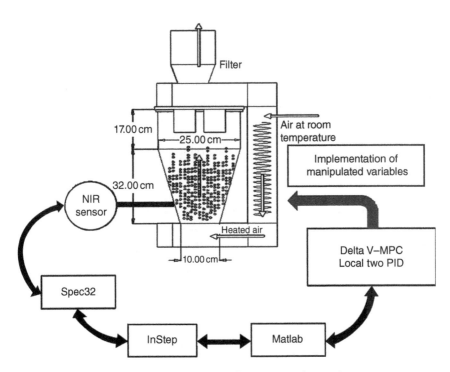

Figure 13.4 Diagram of the system integration for sensing and control.

limitation can easily be overcome by integrating technologies with the right design, that is, adequate computational capacity and speed.

13.4 Results and Discussion

13.4.1 Process Model

The first step to implement the MPC solution was the development of the FBD model, which was of the data driven type. Temple *et al.* [8] used data-driven low order transfer functions to model their continuous FBD. Following their approach, it was assumed for this work that the batch fluid bed drying behaves quite similar to the dynamic part of the continuous drying for a step change in one of the inputs. Therefore, the batch case could be modeled with a low order transfer function without the contribution of the inputs of the material entering and leaving the dryer. The model in this case is as follows:

$$W_p(s) = \frac{-K_F}{\tau s + 1} F_{air,i}(s) + \frac{-K_T}{\tau s + 1} T_{air,i}(s) \tag{13.4}$$

where $F_{air,i}$ is the air inlet flow, $T_{air,i}$ is the air inlet temperature, Wp is the powder moisture content as a function of time, and the fraction terms are the transfer functions in Laplace domain. This equation permits the calculation of powder's moisture content for any values of $F_{air,i}$ and $T_{air,i}$.

Table 13.1 includes the parameters for Equation 13.4 fitted to several drying curves with a partial least squares technique. The fitted model parameters were then used to configure the constant matrix K_c in Equation 13.3. A detailed description of the procedure to obtain the drying curves, the fitted parameters, and K_c can be found in Obregón *et al.* [22].

Table 13.1 Nominal values of the FBD and MPC tuning parameters.

Parameter	Nominal values
K_F	−0.0768
K_T	−0.0987
τ	491 s
Q_1	4×10^{-7}
Q_2	4×10^{-6}
R	0.3

13.4.2 Control Performance with Nominal Process Parameters

Table 13.1 also includes the tuning parameters used to configure the MPC algorithm, Equation 13.3. These parameters were obtained using the tuning guidelines of Dougherty and Cooper [20] and some fine-tuning to obtain an acceptable controller behavior. Specifically, the initially computed Q values were further adjusted as to obtain a compromised between aggressive changes in air inlet temperature and speed and drying rate. Too aggressive changes, as a result of large Q values, would cause a too rapid drying, which would have caused the granules to entrap water; too slow changes, corresponding to small Q values, would not be a desirable operation because of extended drying times. In the end, the Q values established tended to produce changes in the manipulated variables that dried at an adequate rate.

Figures 13.5 and 13.6 depict the NIR-predicted moisture content of four different runs under closed loop and their respective airflow rate and temperature profile computed by the MPC, respectively. The slightly differences noticed between the four runs are related to initial moisture content, mean particle size, and particle size distribution of the granulated lactose; see Table 13.2. Specifically, the difference in particle size establishes difference in surface area for energy and mass transfer; therefore it affects the drying rate. The mentioned differences caused variances in the response of the controller to achieve the desired value as will be seen in Figure 13.6.

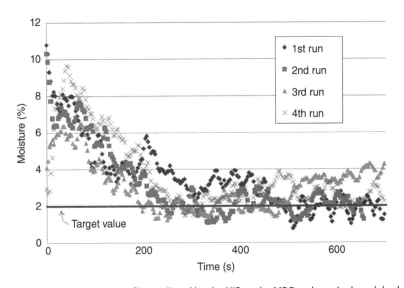

Figure 13.5 Moisture profile predicted by the NIR under MPC and nominal model values.

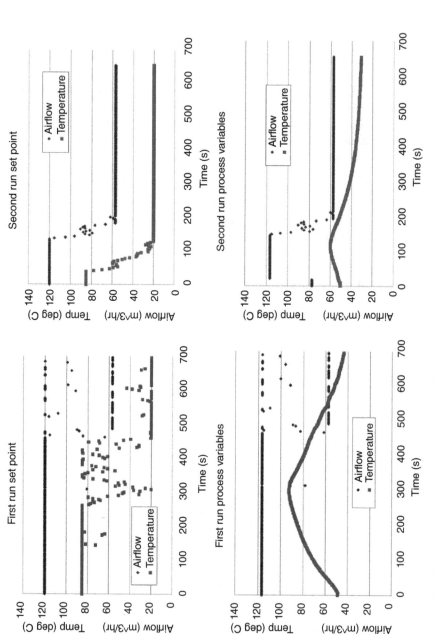

Figure 13.6 Manipulated variables computed by the MPC (top) and its values applied through local PIDs (bottom), nominal model values, run #1, and #2.

Table 13.2 Initial conditions of the granulated lactose.

Sample	Moisture content	Mean diameter, D_{av} (µm)	D_{10} (µm)	D_{90} (µm)
1	10.8	454	361	543
2	10.5	449	362	530
3	10.1	432	330	542
4	10.3	483	391	588

The moisture content set point (desired value) trajectory given to the MPC was a step change from the initial moisture content to 2%. As can be seen in Figure 13.5 for all four runs, the controller configuration was able to drive the moisture content to the set point of 2%. Since this is a batch operation, the drying curve exhibits a decreasing monotonic behavior, since the powders could not gain back humidity from the passing gas at the conditions used.

Figure 13.6 includes two samples of the MPC computed values (top figure) of airflow rate and temperature and its respective real behavior in the bottom figures. All of the values established as set point for the two local PID loops fell within the acceptable operating space. As can be seen, the patterns applied by the controller agreed with the theory. At the beginning during the constant evaporation rate period, the system maintained the highest possible airflow and temperature to promote a faster evaporation rate. This was similar to the open loop operation. After that, the system started to decrease those values since the drying at this point is mainly governed by the diffusion phenomenon. As such, the drying rate could not be improved by maintaining high values of the two manipulated variables.

These samples demonstrate how the MPC algorithm manipulated the changes of each input to drive the moisture content to the desired value, without violating the physical constraints, despite the differences in the physical characteristics of the granulated blend. The configuration of the MPC with the process model, the constraints, and the tuning parameters permitted the MPC computing, using predicted behavior, only feasible operating conditions of the inputs that more importantly ensured the desired quality.

In this example, a step change in set point (moisture content) was established though other types of profiles can be used too. For example, the set point profile could be a time-dependent profile of the drying rate to ensure the granules were dried at a rate that would avoid, for example, caking or entrapment of solvent inside the granule, or too much attrition. These QbD requirements could also be implemented in real time with strategies such as MPC because the methodology can readily accommodate both state and control variable constraints.

13.4.3 Control Performance with Non-nominal Model Parameters

Previous works with model-based control strategies have demonstrated the strong dependence of the strategies upon the accuracy of the models to predict the controlled variable. The next set of results demonstrate that even if there were a slight change in the formulation, which would be the same as to use different model parameters, the integrated system would provide adequate control. In this experiment, the model and the tuning for the MPC algorithm were modified with the parameters in Table 13.3. The performance of the MPC with the nominal model parameters, shown earlier, was compared with the one with the non-nominal parameters. For comparison purpose, the same type of powder system used in previous experiments was utilized here.

Three runs were performed with the results of moisture content prediction and control depicted in Figure 13.7, while the MPC computed values of airflow rate and temperature are depicted in Figure 13.8. The initial conditions of the three runs for the non-nominal case are described in Table 13.4.

Comparing Figure 13.5 with Figure 13.7, it can be seen that the system was able to achieve the desired set point of 2% moisture content in approximately 5 min in both cases, except in run 3 of the non-nominal case. However, there is a difference in the dynamic part of the curves. This difference was the result of the controller behavior, which favored, especially for the non-nominal case #3, higher temperatures, and lower airflow rates. Figures 13.6 and 13.8 show similar response of the implemented values of airflow rate and temperature to the set points established by the MPC.

Figure 13.8 corresponds to the airflow rate and temperature for run #2 of the non-nominal parameters, where it can be seen that the moisture content increased around 300 s. The same behavior is noticed in the temperature profile where an increase in temperature is seen around 300 s. The MPC algorithm increased the temperature at that point using the process model and

Table 13.3 Non-nominal values for the FBD and corresponding MPC tuning parameters.

Parameter	Non-nominal value
K_F	−0.05
K_T	−0.04
τ	0.3
Q_1	2.43×10^{-5}
Q_2	1.15×10^{-5}
R	0.3

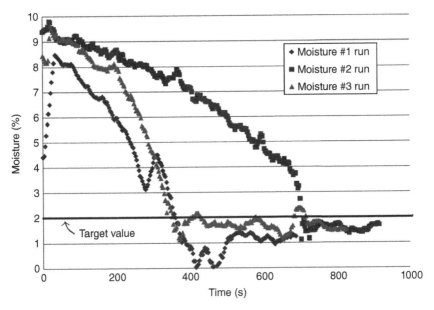

Figure 13.7 Moisture profile predicted by the NIR under MPC for non-nominal model values.

configuration determined in order to achieve the desired value despite the difference in the powder's physical characteristics. This is another demonstration that the quality can be achieved with MPC without violating any physical constraints applying the entire design space despite the changes in operating conditions or system's characteristics.

It is worth noticing that the MPC tuning parameters, the **Q** values, were different for both cases. This was necessary due to the characteristics of the granulated lactose. In the latter case, the granules were slightly bigger with a mean diameter of 532 (μm), which required a more aggressive controller performance to dry the granulated material as fast as in the nominal case. In the practice, commercial control systems are designed to provide the degree of freedom of tuning the control algorithm in real time. Depending on the controlled process behavior, it is possible to adjust the tuning parameter during the manufacturing so that the control algorithm can work better inside the design space in achieving the desired quality. For example, if a slow drying is noticed, the **Q** values could be increased to turn the controller into a more aggressive one. However, this must be done very cautiously and by a very experienced professional. If the wrong values of **Q** are given, the controller would operate outside the design space and the product will end up with the wrong quality.

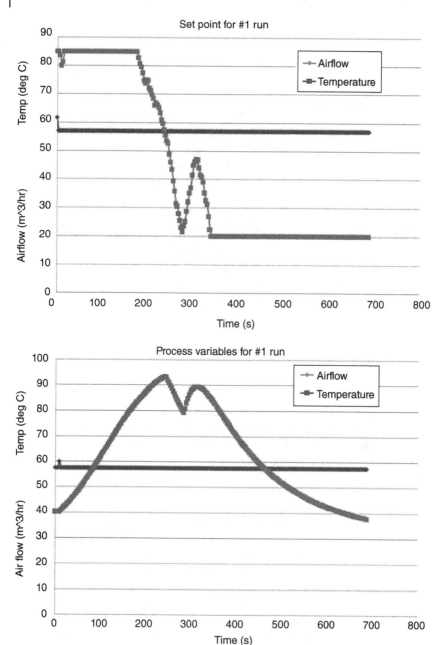

Figure 13.8 Manipulated variables computed by the MPC (top) and process its values applied through local PIDs (bottom), non-nominal model values, run #1.

Table 13.4 Initial conditions of the granulated lactose for the non-nominal case.

Sample	Moisture content	Mean diameter, D_{av} (µm)	D_{10} (µm)	D_{90} (µm)
1	10.1	447	377	517
2	10.3	463	358	545
3	9.8	452	375	527

13.5 Continuous Fluidized Bed Drying

In the continuous drying case, Figure 13.9, the material to be dried enters and leaves continuously the dryer vessel, in contrast to the batch case where all the material is fed to the vessel at once.

Additional differences include the following: (i) system reaches steady state in powder moisture content, (ii) less material can be processed per unit time, (iii) more material can be sensed percentagewise, (iv) a constant set point is typically established, and (v) the model has a different structure. Regarding the model, it contains the same inputs as in the batch case plus the inputs related to the material entering and leaving the unit. A generalized form of Equation 13.1 for the continuous case is shown in Equation 13.5:

$$
W_{p,o}(s) = \frac{-K_F}{\tau s + 1} F_{air,i}(s) + \frac{-K_T}{\tau s + 1} T_{air,i}(s) + \frac{K_m}{\tau s + 1} F_{p,i}(s)
$$
$$
+ \frac{K_W}{\tau s + 1} W_{p,i}(s) + \frac{K_Y}{\tau s + 1} Y_{A,i}(s)
$$

(13.5)

where the new inputs are $F_{p,i}$, powder inlet flow; $W_{p,i}$, powder inlet moisture content; and $Y_{A,i}$, air inlet humidity.

One more advantage of the continuous operation over the batch one is the number of degree of freedom the MPC controller would have to ensure the typically constant desired quality. For example, the MPC algorithm could manipulate the air inlet temperature and flow rate, as in the batch case, plus the powder inlet flow ($F_{mat,i}$) to ensure the final desired moisture content is achieved. In general, the MPC will consider all the interactions between the controlled variable (final moisture content) and selected input variables through the given model; therefore it will manipulate at the same time all those selected inputs in the optimum way as to fulfill the control objective. Contrary to the batch case, in this case the optimum values will be maintained constant as long as the controlled variable is at the target value. These optimal profiles of the inputs cannot be achieved by single PID controllers since its structure does not allowed inclusion of that type of information; therefore the control profile will always be a suboptimal one.

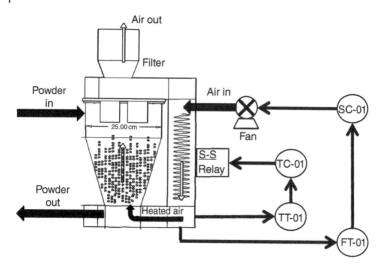

Figure 13.9 A continuous fluid bed dryer conceptual schematic.

13.6 Control Limitations

The values in Tables 13.2 and 13.4 are typical for normal operations; however disturbances can occur in some of the parameters like the initial moisture content or even the air inlet humidity. Depending on the size of the disturbance, the controller actions could be handicapped to achieve the target set point. For example, if the moisture content of the granulated material is 20% or above, the material will be too heavy to be fluidized. Even if the air speed is increased to the maximum capacity of the fan, the material will not fluidize, decreasing the surface area for energy and mass transfer and thus dropping the drying rate to a negligible one. Moreover, the high speed will cause channeling in the wet material ensuring that some material will not dry at all. Therefore, this must be considered during the establishment of the design space, as in Figure 13.3, to avoid operations that are destined to fail.

Another limitation can arise if the air inlet humidity is too high. The consequence is a large reduction in the capacity of the air to evaporate the water in the powders. In the continuous case, either the residence time (time spent inside of the dryer) is not enough to dry the powders as desired or it has to be increased so much that the powder flow rate is too small for an adequate operation. In this case, the controller is limited on what it can do to achieve the desired behavior; therefore the air inlet humidity must be considered too in the design space.

13.7 Conclusions

An FBD was controlled with MPC operating inside the desired design space to consistently obtain the desired moisture content. By tuning properly the control algorithm with knowledge about the process, the desired quality level was achieved consistently. A real control was obtained using an NIR sensor and the advance control strategy MPC with a low order transfer function as the process model and constraints in the inputs. The MPC strategy controlled the FBD even when non-nominal model parameters were used in the MPC configuration. In this case, the MPC was fine-tuned to ensure the adequate the control. This case can be interpreted as a disturbance rejection case demonstrating the robustness of the control algorithm. Additional disturbances, such as large powder initial moisture content or air inlet humidity, could be studied to expand the robustness of the algorithm.

The implementation of the QbD framework to a continuous FBD to achieve the desired moisture content would be similar to the batch case. It would manipulate the common inputs and constraints plus it could manipulate other inputs, especially those related to the material entering and leaving, and other constraints. The major advantage, though, is the requirement of maintaining the desired quality at a steady state (constant value).

Acknowledgment

The authors appreciate the support of Pfizer Co., the NSF-ERC for Structured Organic Particulate System (EEC-0540855), and INDUNIV, a local consortium of industry, government, and academia.

References

1 Liu Q, Bakker-Arkema FW. A model-predictive controller for grain drying. Journal of Food Engineering. 2001;49:321–326.

2 Davidson V J, Li X, Brown RB. Forced-air drying of ginseng roots: 2. Control strategy for three-stage drying process. Journal of Food Engineering. 2004;63:369–373.

3 Nagaya K, Li Y, Jin Z, Fukumuro M, Ando Y, Akaishi A. Low-temperature desiccant-based food drying system with airflow and temperature control. Journal of Food Engineering. 2006;75:71–77.

4 Yuzgeç U, Becerikli Y, Türker M. Nonlinear predictive control of a drying process using genetic algorithms. ISA Transaction. 2006;45(4):589–602.

5 De Temmerman J, Dufour P, Nicolaï B, Ramon H. MPC as control strategy for pasta drying processes. Computers & Chemical Engineering. 2009;33(1):50–57.

6 Regina-Luz G, dos Santos-Conceição WA, de Matos-Jorge LM, Paraíso PR, Gonçalves-Andrade CM. Dynamic modeling and control of soybean meal drying in a direct rotary dryer. Food and Bioproducts Processing. 2010;88:90–98.

7 Zacour BM, Drennen JK III, Anderson CA. Hybrid controls combining first-principle calculations with empirical modeling for fully automated fluid bed processing. Journal of Pharmaceutical Innovation. 2012;7(3–4):140–150.

8 Temple SJ, van Boxtel AJB, van Straten G. Control of fluid bed tea dryers: controller performance under varying operating conditions. Computer and Electronics in Agriculture. 2000;29:217–231.

9 Atthajariyakul S, Leephakpreeda T. Fluidized bed paddy drying in optimal conditions via adaptive fuzzy logic control. Journal of Food Engineering. 2006;75:104–114.

10 Tasirin SM, Kamarudin SK, Ghani JA, Lee KF. Optimization of drying parameters of bird's eye chilli in a fluidized bed dryer. Journal of Food Engineering 2007;80:695–700.

11 Wang HG, Senior PR, Mann R, Yang WQ. Online measurement and control of solids moisture in fluidised bed dryers. Chemical Engineering Science. 2009;64:2893–2902.

12 Köni M, Yüzgeç U, Türker M, Dinçer H. Optimal quality control of baker's yeast drying in large scale batch fluidized bed. Chemical Engineering and Processing. 2009;48:1361–1370.

13 Mujumdar AS. Handbook of Industrial Drying (3rd edition). Boca Raton, FL: Taylor & Francis Group, 2007.

14 Benítez J. Principles and Modern Applications of Mass Transfer Operations (1st edition). New York: John Wiley & Sons, Inc., 2002.

15 Alden M, Torkington P, Strutt ACR. Control and instrumentation of a fluid bed dryer using temperature difference technique: I, development of a working model. Powder Technology. 1988;54:15–25.

16 Chandran AN, Rao SS, Varma YBG. Fluidized bed drying of solids. AIChE Journal. 1990;36:29–38.

17 Robinson JW. Improve dryer control. Chemical Engineering Progress. 1992;88(12):28–33.

18 Ogunnaike BA, Ray WH. Process Dynamics, Modeling, and Control. New York: Oxford University Press, 1994.

19 Seborg DE, Edgar TF, Mellichamp DA. Process Dynamics and Control (2nd edition). Hoboken: John Wiley & Sons, Inc, 2003.

20 Dougherty D, Cooper D. A practical multiple model adaptive strategy for multivariable model predictive control. Control Engineering Practice. 2003; 11:649–664.

21 Roggo Y, Chalus P, Maurer L, Lema-Martinez C, Edmond A, Jent N. A review of near infrared spectroscopy and chemometrics in pharmaceutical technologies. Journal of Pharmaceutical and Biomedical Analysis. 2007;44:683–700.

22 Obregón L, Quiñones L, Velázquez C. Model predictive control of a fluidized bed dryer with an inline NIR as moisture sensor. Control Engineering Practice. 2013;21(4):509–517.

14

Knowledge Management in Support of QbD

G. Joglekar, Gintaras V. Reklaitis, A. Giridhar, and Linas Mockus

Davidson School of Chemical Engineering, Purdue University, West Lafayette, IN, USA

14.1 Introduction

In the past decade a paradigm shift has taken place in pharmaceutical development, the set of activities that encompasses designing a drug product and its associated manufacturing processes to result in a product that consistently and reliably achieves intended performance targets. The "quality by design" (QbD) approach described in the ICH guidelines (ICH Q8 [1], ICH Q9 [2], and ICH Q10 [3]) aims to replace empirical methods in drug and process development by a science and risk-based approach. In addition to a deeper scientific understanding of products and processes, risk management and knowledge management (KM) are considered critical for the implementation of this new paradigm.

During pharmaceutical development, significant amounts of information and data are generated. However, they have to be handled, managed, reused, and shared over the entire life cycle of a drug product. The use of KM during the entire product life cycle is emphasized by the ICH Q10 guidelines, defining it as a "systematic approach to acquiring, analyzing, storing, and disseminating information related to products, manufacturing processes and components." Holm *et al.* [4] defined the general objective of KM as "getting the right information to the right people at the right time," such that the information can be used more effectively. Thus, KM—in addition to risk management—is a key instrument for helping pharmaceutical product and process developers and manufacturers to implement QbD according to the ICH guidelines. However, as usual the ICH guidelines neither recommend nor describe any KM methods or tools for its implementation. Preferably, application of ICH Q8 elements starts in the early design phase of a drug product where both patient needs and process design are considered. During the design phase, it is important to determine the critical quality attributes (CQAs), to identify critical process parameters (CPPs) and critical material attributes (CMAs), and to understand how the process parameters and material attributes affect the CQAs. The relationship between process inputs (material attributes and process parameters) and the CQAs constitutes the design, which, when integrated with an appropriate process control strategy, leads to greater operational flexibility with reduced regulatory filing requirements. The QbD approach also plays a key role in the transition from batch to continuous pharmaceutical manufacturing [5].

There are three pillars for QbD implementation: change control, KM, and risk management. In this chapter we will focus on the KM pillar. Development knowledge supports filings and is used by regulators to understand the basis for the manufacturing process and its control strategy, including the rationale for the selection of CPPs and CQAs, and for the design space. As shown in Figure 14.1, KM is one of the enablers for the pharmaceutical quality system.

The knowledge management component of QbD includes both experimental and modeling aspects. An overview of quantitative models used in the drug product and process development is given by Chatterjee *et al.* [6]. The models may be broadly classified into mechanistic or first-principles models, empirical models, and semiempirical or hybrid models. In order to efficiently capture the knowledge about the products and processes, it is imperative to link experimental data and predictions with suitable models.

Under the QbD paradigm, mathematical models are important for capturing pharmaceutical process understanding and can be used effectively throughout

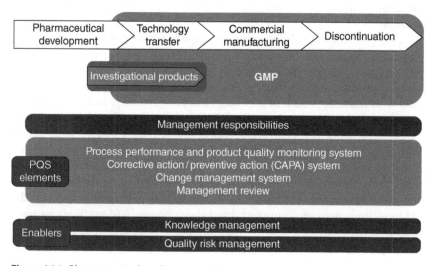

Figure 14.1 Pharmaceutical quality system [3].

development and manufacturing, including process design, scale-up, process monitoring, control, and continual improvement. Use of models can support efficient development and implementation of a robust process that ensures consistent manufacture of desired quality product. Although many such models have been implemented in pharmaceutical process development and manufacture, by and large these modeling approaches are still evolving and more understanding is expected to be garnered in the coming years.

In this chapter we outline a novel approach for systematically managing the knowledge gained during product and process development in the drug substance and drug product domains. The proposed approach also can be easily extended to drug product packaging.

First, the knowledge hierarchy is discussed, followed by a review of various tools for information and KM currently used in pharmaceutical process development. The conceptual framework of the proposed approach is discussed after that. The details of the proposed workflow-based framework are illustrated with the use of industrially motivated case studies. The remaining challenges in the development of the proposed framework are discussed in the last section.

14.2 Knowledge Hierarchy

The data–information–knowledge–wisdom hierarchy (DIKW), commonly known as the "knowledge pyramid," is shown in Figure 14.2. It is one of the fundamental conceptual models in the information and knowledge literature. The data are defined as symbols that represent properties of objects, events, and their environment. They are the product of observations. Information is

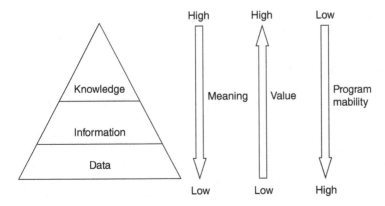

Figure 14.2 The knowledge pyramid.

contained in descriptions, answers to questions that begin with such words as who, what, when, and how many. Information is inferred from data.

While the distinction between data and information is well understood, the definitional statement on knowledge is often much more complex. Some of the broad definitions of knowledge are:

- Knowledge is information, to which is added expert opinion, skill, and experience, which can be used to aid in decision making.
- Knowledge is information that has been organized and processed to convey understanding, experience, accumulated learning, and expertise as they apply to a current activity.
- Knowledge is information from the human mind and includes reflection, synthesis, and context.

Knowledge exists along a continuum between tacit knowledge and explicit knowledge. Tacit knowledge refers to personal knowledge embedded in individual experience, while explicit knowledge refers to tacit knowledge that has been documented. We report on concepts and structures with the goal of the development of a framework for managing explicit knowledge.

A study reported by Junker *et al.* [7] highlights the efforts required to develop a coherent and comprehensive strategy for KM. An illustrative business case showed that for an example group of 50 staff, the total annual cost of the knowledge losses arising from some simple yet common circumstances, such as time wasted in searching but not finding information, time spent in looking for guidance on published values, and so on, ranged from $2 to 3 million. In addition, the authors have developed a conceptual design of a KM system that would fulfill the needs of a QbD approach for drug product development. The two key conclusions of the study were the following:

- A workflow-based framework could provide a solid foundation for tying together content repositories.
- There is a need to build tools to improve information linkages and interfaces to provide access across all functions and internal/external network sites.

14.3 Review of Existing Software

Systematic recording of data has been practiced in the pharmaceutical industry since the creation of a regulatory framework for drugs, first because of its importance in protecting intellectual property (IP) and second because of its requirement to support regulatory filings. It started with laboratory notebooks with handwritten entries and then progressed to recording information electronically. The commercial software packages for this purpose fall into one of the three categories: e-lab notebooks (ELN), laboratory information management systems (LIMS), and scientific data management systems (SDMS).

The first generation of ELN systems provided basic data capture capability used mainly for IP documentation and retention. They are closest to pure paper replacement. Separate applications were used for data aggregation, analysis, and reporting. Specific ELN solutions, which emerged from first-generation ELNs, enhanced user's efficiency by providing new tools previously unavailable, such as searchable reaction database, automated stoichiometry, automation of standard procedures, etc. Subsequently, domain-specific solutions were added to provide additional capabilities. These enhancements improved the productivity of end users. The current generation of tools addresses the needs to manage structured data and automation of the laboratory's information workflow as addition to ELN. There is a convergence of functionalities seen in ELN, LIMS, and even SDMS.

LIMS has been particularly successful in discovery research for standardized processes such as high-throughput screening. Few commercial LIMS products have the flexibility the researchers need to adapt to frequent changes in protocols and study designs. Even fewer products include the functionality for experiment design and documentation found in an ELN. Nevertheless, LIMS manages structured data, tracking, and automation tasks very well. Some of the popular LIMS in use are STARLIMS [8] and BIOVIA LIMS [9].

The BIOVIA corporation also provides a product named Discoverant [10], an enterprise manufacturing intelligence software platform that enables manufacturers to produce regulated and specialty products more efficiently. Discoverant provides on-demand access, contextualization, analysis, and reporting for all manufacturing, quality, and process development data. The four main areas it addresses: quality, technical operations, data analysis, and process development. It also facilitates managing contract manufacturing organizations (CMOs) through improved data sharing.

None of the existing software products provide a framework for capturing and sharing knowledge and its provenance across functions in a way stipulated and envisioned in the ICH guidelines. There is a tremendous potential for developing such a framework, which will naturally become the backbone of a QbD methodology.

14.4 Workflow-Based Framework

We have been advancing a workflow-based framework for managing explicit knowledge. This framework, first and foremost, has potential applications in all phases of pharmaceutical process development and can play a crucial role in technology transfer and continuous improvement of the knowledge base.

A workflow captures the operational aspect of a work procedure: how tasks are structured, who performs them, what their relative order is, how they are synchronized, how information flows to support the tasks, and how tasks are tracked. Workflows can be constructed at almost any stage of development and to any degree of complexity. The complexity varies based on the needs of the user community. There are four main workflow types: business, scientific, experimental, and manufacturing recipe.

Workflow-based KM systems have been developed for business and scientific workflows, while the LIMS have been used for managing experimental data but not for managing experimental procedures. Workflow-based systems for managing manufacturing recipes, particularly in the process development setting, have not been available commercially. However, several automation systems have been developed for implementing recipes for batch processes, for example, SIMATIC BATCH [11] by Siemens and System 800xA [12] by ABB.

14.4.1 Scientific Workflows

A scientific workflow is a formal description of a process for accomplishing a scientific objective, usually expressed in terms of tasks and their dependencies. Scientific workflows can be used during several different phases of a larger science process, that is, the cycle of hypothesis formation, experiment design, execution, and data analysis. Scientific workflows can include steps for the acquisition, integration, reduction, analysis, visualization, and publication of scientific data. They constitute knowledge artifacts or "recipes" that provide a means to automate, document, and make repeatable a scientific process.

The primary task of a scientific workflow system is to automate the execution of such workflows. These systems may additionally support users in the design, composition, and verification of scientific workflows. They also may include support for monitoring the execution of workflows in real time, recording the processing history of data, planning resource allocation in distributed execution environments, discovering existing workflows and workflow components, recording the lineage of data and evolution of workflows, and generally managing scientific data.

Taverna [13] is an open-source and domain-independent workflow management system (WFMS)—a suite of tools used to design and execute scientific workflows and aid in silico experimentation. It has been applied in bioinformatics, chemistry, data and text mining, and engineering. For example, Taverna allows Jet Propulsion Lab (JPL) scientists to model how mission design objectives impact science. One of the challenges faced by the research team is the integration of various models that are used to explore the science behind the JPL's missions. These models have been developed independently and run on different platforms, as a consequence of which their integration is not straightforward. JPL's solution is to wrap each model as a Web service and to execute them with a Web service-enabled workflow. Taverna enables dynamic Web Services Description Language (WSDL) introspection and presentation of the WSDL interface as ports of a process component.

The Smart Manufacturing Leadership Coalition (SMLC) has developed a smart manufacturing (SM) platform that marries cloud technologies with real-time manufacturing data and operational requirements, making it possible to build dynamic enterprise data systems, scale IT infrastructure, and manage software applications while managing their resulting actions locally [14]. The platform uses Taverna-based workflows to perform scientific computation that tie data input and data generation.

Cyber–physical systems (CPS) [15] falls in the scientific workflow category. This tool category targets integrations of computation and physical processes. For example, the Ptolemy project has been used in a variety of applications. Similarly, the Pegasus project encompasses a set of technologies that help workflow-based applications execute in a number of different environments including desktops, campus clusters, grids, and now clouds. Scientific workflows allow users to easily express multistep computations, for example, retrieve data from a database, reformat the data, and run an analysis. Once an application is formalized as a workflow, the Pegasus Workflow Management Service can map it onto available compute resources and execute the steps in appropriate order. Pegasus can easily handle workflows with several million computational tasks.

Over the past few years, a considerable amount of work has been done to automate the management of individual patient care in the healthcare industry. Primarily, the treatment of a health incidence is treated as a clinical workflow, with links to all the necessary diagnostics tests and information. The CHIRON project [16] intends to combine state-of-the-art technologies and innovative solutions into an integrated framework designed for an effective and person-centric health management along the complete care cycle.

14.4.2 Business Workflows

Business workflows are mainly concerned with the modeling of business rules, policies, and case management and therefore are often control and activity oriented. By contrast, to support the work of computational scientists, scientific workflows are mainly concerned with capturing scientific data analysis or simulation processes and the associated management of data and computational resources. While scientific workflow technology and research can inherit and adopt techniques from the field of business workflows, there are several sometimes subtle differences, ranging from the modeling paradigms used to the underlying computation models employed to execute workflows. For example, scientific workflows are usually data flow-oriented "analysis pipelines" that often exhibit pipeline parallelism over data streams in addition to supporting the data parallelism and task parallelism common in business workflows.

Workflow modeling is the key part of business process management (BPM) solutions, which have been prevalent in both industry products and academic prototypes since the late 1990s. There are two main formalisms used in WFMS, graph based and rule based. Current industrial standards such as the one developed by Business Process Management Initiative (BPMI) endorse the graph-based formalism as the process definition language. Examples of commercial BPMs that use graphical process definition include SAP NetWeaver [17], Tibco Staffware Process Suite [18], and Ultimus [19].

14.4.3 Comprehensive Workflow-Based Knowledge Management System

The KProMS system developed at Purdue University provides a framework for modeling all four of the workflow types mentioned earlier and managing information resulting from their implementation. In addition, the hierarchical structure of workflows in the work processes can be handled naturally by this framework. The underlying concepts of this framework are described by Joglekar *et al.* [20]. In the following sections we will describe the application of KProMS through two case studies.

14.5 Drug Substance Case Study

We use the semagacestat process described by Burcham *et al.* [21] to demonstrate the proposed workflow-based concepts in the drug substance domain. Semagacestat was a candidate drug for a causal therapy against Alzheimer's disease. The semagacestat process was selected because both experiments and mechanistic predictive models were used throughout the study. The objective of the study was to control the concentration of impurities in the drug substance to ensure that they were below the limits set by the FDA.

14.5.1 Process Description

The key chemical transformations during the synthesis of this API can be summarized by the simplified reaction scheme.

$$A \rightarrow B \rightarrow C$$

The final bond-forming step in the semagacestat process involves two chemical transformations: *N-t*-butoxycarbonyl amine protecting group removal followed by HCl-salt peptide coupling with (*S*)-2-hydroxyisovaleric acid (see Ref. [21] for details of the reaction scheme). The drug substance manufacturing process for semagacestat includes two additional steps, namely, formation of A when aminobenzoazapinone hydrochloride salt is reacted with BOC-protected alanine and the purification of C by selective crystallization followed by particle size reduction and drying; however, the focus is on the impurity formation during peptide coupling.

A number of the required operations, which occur during the final bond-forming step, introduce significant opportunity for generating impurities, that is, process-derived impurities. Components as innocuous as evolved gases, like CO_2, can have a dramatic effect on the impurity distribution in the product. If entrained/adsorbed CO_2 is not removed prior to the coupling step, a symmetrical urea is formed as a side product. This impurity, an unqualified impurity, has very low solubility. The low solubility results in very little rejection during crystallization of the active pharmaceutical ingredient (API). Thus, controlling the formation of urea became a primary focus of the intense multidisciplinary study presented in the referenced paper.

To fully understand and predict impurity formation, it was desired to build a mechanistic model of the reaction system. It was imperative to systematically determine each parameter required for such a complete model. The parameters were obtained experimentally and, in certain cases, by more than one experimental method in order to maximize accuracy and minimize error. As a consequence of this effort, detailed mechanistic models of CO_2 mass transfer, deprotection, and coupling reaction kinetics were developed.

Determining Henry's law constant for carbon dioxide in the reaction matrix was fundamental to the understanding of the mass transfer rate and the reaction rate for the formation of the urea impurity. To ensure accuracy and minimize error in determining these values, two orthogonal methods were utilized at multiple scales.

Mass transfer is inherently scale and equipment dependent. Even with geometrically similar reactor systems, mass transfer coefficients are difficult to predict. In the referenced case study, mass transfer coefficients were experimentally determined at laboratory scale. Ad hoc correlation was used to predict those coefficients at commercial scale.

For the deprotection reaction to occur, compound A must first be solubilized. In the case where the reaction rate is considerably higher than the rate of A dissolution, the rate of mass transfer has the ability to control the overall rate. In the event that the dissolution rate is considerably faster than the reaction rate, the overall reaction is controlled by the reaction rate and the reaction takes place more in the bulk liquid. Based on prior work, mass transfer was known to be rate limiting. As a result it was necessary to account for the solid dissolution mass transfer rate in the model. Reaction rate constants were experimentally determined for the deprotection and coupling reactions.

Some of the crucial studies performed in the initial stages of process development for a pharmaceutical substance are designed to develop better understanding of all the chemical reactions that take place during the synthesis. These include establishing the reaction mechanism, determining the kinetic rate constants, studying the role played by heat and mass transfer, and so on. The knowledge generated during these studies must be managed effectively because it is not only a result of significant investment of resources but more importantly has a long-term use all the way into manufacturing. By drawing on the Burcham *et al.* study, a general scheme based on workflow-based KM is described in the next section.

14.5.2 Workflow-Based Representation of the Semagacestat Study

The team in charge of the study was in the chemical manufacturing and controls (CMC) division of the business. The main objective of the study was to design and develop an impurity control strategy to deliver high quality API semagacestat. The impurities in the final API result from impurities entering with the raw materials and/or generated as reaction products. While the impurities entering with the raw materials can be reduced by tightening the specs, thus transferring the responsibility to the materials vendors, the impurities resulting from undesired side reactions must be controlled by identifying the proper operating regimes.

From the perspective of KM, the project can be represented as a workflow consisting of a series of activities with well-defined objectives. An example of a business workflow for the project is given in Figure 14.3. The workflow (dark gray rectangle) is named *ImpurityControl*, which is the name of the project. It represents a collection of tasks (light gray rectangles) and data nodes (rectangles with solid boundaries) that identify data specified or generated during the execution of that workflow. Each task consists of one or more subtasks (white rectangles). Dotted lines represent information flow between associated icons, while solid lines represent mass flow between associated icons. Typically, a project is organized as a hierarchy of such workflows. At the highest level, shown as Level 1 in Figure 14.3, the major tasks and their relationships are captured.

In this example, the project begins with the establishment of the reaction mechanism of the associated reactions. This work would be typically done by research chemists and may already be completed as part of discovery. In general, the various reactions occurring during synthesis are grouped into *n* reaction systems. The basic assumption is that reactions in a system occur simultaneously when the associated operation is performed. Conceptually, for each reaction system, there is one reactor operation.

A reaction system is a set of reactions describing the associated chemical transformations. Each reaction is represented by a reaction object, an instance

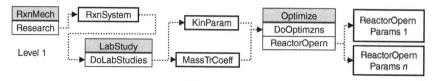

Figure 14.3 Workflow for the project to control impurities in a reaction system.

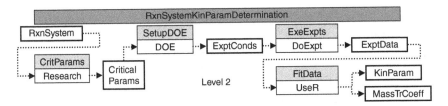

Figure 14.4 Workflow for the laboratory study for reaction kinetics.

of a specific class, which in turn is described by a standard set of attributes, such as chemical species identifiers, stoichiometric coefficients, key component, reaction rate expression, and rate constants. The data model for a reaction could be an ontology or a table in a database. In the KProMS system, which uses MySQL, a reaction is a row in a MySQL database table. A reaction system is defined as a list of IDs from the reaction table corresponding to the associated reactions.

A task object in a workflow has the following attributes: brief description of the purpose, the person responsible for that task, and other details such as where that task is performed and its estimated duration. The DoLabStudies subtask of the LabStudy task in Figure 14.3 invokes the workflow shown in Figure 14.4, a Level 2 workflow, for each reaction system identified to be relevant and important by the Research subtask. The workflow hierarchies accurately represent the distribution of work and responsibility in a typical work environment in that the group performing laboratories studies would be different from the group managing the main project.

The input to the laboratory study workflow in Figure 14.4 is the reaction system to be studied, which is passed by the DoLabStudies subtask of the *ImpurityControl* workflow discussed earlier. The first task is to determine the critical parameters to be investigated in the experimental study, for example, temperature, pressure, starting composition, and so on. This task would be performed by the experts in reaction kinetics. The result of this research is the set of parameters with the range of values to be studied. These constitute the main input to set up a design of experiments. This step is performed by the team with the appropriate expertise. The result of the DOE is a table of conditions for each experiment to be performed in the laboratory. Each row in the table represents one experiment in a laboratory. The DoExpt subtask of the ExeExpts task loops through the table provided to it. For each experiment, it invokes the workflow describing the associated experiment, workflow *RxnSys1Experiment* shown in Figure 14.5. Once the experiment is completed, the results are captured and stored in an organized fashion.

An experiment associated with kinetic study is performed in a laboratory using a Mettler reactor according to the experimental workflow shown in

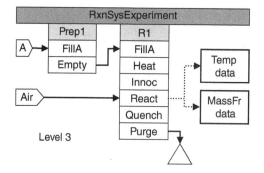

Figure 14.5 Example of experiment workflow.

Figure 14.5, a Level 3 workflow. There are new icons shown that define the materials used in an experiment, for example, the chemicals consumed (pentagon) and materials generated as "output" from experiment (triangle).

The key feature of an experiment workflow is that it models the "recipe" or the standard operating procedure of that experiment. In this example, an experiment for studying the kinetics of a reaction consists of two tasks. In task Prep1, a predetermined amount of material A is filled into a suitable unit. There may be additional processing required, such as analyzing the material, preheating it to the desired temperature, and so on. The material is then transferred into the reactor used for task R1. The sequence of subtasks defining that task is implemented in the sequence shown in the workflow diagram and according to the conditions defined by the recipe. At the end, the material is disposed during the Purge subtask. Various kinds of data collected during a subtask are represented by the data nodes. As an example, in Figure 14.5 during the React subtask, the temperature profile and concentration profiles are collected via online instruments installed with the reactor.

The UseR subtask of the workflow in Figure 14.4 points to a computational workflow. A computational workflow is similar to an experiment workflow, except that instead of implementing a laboratory procedure, a set of mathematical tools are used according to the "recipe" required by those tools. An example of a computational workflow for extracting kinetic parameters and mass transfer coefficient from a set of experiments is shown in Figure 14.6. Suppose that the statistical analysis tool R [22] is used for the data fitting application in this case. The first step to using R in this example would to develop a C function, which is called by R to model the concentration and temperature dynamics of a stirred tank reactor, like the one used in the experiments. Of course, this task may be done only once, that is, the resulting model could be written in such a way that it can be used in similar data fitting situations. The result of this activity could be a compiled library file (a .dll file), which provides the values expected by R. The first subtask, named DevRModel, of the DataFitting task represents the effort in developing a program or a script that

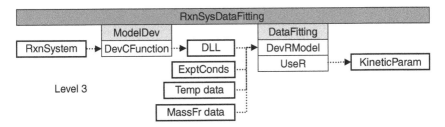

Figure 14.6 Example of computational workflow.

preprocesses the data used in the data fitting exercise as per the requirements of R, specifies numerical parameters such as tolerances, and invokes the appropriate functions in R. The UserR subtask executes the program, and on successful completion, the kinetic parameters and mass transfer coefficient are generated. The computational workflow ties together a wide variety of data, for example, the RxnSys and ExpConds defined in the main workflow and the data generated during the experiments. Similarly, the results are communicated to the main workflow and are accessible from that workflow. In fact, it is the shared data that binds the various types of workflows.

The additional important advantage of the workflow-based approach is that it facilitates communication between all of the members of the team involved with the project. As is clear from the example, each person responsible for implementing tasks and workflows could be different. The workflow approach also provides a clean mechanism to track the progress of a project, because the current status can be maintained with each subtask, task, and workflow.

14.5.3 Using Workflows

In this section, the mechanics of using the workflow constructs are illustrated.

When an *ImpurityControl* project needs to be initiated, an instance of the template shown in Figure 14.3 is created. Since it is a business workflow, a notice is sent to initiate all independent tasks, in this case the RxnMech task. In addition, placeholders are created for all data items identified by the metadata blocks in the workflow. Once the person leading the RxnMech task completes the Research subtask, the instance of the RxnSystems data node is populated with appropriate data, which in turn initiates the LabStudy task. Note that logically the Optimize task cannot be initiated yet because the kinetic parameters and mass transfer coefficients have not been determined yet.

The DoLabStudies subtask scans through all the reaction systems created and for each system invokes the RxnSystemKinParamDetermination workflow. The selection of a workflow for each reaction system is done by the person responsible for the LabStudy task. When this is completed, the KinParam and MassTrCoeff data nodes contain references to the corresponding data nodes created for each instance of the RxnSystemKinParamDetermination workflow. Note that the generation of the kinetic parameters and mass transfer coefficient for one reaction system requires the completion of the entire hierarchy of workflows defined in Figures 14.3, 14.4, 14.5, and 14.6. The Optimize task can be initiated only after all the reaction systems have been studied.

14.6 Design Space

The concept of design space is a key component of QbD implementation for pharmaceutical products. ICH Q8(R2) defines design space as "the multidimensional combination and interaction of input variables (e.g., material attributes) and process parameters that have been demonstrated to provide assurance of quality." Working within the design space is not considered as a change. Movement out of the design space is considered to be a change and would normally initiate a regulatory post-approval change process. It is understood that the design space is proposed by the applicant and is subject to regulatory assessment and approval [23]. Several case studies have been reported highlighting the use of process monitoring techniques and process analytical technology (PAT) to provide information on deviations from the established design space of specific unit operations [24]. A discussion of the possibility of extending the techniques to establish a design space for an entire process has been included, but there is no concrete roadmap presented for how to achieve that goal. In the following section, we suggest a conceptual design of such a roadmap and illustrate the key building blocks in this design.

14.6.1 Design Space Example

As an example, consider a tablet manufacturing process using wet granulation and operating in batch mode as shown in Figure 14.7. The four main unit operations in the process are wet granulation, fluidized bed drying, blending, and tablet compression. The objective is to determine the design space for this process.

The first step in determining the design space is to identify the CQAs of the product. For instance, for tablets important quality attributes include the drug loading, disintegration time (dissolution time), and tablet strength/hardness. Often, the disintegration time can be correlated with some other measurable

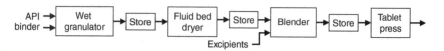

Figure 14.7 Diagram of manufacturing process.

properties of the tablet. The specifications on the CQAs are either fixed target values or ranges on the attribute values, for example, disintegration time between (min, max) seconds. Some of the CQA ranges are established from very detailed in vivo pharmacokinetic studies during clinical trials. For determining the design space, it is assumed that the CQAs are known. The determination of design space consists of establishing the ranges for the CPPs and the CMAs for all the materials passed between various unit operations.

14.6.2 Systematic Approach to Determining Design Space

In principle, in order to determine the design space experimentally, the complete set of CPPs of all unit operations must be considered in the DOE. This set may become very large, requiring an excessively large number of experiments. A systematic approach is necessary to facilitate reliable determination of the design space. Such an approach should incorporate other methodologies such as the use of models; use of methods, such as sensitivity analysis, to identify variables with biggest impact; and the use of feedback control for manufacture of drug product. When CMAs can be actively monitored and feedback control applied to the CPP, then variations in the environment or input materials can be counteracted by new values of the CPP (even values outside of a design space that represents prior experience) to keep the CMA within desired limits. Variables of biggest impact may be determined via risk assessment, previous knowledge, and/or designed experiments (e.g., via Plackett–Burman designs).

Since the finished product must satisfy the CQAs, the determination of design space could begin with the last unit operation of the process and progress upstream one unit operation at a time. Thus, for the tablet compression process given earlier, based on the CQAs on the tablet, the CPPs of the tablet press and the CMAs of the material from the upstream blending step could be determined.

In the next section, we use a simple example to outline a systematic approach for determining the design space for a single unit operation using in silico methods. This approach can be extended for determining the design space for an entire process, which may use experimental as well as in silico methods. As the complexity increases, the amount of information generated increases exponentially, which necessitates the use of an advanced KM system such as KProMS.

14.6.3 Workflow-Based Approach to Design Space Development

We consider a single operation—a tablet press. Let us assume that the tablet press has two CPPs, dwell time and compression force (pp1 and pp2), and the incoming mixture has two CMAs, particle size distribution and flowability (ma1 and ma2). The two CQAs are disintegration time and strength (cq1 and cq2). Also, let us assume that we have two models, one empirical and one theoretical to simulate the tablet press operating at steady state. Both models provide

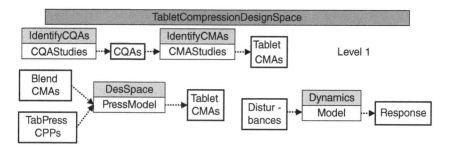

Figure 14.8 Define design space for tablet compression.

the relationship between the input values of two CPPs and two CMAs and the output values of two CQAs. In addition, we assume that a dynamic model for the tablet press is also available.

As an example, a Level 1 workflow for determining the design space for a single operation, tableting, is given in Figure 14.8. The IdentifyCQAs task represents the steps required to identify the CQAs and their ranges, for example, the pharmacokinetic studies mentioned earlier. The IdentifyCMAs task represents the steps required to translate CQAs to some measurable properties of a tablet and associated ranges for the property values. The CQAStudies and CMAStudies subtasks may invoke workflows for specific studies, which would be at Level 2. These studies may require analysis of experimental results and mathematical modeling if appropriate. The results of the Level 2 workflows are captured in the CQAs and TabletCMAs data nodes. If the CQAs cannot be translated into CMAs, appropriate experiments must be performed using tablet samples to determine whether the quality criteria are satisfied.

The DesSpace task is the key step in determining the design space for the tableting operation. The main objective of this task is to determine the range of values for each CMA of the feed blend for which a combination of tablet process parameters exists that will produce a tablet within the allowable ranges for the specified TabletCMAs. Typically, the design space would be determined using expensive and time-consuming experimentation using just the tablet press. However, since feed blend is an intermediate, to produce it the upstream stage must be included, which in turn cascades upstream to include all process operations. For purposes of this discussion, we assume that there exists a simulation model that can be used to support the development of the design space. Since typical unit operations simulation models use description of material inputs (composition, PSD, etc.) and operating parameters to predict output streams, a trial-and-error procedure could be used to determine the design space. A simple simulation model-based approach would be to explore a grid constructed using ranges on feed blend CMAs and tablet process parameters. For each combination in the grid, the model would be executed and the results checked to determine if the computed tablet CMAs are within allowable range. Those cases that do not produce tablets within specified ranges of CMAs would be rejected. The resulting feed blend CMA ranges could serve to define the design space for the given tableting operation.

Another important aspect of determining the design space of a unit operation is establishing controllability of that operation. A dynamic simulation model of the unit operation can be very useful for such controllability studies. In this example, the controlled variables are the tablet CMAs. The manipulated variables and the process control system design could be determined using the process dynamics model. Establishing controllability is critical for continuous manufacturing processes.

The ranges on the CMAs of the powder coming out of the upstream blender determined from this workflow become the "CQAs" for determining the design space for the blender. Similar Level 1 and 2 workflows can be developed using the appropriate models for the blending process. Such a scheme can be repeated successively to each upstream unit operation until the CMAs on all the raw materials coming into the process are established.

After establishing design space for each unit operation, the design space for the complete process can be constructed. The direct superposition of design spaces of each unit operation is not appropriate since the processes and the models describing them may be nonlinear. Thus simulation of all process models will be required to determine input material CMAs so that drug product CQAs are satisfied. It should be noted that although material flow is from left to right, the process of defining the design space goes in reverse order.

An accurate, detailed model for a unit operation can be used very effectively in determining the design space for that unit operation. In the tablet compression example discussed earlier, if a model were available that relates all operating parameters and material properties of the input material to the tablet, then in silico experiments can be performed to determine the design space for the tablet press. If models are available for all unit operations in the process, then the sequential approach described earlier can be applied to determine the design space for the entire process. If a reliable and accurate simulation model for the entire process can be constructed either through the use of a process simulator or by sequentially running individual models, then parametric studies can be performed to define the operating region for each unit operation in the process, thereby defining the design space for the process. However, due to the inherent inaccuracies in the underlying models, there always is an uncertainty associated with the quality of the results. Therefore, model-based design space must be validated with experiments. However, the model-based approach will certainly help in reducing the number of experiments required for validation and will provide a sound basis for justification during the approval process. Ierapetritou *et al.* [25] provide an overview of models for a wide range of operations and varying complexity used in drug product development and manufacturing. Boukouvala *et al.* [26] describe a purely experiment-based approach for determining design space.

14.6.4 Drug Product Case Study

To illustrate the proposed approach for determining the design space in the drug product domain, the drug manufacturing process for the production of gabapentin tablets will be utilized. The series of unit operations involved is shown in Figure 14.7. While experiments in the drug substance domain are focused on reaction kinetics, the effect of process variables on the drug product and intermediates is explored in the drug product domain.

Gabapentin is an anticonvulsant that is used for preventing seizures and for treating postherpetic neuralgia. The stability of gabapentin is complicated by the formation of a toxic degradation product, gabapentin-lactam, which must be maintained at sufficiently low levels [27]. The generation of the lactam is dependent on manufacturing conditions, particularly stresses induced by compression or heating. As in the drug substance case study, the minimization of this impurity is key.

The sequence of processing steps and the interactions of CPPs and CMAs are shown in Figure 14.9. CPPs for various steps and the CMAs associated with various equipment and streams are given in Table 14.1. For further details, see Ref. [28].

Typically, the CPPs and the CMAs constitute the independent variables for the DOE for experimental or simulation studies.

A conceptual approach for determining the design space for the gabapentin process, outlined in this section, is based on the use of both simulation and experimental studies. The proposed approach has the potential for producing a robust and reliable design space in a time-efficient manner. Since the entire activity will generate large amount of information and knowledge, a workflow-based approach is well suited for representing and managing the associated activities as well as the information resulting from those activities. The first phase would be the simulation-based determination of the design space, followed by experimental verification of the identified operating regimes.

The top-level workflow for the first phase is shown in Figure 14.10. The design space determination is done sequentially, one stage at a time starting

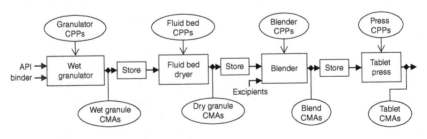

Figure 14.9 Critical process variables and critical material attributes.

Table 14.1 CPPs for processing steps and CMAs for process materials.

Processing step	CPPs	Stream	CMAs
Wet granulation	Water content Wet massing time Spray rate Impeller speed	Wet granules	Median particle size
Fluid bed drying	EEF End moisture target	Dry granules	Median particle size LOD Bulk density
Blending	Blending speed Fill ratio	Final blend	Hydrophobicity Impedance Segregation Compressibility Flowability Cohesion AIF
Tableting	Compression force	Tablets	Tablet weight Hardness Disintegration time Dissolution time

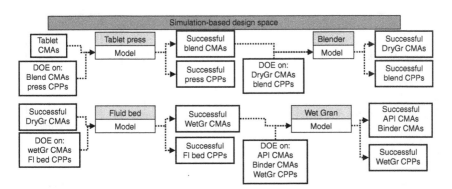

Figure 14.10 Workflow for the simulation-based determination of design space.

with the last stage, namely, tableting. We assume that the tablet CMAs are known, that is, ranges of values on critical tablet attributes for are available and a suitable tableting press model is available. A DOE on CMAs of material coming into the tableting machine from the blender, the blend CMAs, and tableting press CPPs is used as the input to a tablet press model. Those combinations of input values of blend CMAs and press CPPs that result in "good" tablets define the simulation-based design space for the tableting operation and constitute the information in the successful blend CMAs and successful press CPPs data nodes.

After determining the design space for the tablet press, the next step is to repeat the steps for the blender. A DOE for the blender based on the CMAs of material coming into the blender from fluidized bed dryer, and blender CPPs, is the input to a blender model. Those combinations of input values of fluid bed dryer CMAs and blend CPPs that result in blender output CMAs within the successful blend CMAs computed from the previous step define the simulation-based design space for the blend operation and constitute the information in the successful dryer CMAs and blend CPPs data nodes. Repeating these steps for the fluid bed dryer and wet granulator models results in simulation-based design space for the entire process.

The next phase consists of experimentally confirming the validity of model-based results. The experiments are performed using the materials and equipment that are the same as or similar to the actual process. The sequence of experiments is in the order of the material flow, thus beginning with wet granulator and ending with tablet press.

The workflow for the overall sequence of experiments is given in Figure 14.11. The DOE on the first stage is based on the API CMAs, Binder CMA, and wet granulator CPPs, which are within the space determined from simulation-based

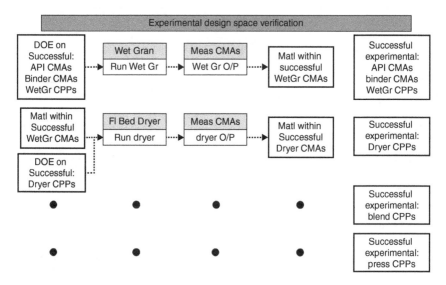

Figure 14.11 Workflow for the experimental verification of the design space.

phase. One wet granulator run is made for each combination generated by the DOE, and the CMAs of the material produced by the wet granulator are measured. If the CMAs are within the successful wet granulator CMA space determined in phase 1, then that combination is deemed as successful and is added to the set in the data node Successful Experimental API CMAs, Binder CMAs, and WetGr CPPs. Designate the number of successful combinations by N_{Wgr}.

For the fluidized bed dryer stage, the experiments are performed using the material generated by the wet granulator. For each successful material, the DOE is based on the Dryer CPPs. Again after a dryer experiment, the CMAs of the dryer output are measured. If the CMAs are within the successful dryer CMA space determined in phase 1, then that combination is deemed as successful and is added to the set in the data node Successful Experimental Dryer CPPs. Designate the number of successful combinations for the first successful wet granulator material by $N_{1,Fbd}$.

Similarly, the material generated by the fluidized bed dryer is used in the experiments with the blender. Let the number of successful blender experiments for the first successful dryer material be represented by $N_{1,1,Bl}$. The successful blender material is used in the experiments on the tablet press. Finally, we assume that the number of successful tablet press experiments for the first successful blend material is represented by $N_{1,1,1,Tp}$.

The total number of successful combinations across all experiments is

$$\sum_{i=1}^{N_{Wgr}} \sum_{j=1}^{N_{i,Fbd}} \sum_{k=1}^{N_{i,j,Bl}} N_{i,j,k,Tp}$$

The data in the nodes on the right in Figure 14.11 defines the design space for the process.

The systematic approach described earlier has two main advantages over the purely experimental approach. Without the insights gained from using mathematical modeling, the search space and the number of experiments performed in the purely experimental approach tend to be much larger, consequently more time consuming and expensive. Often the time available to determine the design space is limited. Due to the constraints on time, the search spaces are often reduced, thereby compromising the utility and power of the design space. Secondly, a purely experimental approach does not leverage the understanding of the physical phenomena embedded in models. As the understanding improves, the predictive capabilities of models also improve, thereby reducing the number of experiments required to verify the design space. The time required to perform simulation runs is insignificant compared with running actual experiments. In the ideal case where the models are perfect, there is no need for experimentally confirming the design space generated from the first phase—of course that is quite a stretch goal.

As evident from the combinatorial aspect, the determination of the design space generated large amount of data. A workflow-based system, such as KProMS, provides the complete context for the associated activities and the framework for organizing the data that is generated.

14.7 Technical Challenges

Conceptually, a workflow-based framework such as KProMS has the required functionality to implement all aspects of the QbD methodology. First, it provides a visual-graphical model to represent all types of workflows that involve material as well as information processing. Second, the framework allows logical and accurate representation of the hierarchical nature and inter-mixing of all processes associated with product development, manufacturing, and continuous process improvement. Third, all information is represented explicitly, that is, each piece of information has a unique place and address in the information repository as well as its identifying metadata and is fully machine accessible.

Due to the sheer volume of information generated and the complexities of the relationships between the processes and data, there are several technical challenges to developing a truly complete KM system for QbD. Some of the important ones are:

1) Human–machine interaction (HMI) design issues
2) Extraction for operational data
3) Collection of tacit knowledge

The key aspects of these challenges are discussed in this section.

14.7.1 Human–Machine Interaction Design

The organization of workflows as a hierarchy of multiple levels helps during recording of information or knowledge creation. The generation of various instances at each level is typically dictated by the actions at the immediately higher level. Also, the responsibilities and assignment of resources to each action are clearly defined at each level. Therefore, a workflow instance can be completely executed, and the results can be communicated to the higher level at the end of its execution.

However, the exploration of data in the knowledge repository can become a challenging task. Although each piece of data has a unique address, before using that data, it is necessary to identify the unique address. Again, if the path to that data is based on prior knowledge of the process or the hierarchy is known, one can specify the address, and the KProMS system provides context-sensitive menus to facilitate the identification. However, often the explorations or queries are general in nature. For example, the user may want to know all the processes in which a particular chemical species was used, or all the processes in which a particular piece of equipment was used, and so on. One solution to address such queries is to build a library of predefined queries, which provide the required answer. Some of these queries can be created based on the input from domain experts and discussions with the end users, with a full understanding that this would be an ongoing task. Another solution is to provide a user interface through which a user can formulate queries without being a database expert. This can be achieved to some extent by presenting predefined views of some of the tables in the repository.

14.7.2 Extraction of Operational Data

The key aspect of QbD is continuous process improvement. This can be realized in a variety of ways, for example, by batch-to-batch optimization, by extraction of information on near misses that can be used in exceptional event management, and by improvements to the design space definition. To successfully implement any of these activities, access to operational data is necessary. A workflow for the manufacturing recipe can be the first tool in facilitating access to operational information.

Operational information is typically stored in process historians, which are provided as components of process automation systems. Traditionally, the historian data have been used for problem resolution in order to conduct investigations of batch failures, such as processing deviations, which result in off spec products. Also, typically process historians are used in continuous manufacturing to record real-time data. However, the extraction of information from a historian is a tedious and time-consuming activity and is undertaken only when required. Commercial products, such as the PI system [29], can provide the ability to simultaneously record selected variables from the set of historian variables as batch files at predefined frequency, for example, one file per day. The data in these files becomes available for analysis of the process data. KProMS can be used to provide complete context to the data thus captured. Thus, each day's production would be an instance of the manufacturing recipe, the metadata for the recorded data would be defined in a corresponding data node in the workflow, and each variable would be a column in the data file stored with that instance. This provenance can be used for extracting any information for the desired analysis mentioned earlier. Of course, the steps in a specific analysis would constitute a scientific workflow. The interested reader is directed to [30] for an example application involving the use of KProMS to support the operation of a dropwise manufacturing system for pharmaceutical solid oral dosage.

14.7.3 Collection of Tacit Knowledge

A KM system is capable of formalizing only explicit knowledge. A considerable amount of knowledge is experiential and therefore tacit. The most challenging task of a KM system is to convert tacit knowledge to explicit knowledge. Tacit knowledge covers a wide range of end uses. For example, it would be something as simple as how an expert handles a specific abnormal situation or as complex as how an expert makes decisions about formulating a product. In general, the steps in converting tacit knowledge to explicit knowledge are the following:

1) Conversation with subject matter experts (SMEs) to identify the steps they go through to solve a problem or resolve a situation
2) Conversion of the steps into a workflow or a hierarchy of workflow
3) Analysis and testing of the resulting workflows

By formalizing the tacit knowledge through a set of workflows, the knowledge can be recorded permanently and thus become available for future use. However, the process itself is time intensive and requires disciplined use of workflows to record not only what was done but also the rationale for decisions made and, subsequently, the outcomes of those decisions, both successes and failures. At that point, the workflow system becomes more than a structured mechanism for accumulating and sharing information but a true knowledge repository.

14.8 Conclusions

In this chapter an approach to managing the data, information, and knowledge required to support the QbD concept and to facilitate the development of the design space for a product is outlined. The key construct is that of a workflow, which can in visual-graphical form capture both the sequence of activities that must be carried out in developing a product and the data generated with each activity together with its associated metadata. It is the latter features that make the workflow construct more than a simple activity chart generator. That functionality is provided by any of a number of graphics-based project planning tools. Although the potential uses of this type of tool to support both drug substance and drug product development were the primary focus of this chapter, there is valuable functionality that a workflow model can also provide in support of manufacturing operations.

Going forward, the most important challenge in KM is to facilitate the capture of tacit knowledge. We believe that the routine construction of workflow-based models and their use to support all activities carried out by development and manufacturing teams can set the stage for not only recording activity outcomes but also decision processes. The consequences of those decision processes can really only be assessed if the full data provenance is available. When assessment is combined with provenance, then the true foundation is set for knowledge creation.

References

1 ICH Q8 (R2), ICH Harmonised Tripartite Guidelines—Pharmaceutical Development, 2009.

2 ICH Q9, ICH Harmonised Tripartite Guidelines—Quality risk management, 2005.

3 ICH Q10, ICH Harmonised Tripartite Guidelines—Pharmaceutical Quality System, 2008.

4 Holm, J., Olla, P., Moura, D., Warhaut, M. Creating architectural approaches to knowledge management: an example from the space industry. J. Knowl. Manag., 10(2), 36–51, 2006.

5 Lee, S.L., O'Connor, T.F., Yang, X., Cruz, C.N., Chatterjee, S., Madurawe, R.D., Moore, C.M.V., Yu, L.X., Woodcock, J. Modernizing pharmaceutical manufacturing: from batch to continuous production. J. Pharm. Innov., 10(3), 191–199, 2015.

6 Chatterjee, S., Moore, C.M.V., Nasr, M.M., An overview of role of mathematical models in implementation of quality by design (QbD) paradigm for drug development and manufacturing. In: Reklaitis, G.V., García-Munoz, S., Seymour, C. (eds.), Comprehensive Quality by Design for Pharmaceutical Product Development and Manufacture. John Wiley & Sons, Inc., Hoboken, 2017, pp. XX–XX.

7 Junker, B., Maheshwari, G., Ranheim, T., Altaras, N., Stankevicz, M., Harmon, L., Rios, S., D'anjou, M., Design-for-Six-Sigma to develop a bioprocess knowledge management framework. PDA J. Pharm. Sci. Technol. 65, 140–165, March/April 2011.

8 STARLIMS, http://www.starlims.com (accessed on April 12, 2017).

9 BIOVIA LIMS, http://accelrys.com/products/pdf/pmc/biovia-lims-ds.pdf (accessed on April 12, 2017).

10 BIOVIA Discoverant, http://accelrys.com/products/process-production-operations/biovia-discoverant/ (accessed on April 12, 2017).

11 http://w3.siemens.com/mcms/process-control-systems/en/distributed-control-system-simatic-pcs-7/simatic-pcs-7-system-components/batch-automation/pages/batch-automation.aspx (accessed on April 12, 2017).

12 http://new.abb.com/control-systems/system-800xa/800xa-dcs/batch-control (accessed on April 12, 2017).

13 Taverna, http://www.taverna.org.uk/ (accessed on April 12, 2017).

14 Davis, J., Edgar, T., Graybill, R., Korambath, P., Schott, B., Swink, D., Wang, J., Wetzel, J., Smart manufacturing. Annu. Rev. Chem. Biomol. Eng., 6, 7.1–7.20, 2015.

15 Cyber Physical Systems (CPS), http://www.nist.gov/cps/ (accessed on April 12, 2017).

16 CHIRON project of Medvision360 B.V., http://www.medvision360.com/en/CHIRON (accessed April 21, 2017).

17 SAP Netweaver, http://scn.sap.com/community/netweaver (accessed on April 12, 2017).

18 TIBCO BPM, http://www.tibco.com/products/automation/business-process-management/activematrix-bpm/business-studio (accessed on April 12, 2017).

19 Ultimus enterprise solutions, http://www.ultimus.com/ (accessed on April 12, 2017).

20 Joglekar, G.S., Giridhar, A., Reklaitis, G.V., A workflow modeling system for capturing data provenance. Comput. Chem. Eng., 67, 148–158, 2014.

21 Burcham, C.L., LaPack, M., Martinelli, J.R., McCracken, N., Using quality by design principles as a guide for designing a process control strategy. In: Reklaitis, G.V., García-Munoz, S., Seymour, C. (eds.), Comprehensive Quality by Design for Pharmaceutical Product Development and Manufacture. John Wiley & Sons, Inc., Hoboken, 2017, pp. XX–XX.

22 R Core Team. R: A language and environment for statistical computing. R Foundation for Statistical Computing, Vienna, Austria, 2013. http://www.R-project.org/ (accessed on April 12, 2017).

23 Rathore, A.S., Roadmap for implementation of quality by design (QbD) for biotechnology products. Trends Biotechnol., 27(9), 546–553, 2009.

24 Rathore, A.S., Saleki-Gerhardt, A., Montgomery, S.H., Tyler, S.M. Quality by design: industrial case studies on defining and implementing design space for pharmaceutical processes (Part 2). Mod. Med., January 1, 2009. http://www.modernmedicine.com/modernmedicine/article/articleDetail.jsp?id=575066 (accessed on April 12, 2017).

25 Ierapetritou, M.G., Ramachandran, R. (Editors). Process Simulation and Data Modeling in Solid Oral Drug Development and Manufacture. Springer, New York, 2016.

26 Boukouvala, F., Muzzio, F.J., Ierapetritou, M.G. Design Space of Pharmaceutical Processes Using Data-Driven-Based Methods. J. Pharm. Innov., 5, 119–137, 2011.

27 Zong, Z., Desai, S., Barich, A., Huang, H.-S., Munson, E., Suryanarayanan, R., Kirsch, L.E. The stabilizing effect of moisture on the solid-state degradation of Gabapentin. AAPS PharmSciTech, 12(3), 924–931, 2011.

28 Mockus, L., Lainez, J., Reklaitis, G., Kirsch, L.E. A Bayesian approach to pharmaceutical product quality risk quantification. Informatica, 22(4), 537–558, 2011.

29 The PI System, http://www.osisoft.com/pi-system/ (accessed April 12, 2017).

30 Icten, E., Joglekar, G., Wallace, C., Loehr, K., Sacksteder, J., Girdhar, A., Nagy, Z.K. and Reklaitis, G.V. A knowledge provenance management system for a dropwise additive manufacturing system for pharmaceutical products. I&EC Res., 55(36), 9676–9686, 2016.

Index

a

acetaminophen (APAP)   103
acetylsalicylic acid (ASA)   265, 267,
    269, 270
acid-base neutralizations   61
acousto-optic tunable filter (AOTF)-NIR
    process spectrometer   237
active pharmaceutical ingredients (APIs)
    61, 103, 125, 262
  impurities in   130, 131
  PAR   175
  RSS stereoisomer of   136
  tablet active film coating process
    coating liquid sprayed   196
    EEF   227
    inert core tablet   194
    RSD   204
    water-soluble and insoluble   194
Aeromatic AG STREA 1 model   344
aggregation kernel   331
Albemarle Corporation   237
analysis of variance (ANOVA)   103
APIs *see* active pharmaceutical
    ingredients (APIs)
aromatic chloride (ArCl)   61
Arrhenius Accelerated Stability Model
    302–307
Arrhenius constants   154, 157
automatic process control   5

active pharmaceutical ingredient   26
bioavailability and tablet stability
    26, 27
control strategy   29, 31
CSD and polymorphic form   28
feedback control system
  antisolvent addition profile   39, 40
  L-glutamic acid   36–39
  Merck compound   35–36
  predetermined temperature   39, 40
kinetics modeling   40–42
microfluidic chips and platform   29, 30
pharmaceutical crystallization   31,
    42–44
pharmaceutical development   25–26
polymorphs   26, 27
protein and pharmaceutical
    crystallization   29, 30
solvatomorphism/
    pseudopolymorphism   26
strategy design   31–34

b

Bayesian posterior predictive
    distributions DS   61, 65, 67
Bayesian treatment, lyophilized
    small molecule parenteral   6–7
  Calling WinBUGS from R   314
  data used for analysis   312–313

Comprehensive Quality by Design for Pharmaceutical Product Development and Manufacture,
First Edition. Edited by Gintaras V. Reklaitis, Christine Seymour, and Salvador García-Munoz.
© 2017 American Institute of Chemical Engineers, Inc. Published 2017 by John Wiley & Sons, Inc.

Bayesian treatment, lyophilized
   small molecule parenteral (*cont'd*)
  design space definition 307
  pseudo-zero-order degradation
    process 302–307
  shelf life 307–309
  sodium ethacrynate 302
  solid oral dosage forms 302
biconvex tablet 284
blending unit operation
  APIs 262
  ASA 265, 267
  blend homogeneity 272, 275
  blending quality 275
  content uniformity 262
  CQAs 261, 265
  design space 275, 276
  DEM simulations 267–268, 274
  double cone mixers 262–263
  FMEA 265, 266
  granular flows 262
  Lacey's index evolution 272, 274
  lactose 267
  objectives 264
  PEPT 263
  powder blending 263
  prior knowledge 262
  process characterization experimental
    design 268–271
  risk assessment 263–264
  RSD evolution 272, 273
BRugs Bayesian statistical package 63
BRugsFit function 63

c

CFD *see* computational fluid dynamics
  (CFD) simulations
chemistry/chemical manufacturing and
  controls (CMC) 125, 370
CHIRON project 367
Claritin_D^{TM} 194
coefficient of restitution (COR) 269
Colorcon Inc. Tablet samples 212
compartmentalized population balance
  model 322–323
computational fluid dynamics (CFD)
  simulations 259

crystallization kinetics 41
tablet coating process
  coating film treatment 283–284
  design and characterization
    284–286
  discrete droplet method 281–282
  droplet–wall interaction 282–283
  software package AVL FIRE
    v2009 281
content uniformity model 262
  fractional residence time 199–201
  mass coated-per-pass
    distribution 197
  model parameters
    process parameters and measurable
      conditions 204
    spray zone width 208–212
    tablet number density 207–208
    tablet velocity 204–207
  pan coating 197
  principles of 198–199
  RSD model 201–203, 215–219
  RTD theory 198
  spray zone 197, 198
  surface renewal theory 197, 198
  total residence time 199–201
continuous fluidized bed drying 355, 356
continuous granulation process 319
continuous stirred-tank reactor
  (CSTR) 201–203
contract manufacturing organizations
  (CMOs) 365
control algorithms 341
Control Development Incorporated 345
control strategy, PQLI 97
coprecipitation process 6
  alcohol-water binary liquid mixing
    process 242–243
  data analysis methods
    PCA 239
    process trajectory 239
    singular points (SPs) 239–240
  dissolution and absorption of
    drugs 235
  formulation development 235
  multivariate statistical modeling
    techniques 236

naproxen   236–239, 241–242
NIR absorbance-wavelength-process
    time, 3D map of   244–245
online turbidity monitoring   248–250
PAT approach   236
process signature identification   245,
    247–248
product characterization   235
real-time integrated PAT
    monitoring   243–244
real-time process monitoring data   237
3D map of NIR absorbance-wavelength-
    process time   244–247
cosmetic coating   193
coupling kinetics reaction   159–165
criticality, PQLI   96–97
critical material attributes (CMAs)   362,
    378–381
Critical Path Initiative   55
critical process parameters (CPPs)   270,
    294, 362, 378–381
critical quality attributes (CQAs)   13, 258,
    261, 265, 294, 301, 362, 375–376
crystal engineering   235
crystal size distribution (CSD)   28, 42, 44
current good manufacturing practice
    (cGMP)   55
cyber-physical systems (cps)   367
cyclopropylamine (CPA)   61

d
data-information-knowledge-wisdom
    hierarchy (DIKW)   363
degradation pathways   150, 151
DeltaV from Emerson Process
    Management   345
deprotection chemistry process   177
deprotection reaction kinetics   151–157
Design Expert^{TM} 56
design of experiment (DOE)   15, 103,
    221, 222, 258–259
design space (DS)
    Bayesian treatment, lyophilized small
        molecule parenteral   6–7
    Calling WinBUGS from R   314
    data used for analysis   312–313
    design space definition   307

pseudo-zero-order degradation
    process   302–307
shelf life   307–309, 315–316
sodium ethacrynate   302
solid oral dosage forms   302
definition   95
knowledge management
    CQAs   375
    drug product domain   378–382
    systematic approach   375
    tablet manufacturing
        process   374–375
    tablet press   375–378
in pharmaceutical development   5
    Bayesian framework   99
    black-box process   96
    chromatographic analytical
        methods   98
    continuous mixer   102–107
    control strategy   102
    feasibility and flexibility
        process   98, 100
    first-principles model   96
    flexibility analysis   100
    fluidized bed granulation   98
    knowledge space   96, 97
    normal operating ranges   96, 97
    optimal process design   100
    PQLI   96–97
    process systems engineering
        community   98
    robustness   101
    roller compaction   107–119
    standard deviation (SD)   101
    statistics in   99
    stochastic flexibility   99–100
quality product
    advantage   176
    confirmation of   186
    data analysis   183–186
    multifactor experiment
        simulation   175, 183–186
    screening analysis   186
    total impurity level   178, 182–183
tablet active film coating process
    DOE factors   221, 222
    EEF   220, 221

design space (DS) (*cont'd*)
  process map 221
  product critical quality attributes 220
  RSD model 222–225
  scale-up parameters 226–229
  spray zone width 222, 223
  tablet velocity 223
  troubleshooting 228
Design Space Task Team 58
desirability functions 57
diastereomer of 1 (SR-1) 138
1-(3-(dimethylamino)propyl)-3-ethylurea
    (EDCU) 130
discrete droplet method (DDM) 281–282
discrete element method (DEM) 259
double cone blender 262–263, 268, 269
drug substance quality 125
drum granulator

e

e-lab notebooks (ELN) 364, 365
Elixir sulfanilamide 2
enantiomer of 6 (ent-6) 138, 139
environmental equivalency factor
    (EEF) 220
equivalent tablet surface area (ETSA) 217
1-ethyl-3-(3-dimethylaminopropyl)
    carbodiimide (EDCI) 130–133
Eudragit L100 *see* naproxen
Eudragit L30D-55 287

f

failure mode effects analysis (FMEA)
    13, 265, 266
failure mode effects and criticality
    analysis (FMECA) 13
fault tree analysis (FTA) 13
fixed-dose combination (FDC) 193
flexibility index (FI) 100
fluidized bed dryer (FBD)
  constant rate evaporation 341
  continuous fluidized bed drying
    355, 356
  control limitations 356
  desired moisture content 341
  diffusion-limited regime 341
  materials and methods 344

MPC strategy
  implementation 346–348
  nominal process
    parameters 349–351
  non-nominal model parameters
    352–355
  parameters 348
  process control strategy 342–344
  PID control law 340, 342
  process model 348
Food and Drug Administration (FDA)
    2, 55, 126
Food, Drug, and Cosmetic Act 2
Fourier transform infrared (FTIR) 126
Froude number 269

g

gabapentin 378
gas-liquid mass transfer 140
gas sparging systems 147
gas streams 143
genetic algorithm (GA) 340
good manufacturing practice (GMP) 260
granular flows 262
granulation process 7
  continuous 319
  control-loop performance 319, 320
  novel process design
    identification of kernels 331
    integrated systems 330
    liquid binder properties 330
    proposed design and control
      configuration 331–335
    viscosity and surface tension
      effects 330–331
  optimal control-loop pairings
    multiple-input multiple-output
      system 327
    relative gain array 327–328
  population balance model
    aggregation phenomenon 321
    breakage kernel 321
    compartmentalized 322–323
    contact forces 320
    microscale properties 320
    population density function 320
    primary particles 322

size-dependent aggregation
  kernel   321
product specification   319
simulation and controllability   320,
  323–327
single-compartment model   320

h

hazard analysis and critical control points
  (HACCP)   13
hazard operability analysis (HAZOP)   13
Henry's law coefficient   160
high dimensional model representation
  (HDMR)   103–106
high performance liquid chromatography
  (HPLC)   126
human-machine interaction design
  (HMI)   382–383
2-hydroxyisovaleric acid (HOVal)   130,
  131, 136

i

ibuprofen coprecipitates with Eudragit
  S100   235
ImpurityControl workflow   370, 371
industrial granulation processes   319
in-line NIR spectroscopy   242–243
integrated enzymatic and chemo-
  enzymatic processes
  characteristics   72
  design problem
    decomposition-based solution
      approach   75
    mathematical formulation   73–75
  ISPR   72
  membrane bioreactors   72
  methodology
    decomposition strategy   79–80
    dynamic model   78
    superstructure   77–78
    workflow   76–77
  Neu5Ac synthesis
    conventional process scheme   80, 81
    data/information collection/
      analysis   81–82
    feasible separation techniques   82, 84
    problem definition statement   81

reactions   80
structural constraints   84–86
superstructure   82, 83
OPR-CRYST   84, 86, 88–90
integrated real-time PAT process
  monitoring strategy   243–244
Integrated simultaneous product
  recovery (ISPR)   72
International Council for Harmonization
  (ICH)   2, 3
  guidelines thresholds   3, 125, 126
International Council for Harmonization
  (ICH) Q8
  Annex   58
  design space   5
    computer graphics   57
    Design Expert   57
    experimental design   57
    JMP   57
    mechanistic model   62–64
    multiple-response configuration
      56, 57
    OMRs   56–59
    pharmaceutical tableting process
      61, 64–68
    predictive distribution
      approach   56, 59–61
    response surface methodology   58
    response surface optimization   57
    statistical applications   57
    surface plot   58, 59
    sweet spot   57
  guidance   55
  R2   12–20
    CQA   13
    design space   15–19
    QTPP   12
    Pharmaceutical Development   3
International Council for Harmonization
  (ICH) Q9
  Quality Risk Management   3
  risk assessment   13–14
International Council for Harmonization
  (ICH) Q10
  control strategy   19–20
  guidelines   361
  Pharmaceutical Quality Systems   3

International Council for Harmonization (ICH) Q11
  Development and Manufacturing of Drug Substances   3
International Society for Pharmaceutical Engineering (ISPE)   96

j

Jet Propulsion Lab (JPL) scientists   366
JMP®   56
Johanson's rolling theory   108
Johanson's theory model   110

k

Kefauver–Harris Drug Amendment   2
knowledge management (KM)   7
  BIOVIA corporation   365
  CQAs and CPPs   362
  design space
    CQAs   375
    drug product domain   378–382
    systematic approach   375
    tablet manufacturing
      process   374–375
    tablet press   375–378
  ELN   364, 365
  knowledge hierarchy   363–364
  KProMS
    extraction for operational
      data   383
    HMI   382–383
    tacit knowledge collection   383–384
  LIMS   364, 365
  pharmaceutical quality system   362, 363
  semagacestat process
    Alzheimer's disease   368
    process description   368–370
    RxnMech task   373–374
    workflow-based
      representation   370–373
  workflow-based framework
    business workflows   367–368
    KProMS system   368
    scientific workflows   366–367
  KProMS system
    extraction for operational data   383
    HMI   382–383

tacit knowledge collection   383–384
workflow-based system   382
Kriging   103–106

l

laboratory information management systems (LIMS)   346, 364
Lacey's index   271, 272
lattice Boltzmann methods   259
L-glutamic acid   36–39
linear spring-dashpot model   267, 268
loratadine   194, 195
lyophilized small molecule parenteral   6–7
  Calling WinBUGS from R   314
  data used for analysis   312–313
  design space definition   307
  pseudo-zero-order degradation
    process   302–307
  shelf life   307–309
  sodium ethacrynate   302
  solid oral dosage forms   302

m

Mahalanobis distance (MD)   240
Markov Chain Monte Carlo (MCMC)
    techniques   60
mass transfer
  coefficient   145–149
  CO_2 mass transfer model
    CO_2 removal model   141–142
    gas-to-liquid mass transfer   140
    Henry's law constant, determination
      of   142–145
    ideal gas law   142, 143
    production scale   141
    reactor schematic   141
MassTrCoeff data nodes   373
mathematical models   4–5
  empirical models   10–12
  internal metrics and external
    validation   21
  mechanistic models   9–12
  monitoring model performance   22
  process characterization models   21
  process monitoring and control   22
  semi empirical/hybrid models   11, 12
  supporting analytical procedures   22

mechanistic models   126
metastable polymorphs   27
MIMO granulation process   336
model predictive control (MPC)
  implementation   346–348
  nominal process parameters   349–351
  non-nominal model parameters
    352–355
  parameters   348
  process control strategy   342–344
Monte Carlo-based algorithm   278
Monte Carlo procedures   60

n

naproxen
  acute gout   236
  analgesic and antipyretic   236
  binary systems of   241–242
  equipment and instruments   237–239
  gastrointestinal side effects   236
  materials   237
  osteoarthritis   236
  rheumatoid arthritis   236
  ternary system of   244
  3D (NIR spectral number-wavelength-
    NIR absorbance) map   245, 246
nephelometric turbidity units (NTU)
  238, 248, 249
neuraminic acid (Neu5Ac) synthesis
  conventional process scheme   80, 81
  data/information collection/
    analysis   81–82
  feasible separation techniques   82, 84
  problem definition statement   81
  reactions   80
  structural constraints   84–86
  superstructure   82, 83
NIR model 128 Element T.E. cooled
  InGaAs Array   345
Norvir soft gelatin capsules   13
N-t-butoxycarbonyl (BOC)   130, 133, 136
nucleation kernel   331

o

Ohnesorge number (Oh)   282
online turbidity measurements   238
online turbidity monitoring   248–250

OSIER SprayWatch system   208, 209
overlapping mean responses (OMRs)
    method
  computer graphics   57
  Design Expert   57
  experimental design   57
  JMP   57
  multiple-response configuration   56, 57
  predictive distributions   56
  response surface methodology   58
  response surface optimization   57
  statistical applications   57
  surface plot   58, 59
  sweet spot   57

p

pan coating process   197
parametric bootstrapping DS   61, 64, 65
partial least squares technique   348
  peliglitazar   194
pharmaceutical coprecipitation process
    see coprecipitation process
Pharmaceutical Development, annex to
    Q8 (2007)   58
pharmaceutical PAT
  NIRS   236
  PCA-based data analysis and
    modeling   250–252
Pharmaceutical Product Lifecycle
    Management   3
planned registered sequence   126, 129
plug flow reactor (PFR)   201
polymorphs   26, 27
positron emission particle tracking
    (PEPT)   263
powder blending processes   263
predictive distribution   59–61
preliminary hazard analysis (PHA)   13
pressure-rated HEL AutoMATER
    reactor   143
principal component analysis (PCA)
    250–252
probabilistic risk assessment   97
process analytical technology (PAT)
    10, 126
  methods   22
  pharmaceutical community

process analytical technology
(PAT) (*cont'd*)
  coprecipitation process 236
  NIRS 236
  PCA-based data analysis and
    modeling 250–251
  pharmaceutical sector 236
  process monitoring 236
  real-time integrated monitoring
    243–244
Process Analytical Technology
  Initiative 55
process control strategy
  API 126, 128
  cGMP sequence 130
  chemical sequence 130
  CO_2 mass transfer model 140–149
  control strategy 136–140
  design and development flow chart
    126, 127
  elementary reactions steps 130
  impurity formation 131–136
  mechanistic models 130
  optimal processing conditions
    carbon dioxide removal process
      167–174
    degassing temperature 166–168,
      nitrogen flow rate 166
    TRS 166
  process monitoring 126
  product quality
    design space simulation 175–177,
      181–186
    PAR 175–178
    in situ impurity profile 177–178
    urea formation 175, 176
  reaction kinetics
    coupling 159–165
    deprotection reaction 151–157
    dissolution process 157, 159
  semagacestat GMP sequence 129
process trajectory
  chemical and biotech sectors 239
  online NIR real-time process
    monitoring 248
  PCA-based process trajectory 251–252

PCA score plot 241–243, 248
singular points (SPs) 239
Product Quality Lifecycle
  Implementation (PQLI) 96
proportional integral derivative (PID)
  control 342
proven acceptable ranges (PAR) 175–181
pseudoephedrine sulfate 194, 195
pseudopolymorphism 26, 27
pseudo-zero-order degradation
  process 302–307
PTV algorithm 209

q
QbD tripartite 3
quality by design (QbD)
  automatic process control *see*
    automatic process control
  computer simulation methods
    advantages 259
    blending unit operation *see* blending
      unit operation
    coating unit operation 261
    life cycle management 261
    screening tool 260
  drug development and manufacture
    empirical 10–12
    hybrid 11, 12
    mechanistic 9–12
    physical-chemical phenomenon 9
    scientific considerations 20–22
  experimental studies 260
  FBD *see* fluidized bed dryer (FBD)
  fundamental concepts of 257
  GMP environment 260
  implementation of 258
  knowledge management (KM) *see*
    knowledge management (KM)
  overview 1–2
  pharmaceutical industry 2–3
  simulation_ based dossier 260–261
  systematic overview of 258
  tablet coating process 276–294
quality risk management 259
quality target product profile (QTPP) 3,
  4, 12

r

relative gain array 327–328
relative standard deviation (RSD)
    model 265, 270–271
    BFC series coaters 212, 213
    biconvex-shaped tablet 212
    content uniformity
        coating time effect 218–219
        pan load 217–219
        spray zone width 215–216
        tablet number density
            217, 218
        tablet velocity 216–217
    key operational parameters 212
    potency distribution 213, 214
    standard concave tablet 212
    tablet cycles 213
residence time distribution (RTD)
    theory 198
response surface methodology
    (RSM) 103–106
Reynolds-averaged Navier–Stokes
    (RANS) 281
Reynolds number (Re) 282
risk priority number (RPN) 14
risk ranking and filtering 13
roller compaction
    control strategies 113–119
    critical process 109
    deterministic DS mapping approach
        110–111
    dry granulation processes 107
    dust generation of powders 107
    Johanson's rolling theory 108
    Johanson's theory model 110
    material parameters 109
    poor flowability 107
    ribbon density 108–109
    schematic representation of 109
    segregation 107
    stochastic design space 112–114
RSD *see* relative standard deviation
    (RSD) model
RxnSys1Experiment workflow 371
RxnSystemKinParamDetermination
    workflow 374

s

scientific data management systems
    (SDMS) 364
semagacestat process
    Alzheimer's disease 368
    GMP sequence 126, 129, 130
    process description 368–370
    RxnMech task 373–374
    workflow-based representation
        370–373
Smart Manufacturing Leadership
    Coalition (SMLC) 367
sodium ethacrynate 302
sodium salt of hydroxybenzotriazole
    (NaOBt) 130, 131
software package AVL FIRE v2009 281
solvatomorphism 26, 27
spray coating phenomena 281
SprayWatch® imaging system 208, 209
STDEV method 241
stochastic flexibility (SF) 99
sugar-coated tablets 277
surface renewal theory 197

t

tablet active film coating process 6
    API 193, 194
    content uniformity model
        fractional residence time 199–201
        mass coated-per-pass
            distribution 197
        model parameters 204–212
        pan coating 197
        principles of 198–199
        RSD model derivation 201–203
        RTD theory 198
        spray zone 197, 198
        surface renewal theory 197, 198
        total residence time 199–201
    design space
        DOE factors 221, 222
        EEF 220, 221
        process map 221
        product critical quality
            attributes 220
        RSD model 222–225

tablet active film coating process (*cont'd*)
    scale-up parameters   226–229
    spray zone width   222, 223
    tablet velocity   223
    troubleshooting   228
  drug molecule stabilization   193
  FDC tablet formulations   193–195
  pan coating process   196, 197
  peliglitazar   194
  PPAR α/g agonist   194
  RSD model validation, 213–215 *see*
    also relative standard deviation
    (RSD) model
  spray rate   195, 196
  sugar beads   193
  three-layer film coating process   194
  water-soluble compound   194
tablet coating process
  aqueous coating process   280
  aqueous polymer dispersions   277
  via CFD simulations
    coating film treatment   283–284
    design and characterization
     284–286
    discrete droplet method   281–282
    droplet–wall interaction
     282–283
    software package AVL FIRE
     v2009   281
  coating apparatus   278, 279
  coating uniformity   278
  film formation   278
  industrial-scale Bohle BTC   400, 280
  inter-tablet uniformity   278, 280
  intra-tablet uniformity   278, 280
  mask unpleasant odor/taste   277
  Monte Carlo-based algorithm   278
  physical properties   281
  potentially critical input parameters
   286–288

residence time and circulation
  time   278
residence time distribution   278
spray coating   277
spray guns   278
spraying phase   290–292
spray nozzle model   280
spray zone   279
sugar coating   277
time evolution of mean film
  thickness   288, 289
time evolution of RSD   290, 291
trial-and-error approach   277
tacit knowledge   364
Taguchi quality loss concept   97
Thalidomide disaster of   1961, 2
Thermodynamic Analysis of Aqueous
  Film Coating (TAAC)
  program   220
total related substances (TRS)   166

u

urea formation   175, 176

v

Van't Hoff correlation   159

w

web service-enabled workflow   366
Web Services Description Language
  (WSDL)   366
WinBUGS Gibbs sampling
  algorithm   307
WinBUGS Model   312
workflow-based KM systems
  business workflows   367–368
  KProMS system   368
  scientific workflows   366–367
workflow management system
  (WFMS)   366